Advances in Organic Crystal Chemistry

Seiya Kobatake · Hidehiro Uekusa
Editors

Advances in Organic Crystal Chemistry

Comprehensive Reviews 2025 on Crystal Functions

Springer

Editors
Seiya Kobatake
Osaka Metropolitan University
Osaka, Japan

Hidehiro Uekusa
Institute of Science Tokyo
Tokyo, Japan

ISBN 978-981-96-4478-0 ISBN 978-981-96-4479-7 (eBook)
https://doi.org/10.1007/978-981-96-4479-7

This Springer imprint is published by the registered company Springer Nature Singapore Pte Ltd.
The registered company address is: 152 Beach Road, #21-01/04 Gateway East, Singapore 189721, Singapore

Preface

This book is part of a series reviewing recent advances in organic crystal chemistry, a field that has progressed significantly alongside the rapid development of analytical and measurement techniques for solid organic materials. Published periodically (at least every five years), these volumes feature contributions from leading researchers in Japan and internationally. The first volume (2015) systematically summarized progress in diverse aspects of organic crystal chemistry using organic solids and organic–inorganic hybrid materials. The second volume (2020) focused on advances primarily from 2016 to 2020.

In 2025, two volumes are being published to reflect the substantial progress made in the field over the past five years. These volumes address key areas of organic solid-state chemistry, including nucleation and crystal growth, crystal structure design, solid-state reactions, and various physical properties of solid organic materials. One volume focuses primarily on "Crystal Structures," covering nucleation and crystal growth, and the design of crystal structures. This volume, "Crystal Functions," focuses on solid-state reactions, photomechanical crystals, elastic crystals, luminescence, and electrical and magnetic properties.

The chapters in this volume are authored by invited members of the Division of Organic Crystals, the Chemical Society of Japan (CSJ), and prominent international researchers. Founded in 1997, the Division of Organic Crystals comprises leading organic crystal chemistry researchers in Japan. The division holds a biannual domestic conference (the Symposium on Organic Crystals in autumn and the CSJ Annual Meeting in late March) and publishes the *Organic Crystals Division Newsletter* three times a year. A symposium on organic solid-state chemistry titled "Organic Solid-State Chemistry: Advances from Structures to Properties (#ORG021)" will be held at Pacifichem 2025 in Hawaii in December 2025.

This volume offers an exclusive and up-to-date overview of the latest achievements in this interdisciplinary field, with applications across a wide range of scientific and technological disciplines. The chapters are comprehensive and authoritative. We, the editors, express our sincere gratitude to all authors for their valuable contributions to *Advances in Organic Crystal Chemistry: Comprehensive Reviews 2025 on Crystal Functions*. We hope this book will serve as a valuable resource for advanced studies in chemistry, biochemistry, industrial chemistry, and pharmacology.

Osaka, Japan Seiya Kobatake
Tokyo, Japan Hidehiro Uekusa

Contents

Part I
Solid-State Reaction, Photomechanical Crystals, and Elastic Crystals

Chapter 1
Alternating Copolymerizations via Alternating 1D Columnar Molecular Assemblies of *p*-Quinodimethane Derivatives in Solution and Solid States

Takahito Itoh and Mikiji Miyata

Abstract We propose a novel mechanism for alternating copolymerizations of *p*-quinodimethane derivatives in solution. This comes from closely related behaviors of their copolymerizations in solid and solution states. Based on their cocrystal structures, alternating 1D columnar molecular assemblies are formed through stereospecific stacking between acceptor and donor monomers, resulting in very short distances between neighboring reaction centers. Such 1D columns bring about the alternating and stereospecific copolymerizations in solution spontaneously or with radical initiators. We suggest that this polymerization mechanism might be applicable to spontaneous alternating copolymerizations between TCNQ and styrene in solution.

Keywords Alternating copolymerization mechanism · 1D columnar molecular assemblies · *p*-Quinodimethane derivatives · Cocrystal structure

1.1 Introduction

This review deals with an alternating copolymerization mechanism for donor and acceptor monomers in solution as well as solid states. Firstly, you see a historical overview of our research starting from the 1970s. Secondly, we summarize syntheses, various reactions including polymerizations of diverse substituted *p*-quinodimethane derivatives in solution and solid states. Thirdly, successful crystallographic studies

T. Itoh (✉)
Department of Chemistry for Materials, Graduate School of Engineering, Mie University, 1577 Kurimamachiya-cho, Tsu 514-8507, Mie, Japan
e-mail: itoh@chem.mie-u.ac.jp

M. Miyata
The Institute of Scientific and Industrial Research (SANKEN), The University of Osaka, 8-1 Mihogaoka, Ibaraki 567-0047, Osaka, Japan
e-mail: miyata@chem.eng.osaka-u.ac.jp

S. Kobatake and H. Uekusa (eds.), *Advances in Organic Crystal Chemistry*,
https://doi.org/10.1007/978-981-96-4479-7_1

led us to analyze the single-crystal-to-single-crystal transformation in their solid-state copolymerizations, establishing *cis*-conformations of the resultant alternating copolymers. Fourth, we succeeded in acquiring soluble copolymers in solution. ^1H NMR spectroscopic and simulation studies confirmed the *cis*-conformations of the resulting polymers in solution, leading to the establishment of their *cis*-conformations in both solid and solution states. Finally, we propose a novel mechanism for the above-mentioned alternating copolymerizations in solution of *p*-quinodimethane derivatives in comparison with the conventional mechanism.

1.2 Historical Overview for Polymerizations of *p*-Quinodimethane Derivatives

In the mid-1960s to the 1970s, one actively discussed a copolymerization mechanism involving 1:1 charge-transfer complexes formed between donor monomers and acceptor monomers in order to explain the high degree of monomer alternation observed in copolymerizations of vinyl-type monomers [1–4]. At that time, the acceptor monomers were known to be maleic anhydride, alkyl fumarates, fumaronitrile, malenonitrile, *N*-phenylmaleimide, sulfur dioxide, and so on, but they were fewer in number compared to common donor monomers. In addition, these acceptor monomers were restricted within a narrow range of electron-accepting ability.

In the early 1950s, Du Pont chemists started the synthesis of a series of compounds bearing cyano groups as pseudohalogen of tetrafluoroethylene and poly(tetrafluoroethylene) and developed cyanocarbon chemistry [5]. In 1962, Acker and Hertler reported the synthesis of 7,7,8,8-tetracyano-*p*-quinodimethane (TCNQ) [6]. After that, Du Pont chemists actively investigated the electrical conductivities of the metallic and organic cations/TCNQ radical-anion salts and of the charge-transfer complex compounds between TCNQ and aromatic hydrocarbons, amines, and polyhydric phenols [7, 8]. At that time, polymer scientists and researchers regarded TCNQ as an inhibitor and retarder in radical polymerizations, similar to *p*-benzoquinone derivatives and hydroquinone [1, 4].

In 1978, we found that TCNQ copolymerizes spontaneously with styrene in acetonitrile to afford an alternating copolymer, indicating that TCNQ can act as an acceptor monomer (Fig. 1.1) [9]. This finding greatly contributed to the expansion in the variety of acceptor monomers in solution copolymerizations and the range of electron-accepting ability in acceptor monomers. This is because TCNQ had a much stronger electron-accepting ability compared to vinyl-type acceptor ones. Since then, we have investigated the syntheses of novel *p*-quinodimethane derivatives with diverse substituents (CN, COOR, COR, Br, SR) and substitution patterns. Furthermore, based on TCNQ and novel *p*-quinodimethane derivatives, we have developed a charge-transfer complex participation mechanism for alternating copolymerization in solution [10–14]. During the course of this study, we conducted the synthesis of 7,7,8,8-tetrakis(methoxycarbonyl)-*p*-quinodimethane (**1a**(Me)) with much weaker electron-accepting ability compared to TCNQ for the purpose of expanding the range of electron-accepting ability within the family of *p*-quinodimethane derivatives. The **1a**(Me) copolymerized with styrene and vinyl ethers in solution to afford alternating copolymers, similar to TCNQ.

Next we conducted a terpolymerization of the TCNQ-**1a**(Me)-styrene system for a quantitative analysis of the reactivity between the TCNQ-styrene complex and the **1a**(Me)-styrene complex. The compositional analyses of the resulting terpolymers suggested a possible alternating copolymerization between TCNQ and **1a**(Me) [15]. Actually, the copolymerization reaction proceeded between TCNQ and **1a**(Me) in acetonitrile to yield an alternating copolymer. Furthermore, interesting spectroscopic evidence (charge-transfer band) was observed due to some form of donor–acceptor

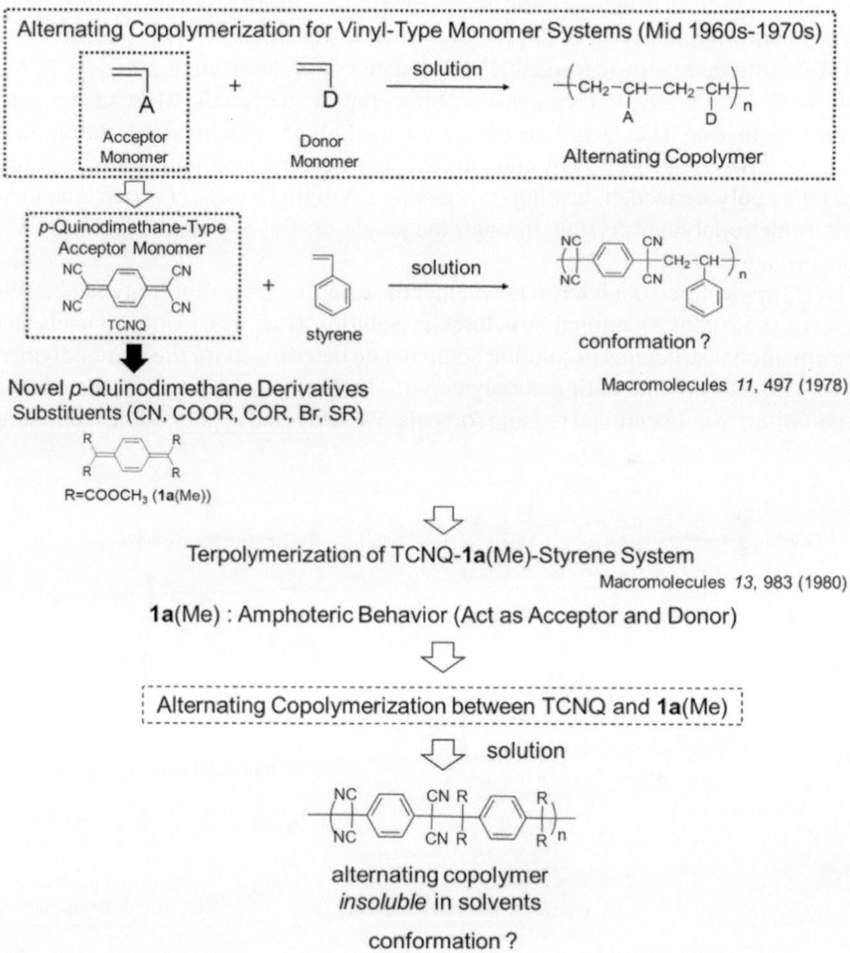

Fig. 1.1 Historical flow of our research in solution states

interaction on mixing **1a**(Me) with TCNQ in acetonitrile. In this way, we discovered a unique behavior of **1a**(Me) in both charge-transfer complexations and copolymerizations, meaning that **1a**(Me) acts as an acceptor monomer toward donor monomers styrene and vinyl ether and acts as a donor one toward acceptor monomer TCNQ [15]. Unfortunately, we did not determine any conformational structures for the alternating copolymer composed of **1a**(Me) with TCNQ because of its insolubility in conventional organic solvents.

TCNQ was known to form only rust-colored prisms crystal on crystallization and show no polymerization in the solid state [6]. Although polymerizations of **1a**(Me) in the solid state were known previously, we found in 2002 that **1a**(Me) affords two polymorphic crystals (yellow needle and yellow prism). The former exhibited no polymerization, but the latter underwent topochemical polymerization through the single-crystal-to-single-crystal transformation to afford a polymer crystal with a *trans*-conformation (Fig. 1.2) [16, 17]. Such a topochemical polymerization of **1a**(Me) reminded us of the potential solid-state copolymerization between TCNQ and **1a**(Me). Actually, 1:1 charge-transfer complex cocrystals were successfully formed by mixing TCNQ and **1a**(Me) at relatively high monomer concentration in acetonitrile and subsequently by allowing to cool to lower temperatures. The resulting cocrystals polymerized on heating or exposing UV light to yield *cis*-conformational alternating copolymer crystals through the single-crystal-to-single-crystal (SCSC) transformation [18].

Next, we explored our interest to whether the resulting crystalline polymers would preserve their conformational structures in solution (Fig. 1.3). Unfortunately, the conformational structures in solution could not be determined for the homopolymers of **1a**(Me) and the alternating copolymers of **1a**(Me) and TCNQ because of their insolubilities in conventional organic solvents. We struggled to get soluble alternating

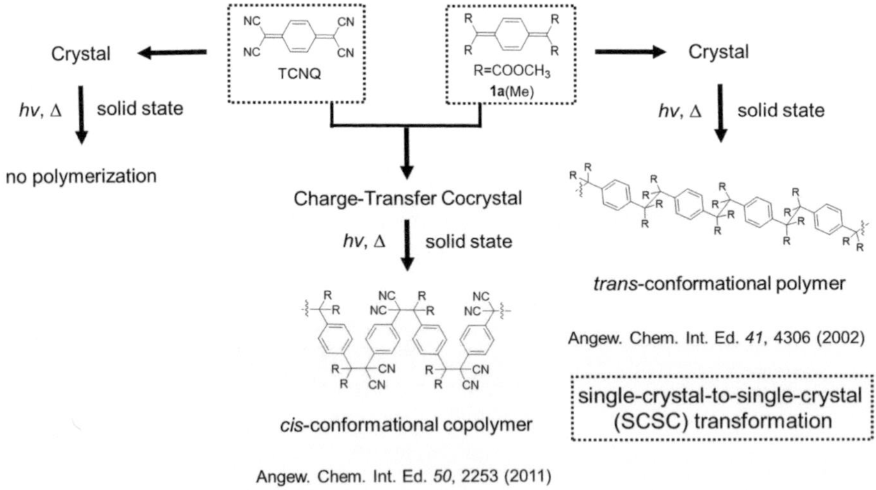

Fig. 1.2 Historical flow of our research in solid states

copolymers and employed *p*-quinodimethane derivatives with longer alkoxy groups such as ethoxy, propoxy, butoxy, and pentyloxy. Unfortunately, all of the resulting alternating copolymers were insoluble in conventional organic solvents. During the studies, we noticed the different speeds of the solid-state copolymerizations. Among them, cocrystals of 7,7,8,8-tetrakis(pentyloxycarbonyl)-*p*-quinodimethane (**1g**(Pen)) and TCNQ proceeded most rapidly. So, we attempted solution copolymerization in acetonitrile [19]. At last, we succeeded in obtaining soluble alternating copolymers in chloroform. ^{1}H and ^{13}C NMR spectral analyses and spectroscopic simulations confirmed a *cis*-conformational structure of the alternating copolymer in solution, leading to the preservation of *cis*-confirmations in solution as well as solid states [19]. This strongly suggested that 1D columnar assembly formation occurs in solution as well as the solid state.

The observation of this behavior has been successful so far only with the combinations of four-ester possessing *p*-quinodimethane derivatives and TCNQ. The 1D

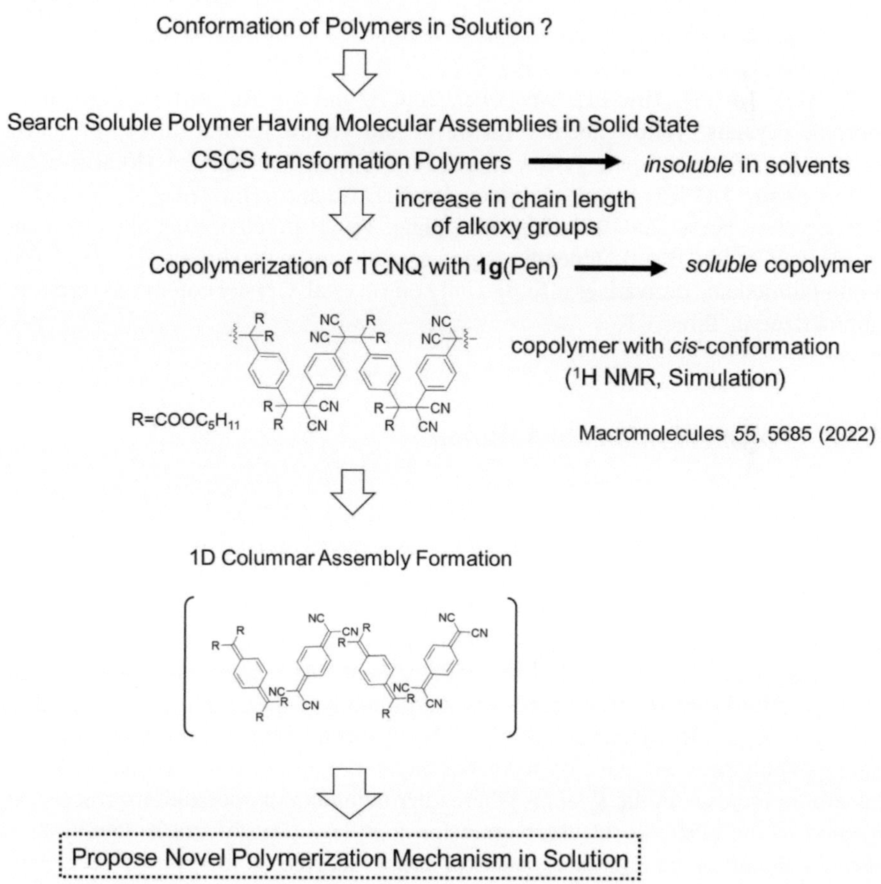

Fig. 1.3 Historical flow of our research in solid and solution states

columnar molecular assembly formation provided us an opportunity to advance the polymerization reaction mechanism we had proposed at the early stage of our research in solution copolymerization involving *p*-quinodimethane derivatives. The important concept is that supramolecular 1D columnar assemblies are formed in solution before the polymerizations. This behavior can be further supported through an approach involving various molecular assembly structures and intermolecular interaction energies in crystals, as described in the subsequent sections.

1.3 Syntheses of *p*-Quinodimethane Derivatives and Their Polymerization Behaviors

1.3.1 Synthesis

We synthesized novel *p*-quinodimethane derivatives with diverse substituents and substitution patterns as shown in Fig. 1.4.

1a(Me), **1d**(^{i}Pr), **1l**(ClEt), **1m**(BrEt), **2c**(Cl), and **8b**(^{t}Bu) provided two polymorphic crystals (yellow needle: **1a**(Me)-α and yellow prism: **1a**(Me)-β; yellow needle: **1c**(^{i}Pr)- α and pale yellow plate: **1c**(^{i}Pr)-β; yellow needle: **1l**(ClEt)-α and yellow prism: **1l**(ClEt)-β; yellow prism: **1m**(BrEt)-α and yellow needle: **1m**(BrEt)-β; pale yellow scaly: **2c**(Cl)-α and yellow plate: **2c**(Cl)-β; red needle: **8b**(^{t}Bu)-α and purple needle: **8b**(^{t}Bu)-β) depending upon crystallization conditions [19–22]. Other *p*-quinodimethane derivatives afforded only one crystal. Crystal colors and forms are summarized in Table 1.1.

1.3.2 Polymerizations in Solution

We investigated homopolymerizations in the presence of 2,2′-azobis(isobutyronitrile) (AIBN) and copolymerizations with styrene (St) and its derivatives (*p*-methylstyrene (MeSt), *p*-methoxystyrene (MeOSt)) in solution for *p*-quinodimethane derivatives except for **1c**(Pr), **1d**(^{i}Pr), **1e**(Bu), **1f**(^{i}Bu), **1g**(Pen), **1h**(Hex), **1i**(Dod), **1j**(Bz), **1k**(BzF_5), **2a**(H), **2b**(F), **2c**(Cl), and **2d**(Br). The results are summarized in Table 1.1.

Almost *p*-quinodimethane derivatives underwent polymerizations in solution with AIBN to afford corresponding polymers. Additionally, *p*-quinodimethane derivatives copolymerized with styrene and its derivatives to afford random copolymers or alternating copolymers. We have been considering copolymerizations involving charge-transfer complexes in the systems where alternating copolymerizations proceed. In relation to the reactivity of charge-transfer complexes, terpolymerization systems such as donor(1)-donor(2)-acceptor and donor–acceptor(1)-acceptor(2) combinations were very useful methods for quantitatively analyzing the reactivity between charge-transfer complexes.

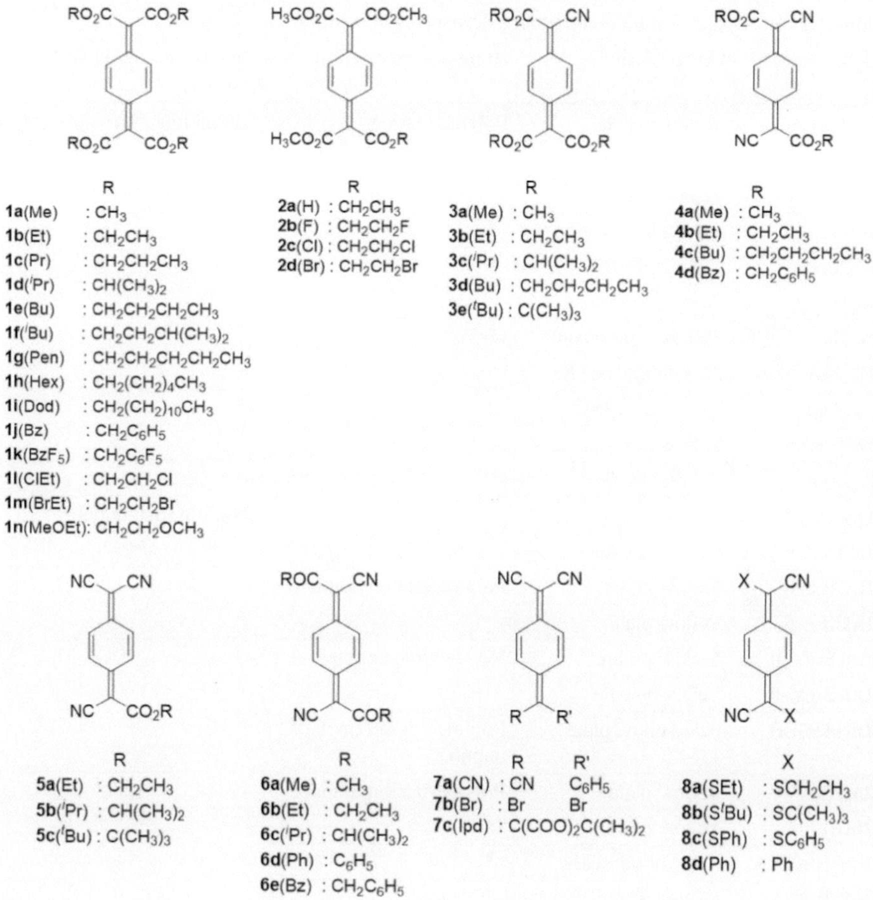

Fig. 1.4 Synthesized *p*-quinodimethane derivatives with diverse substituents and substitution patterns

For the purpose of analyzing the reactivity between the TCNQ-styrene complex and the **1a**(Me)-styrene complex, we conducted a terpolymerization of the TCNQ-**1a**(Me)-styrene system. Unexpectedly, we found that **1a**(Me) can copolymerize alternatingly with TCNQ from the composition analysis of the terpolymers [15]. Actually, the copolymerization reaction proceeded between TCNQ and **1a**(Me) in acetonitrile, resulting in the formation of an alternating copolymer. **1a**(Me) exhibited a unique behavior in copolymerizations, that is, **1a**(Me) acts as an acceptor monomer toward donor monomers such as styrene and vinyl ether and also acts as a donor one toward a stronger acceptor monomer such as TCNQ. We pointed out that the relative gap in electron character between two compounds plays an important role in the determination of their polar character.

Table 1.1 Reactions of *p*-quinodimethane derivatives in solution

Compound	Crystal form	Homopolymerization	Copolymerization with styrenes
1a(Me)-α	Yellow needle	Polymerization	Alternating (St)
1a(Me)-β	Yellow prism		
1b(Et)	Yellow needle	Polymerization	Alternating (St)
1c(Pr)	Pale yellow needle	–[a]	–[a]
1d(^{i}Pr)-α	Yellow needle	–[a]	–[a]
1d(^{i}Pr)-β	Pale yellow plate		–[a]
1e(Bu)	Yellow viscous oil	–[a]	–[a]
1f(^{i}Bu)	Pale yellow needle	–[a]	–[a]
1g(Pen)	Yellow powder	–[a]	–[a]
1h(Hex)	Yellow powder	–[a]	–[a]
1i(Dod)	Yellow needle	–[b]	–[a]
1j(Bz)	Yellow plate	–[a]	Random (St), unpub.[c]
1k(BzF_5)	Yellow needle	Polymerization, unpub.[c]	–[a]
1l(ClEt)-α	Yellow prism	Polymerization, unpub.[c]	–[a]
1l(ClEt)-β	Yellow plate		–[a]
1m(BrEt)-α	Yellow prism	No reaction, unpub.[c]	–[a]
1m(BrEt)-β	Yellow needle		–[a]
1n(MeOEt)	Pale yellow plate	Polymerization (melt), unpub.[c]	–[a]
2a(H)	Yellow plate	Polymerization, unpub.[c]	–[a]
2b(F)	Yellow plate	–[a]	–[a]
2c(Cl)-α	Pale yellow scaly	–[a]	–[a]
2c(Cl)-β	Yellow plate	–[a]	–[a]
2d(Br)	Yellow prism	–[a]	–[a]
3a(Me)	Yellow needle	Polymerization	–[a]
3b(Et)	Yellow needle	Polymerization, unpub.[c]	Random (St)
3c(^{i}Pr)	Yellow needle	Polymerization	–[a]
3d(Bu)	Yellow needle	Polymerization	–[a]
3e(^{t}Bu)	Yellow needle	Polymerization	Alternating (St)
4a(Me)	Yellow needle	Polymerization	–[a]
4b(Et)	Yellow needle	Polymerization	–[a]
4c(Bu)	Yellow needle	Polymerization	Random (St), alternating (MeOSt)
4d(Bz)	Yellow needle	Polymerization	Random (St)
5a(Et)	Unstable[b]	Polymerization	–[a]
5b(^{i}Pr)	Unstable[b]	Polymerization	–[a]
5c(^{t}Bu)	Unstable[b]	Polymerization	Random (St)

(continued)

Table 1.1 (continued)

Compound	Crystal form	Homopolymerization	Copolymerization with styrenes
6a(Me)	Orange plate	Polymerization	Random (St, MeSt, MeOSt)
6b(Et)	Orange plate	Polymerization	–[a]
6c(^{i}Pr)	Orange plate	Polymerization	–[a]
6d(Ph)	Orange plate	Polymerization	Random (St, MeSt, MeOSt)
6e(Bz)	Orange plate	Polymerization	–[a]
7a(CN)	Purple needle	No reaction	Alternating (MeSt)
7b(Br)	Orange needle	Polymerization	Random (St)
7c(Ipd)	Orange needle	Polymerization	Alternating (St)
8a(SEt)	Red needle	Polymerization	Random (St)
8b(S^{t}Bu)-α	Red needle	No reaction	–[a]
8b(S^{t}Bu)-β	Purple needle	No reaction	–[a]
8c(SPh)	Red needle	Polymerization	Random (St)
8d(Ph)	Purple needle	No reaction	Alternating (MeSt)

[a]Not investigated
[b]Not isolated as pure crystals
[c]Unpub.: unpublish result

As to the conformational structures in solution, we were unable to obtain information regarding the conformational structure of the alternating copolymer composed of **1a**(Me) with TCNQ because it was insoluble in conventional organic solvents.

1.3.3 Homopolymerizations in Solid State

We conducted solid-state reactions under photoirradiation and heating for four-ester possessing *p*-quinodimethane derivatives (**1a-n** and **2a-d**), and the results are summarized in Table 1.2.

They exhibited three behaviors (no reaction, dimerization, and polymerization) arising from substituent-responsive reactivities. All of the polymers afforded low-quality polymer crystals, which were not suitable for single-crystal structure analysis. Therefore, we could not determine the conformational structures of the resulting polymers. The conformational structures of the polymer obtained by solid-state polymerizations were determined to have a *trans*-conformation with a high-quality single polymer crystal, a transparent and colorless crystal, of **1c**(BrEt) formed through the single-crystal-to-single-crystal transformation by utilizing ^{60}Co γ-radiation, confirming topochemical polymerizations for *p*-quinodimethane derivatives [17].

The compounds shown in Fig. 1.5 are known as typical monomers for topochemical polymerizations. In 1969, Wegner confirmed the solid-state polymerizations of

Table 1.2 Reactions of *p*-quinodimethane derivatives in solid state

Compound	Crystal form	Photoirradiation[a]	Heating
1a(Me)-α	Yellow needle	No reaction	No reaction
1a(Me)-β	Yellow prism	Polymerization	Polymerization
1b(Et)	Yellow needle	No reaction	No reaction
1c(Pr)	Pale yellow needle	No reaction	No reaction
1d(^{i}Pr)-α	Yellow needle	No reaction	No reaction
1d(^{i}Pr)-β	Pale yellow plate	No reaction	No reaction
1e(Bu)	Yellow viscous oil	–[b]	–[b]
1f(^{i}Bu)	Pale yellow needle	–[b]	–[b]
1g(Pen)	Yellow powder	–[b]	–[b]
1h(Hex)	Yellow powder	–[b]	–[b]
1i(Dod)	Yellow needle	–[b]	–[b]
1j(Bz)	Yellow plate	No reaction	No reaction
1k(BzF_5)	Yellow needle	No reaction	No reaction
1l(ClEt)-α	Yellow prism	Polymerization	Polymerization
1l(ClEt)-β	Yellow plate	Polymerization	Polymerization
1m(BrEt)-α	Yellow prism	Polymerization	Polymerization
1m(BrEt)-β	Yellow needle	No reaction	Polymerization
1n(MeOEt)	Pale yellow plate	Polymerization	Polymerization
2a(H)	Yellow plate	Polymerization	–[b]
2b(F)	Yellow plate	No reaction	–[b]
2c(Cl)-α	Pale yellow scaly	No reaction	–[b]
2c(Cl)-β	Yellow plate	Dimerization	–[b]
2d(Br)	Yellow prism	Dimerization	–[b]

[a]Photoirradiation with a high-pressure mercury lamp (400 W) at a distance of 12 cm and at 30 °C
[b]Not investigated

diacetylene (1,3-diyne) units and established the topochemical 1,4-addition polymerizations of diacetylene-based monomers (1,3-diynes) [23, 24]. The steric requirements of diacetylene-based monomers for topochemical polymerizations are one-dimensional (1D) stacking of 1,3-diyne groups in a translational position around 4.9 Å, a contact distance of 3.5 Å between the reactive carbons of two adjacent molecules, and tilt angles of about 45° of 1,3-diyne unit axes to the stacking axis [25]. The minimum movement of the reactive carbons induced the effective polymerization in the crystal state to afford crystalline poly(diacetylene)s with a *trans*-conformational structure. Extension of the diacetylene (1,3-diyne) monomers to triacetylenes (1,3,5-triynes) was demonstrated successfully by Lauher and Fowler [26]. They achieved the 1,6-addition polymerizations of triacetylenes (1,3,5-triynes) by utilizing the formation of the host–guest cocrystals, where self-assembling host

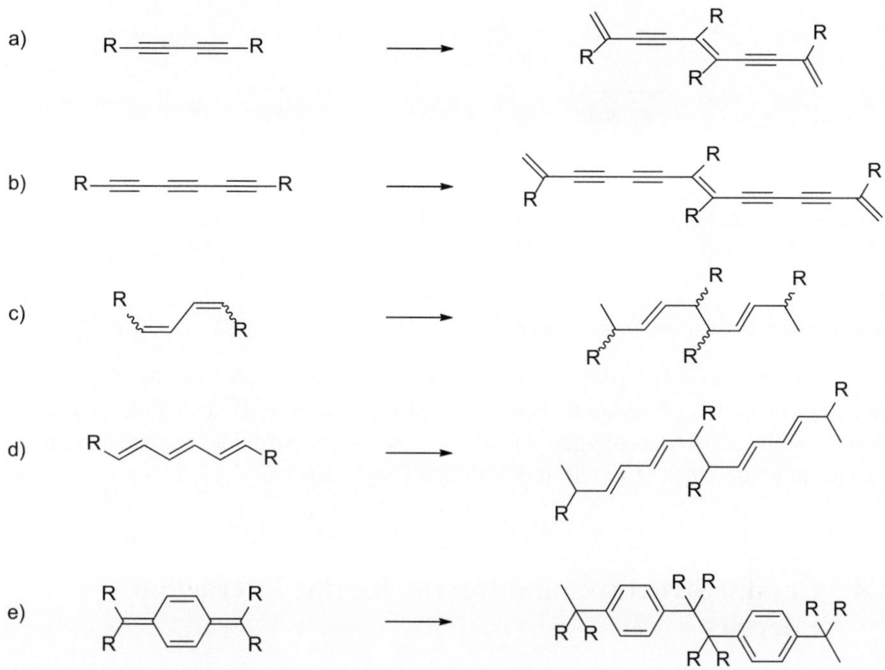

Fig. 1.5 Monomers for topochemical homopolymerizations and their polymers with *trans*-conformations: **a** 1,3-diynes, **b** 1,3,5-triynes, **c** 1,3-dienes, **d** 1,3,5-trienes, and **e** *p*-quinodimethanes

molecules stack with a repeating distance close to 7.4 Å and align the 1,3,5-triynes favorably for the topochemical 1,6-addition polymerizations.

The topochemical 1,4-addition polymerizations of dienes (1,3-dienes) were firstly reported by Tieke [27], and later were explored by Matsumoto et al. in the 1990s [28, 29]. Extension of 1,3-dienes to 1,3,5-trienes was demonstrated by Lauher and Fowler [30]. The steric requirements are found to be close to those of 1,3,5-triynes, where the suitable parameters are (i) a distance of 3.5 Å between the reacting carbon atoms, (ii) a 1D column formation by face-to-face stacking of the triene moieties with the translational distance of 7.2 Å, and (iii) a tilt angle of 34° of the longer axis of the molecules to the 1D stacking axis. *p*-Quinodimethane is regarded as a kind of a *trans–trans* fused triene monomer, which transforms into the phenylene rings after polymerization. Polymerizable monomers shown in Table 1.2 proceed exclusively through *trans*-addition reactions to form polymers with a *trans*-conformational structure.

Solid-state photoirradiations to **2c**(Cl)-β and **2d**(Br) induced topochemical [6+6] photocycloaddition dimerizations to yield exclusively bridge-substituted [2.2]paracyclophanes (1,9-bis(2′-chloroethoxycarbonyl)-1,2,2,9,10,10-hexa(methoxycarbonyl)[2.2]paracyclophane (**9c**(Cl)) and 1,9-bis(2′-bromoethoxycarbonyl)-1,2,2,9,10,10-hexa(methoxycarbonyl)[2.2]paracyclophane

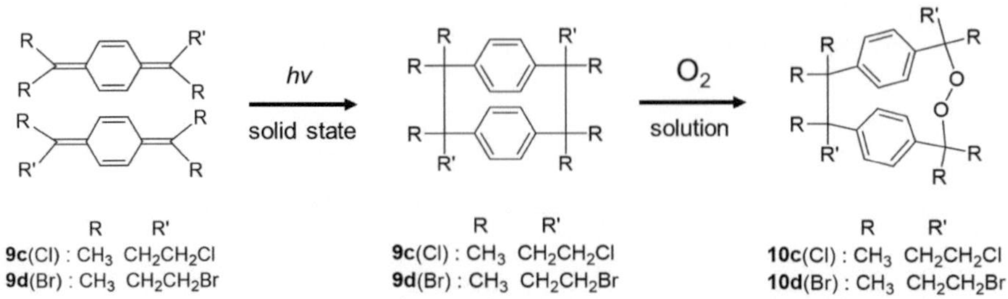

Fig. 1.6 Photocycloaddition dimerizations and reactions of bridge-substituted [2.2]paracyclophanes with molecular oxygen in solution

(**9d**(Br)) [19, 31]. And also, both cyclophanes of **9c**(Cl) and **9d**(Br) reacted slowly with molecular oxygen in solution to form peroxide bridge-substituted [2.2]paracyclophanes (**10c**(Cl) and **10d**(Br)) (Fig. 1.6).

1.4 Crystal Structures and Intermolecular Interaction Energies

1.4.1 Typical Crystal Structures

Three reaction behaviors (no reaction, dimerization, and polymerization) in the solid-state reactions of four-ester possessing *p*-quinodimethane derivatives are expected to be closely associated with the molecular arrangements in those crystals. Therefore, the crystal structures of four-ester possessing *p*-quinodimethane derivatives were determined by X-ray single-crystal analyses, and systematically compared in detail. Here, typical crystal structures and 1D column motifs are shown for **1a**(Me)-α [16], **1a**(Me)-β [16], **1b**(Et) [32], and **2c**(Cl)-β [21] in Fig. 1.7, where one 1D column is surrounded by six identical columns as a common formation of 1D column motifs. In cases of **1a**(Me)-α, **1a**(Me)-β, and **1b**(Et), planar *p*-quinodimethane moieties stack along an identical axis to give a single 1D column. In contrast, in the case of **2c**(Cl)-β, the moieties make dimers which stack along an identical axis to give double 1D columns [19].

1a(Me)-β (space group $P2_1/n$, Fig. 1.7b), **1c**(BrEt) (*P*-1), and **2a**(H) ($P2_1$) form very similar single 1D columns, and three of them undergo topochemical photopolymerizations. On the other hand, **1a**(Me)-α (*P*-1, Fig. 1.7a), **2b**(F) (*P*-1), and **2c**(Cl)-α (*P*-1) undergo no reactions and form single 1D columns which are considerably different from those of **1a**(Me)-β, **1c**(BrEt), and **2a**(H). The former three compounds form 1D columns with a steep inclination and the latter three ones do 1D columns with a gentle inclination, where identical molecules are extended in a front-to-back manner. While, **2c**(Cl)-β ($I4_1/a$, Fig. 1.7d) and **2d**(Br) ($P2_1/c$) make double 1D columns for topochemical dimerization reactions, where dimers are arranged with

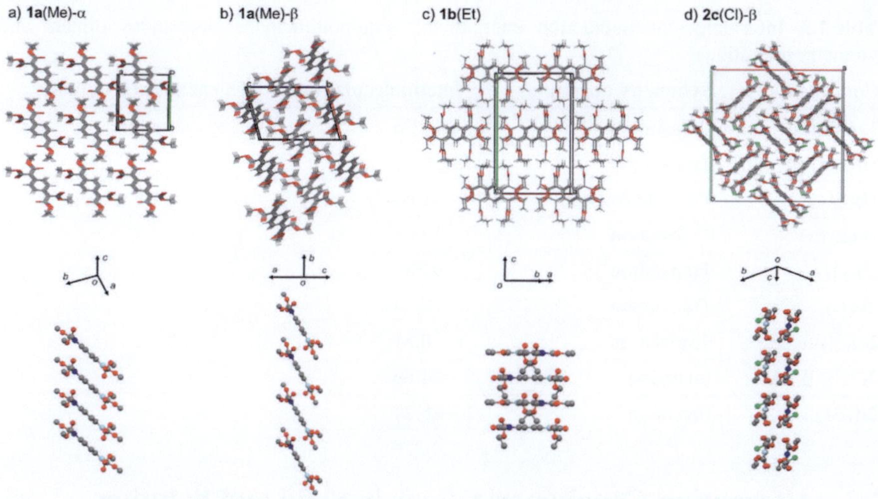

Fig. 1.7 Typical packing diagrams with stick models and 1D columns for crystals of **a 1a**(Me)-α, **b 1a**(Me)-β, **c 1b**(Et), and **d 2c**(Cl)-β. Atoms of carbon, oxygen, hydrogen, and chlorine are colored by gray, red, white, yellow-green, and green

contacts on long-sides of the molecules. The similarity of each 1D columnar motif in the crystals leads to similar reaction behaviors. The molecules in **1b**(Et) (*Iba*2, Fig. 1.7c) are stacked in a front-to-back manner perpendicular to quinoid rings of *p*-quinodimethane derivatives [32].

1.4.2 Intermolecular Interaction Energies

Intermolecular interaction energies among contacting molecules in crystals were evaluated by using the density functional theory (DFT) method for *p*-quinodimethane derivatives such as **1a**(Me)-α, **1a**(Me)-β, **1b**(Et), **1c**(BrEt), **2a**(H), **2b**(F), **2c**(Cl)-α, **2c**(Cl)-β, and **2d**(Br) [33]. The intermolecular interaction energies (kcal/mol) arising from dimer structures in crystals have the largest value among contacting molecules, due to the very close overlapping between two adjacent molecules in the crystals. The intermolecular interaction energies of the dimers were summarized in Table 1.3, together with symmetry operations such as translation operation, twofold rotation operation, and inversion operation.

1b(Et) has the largest negative intermolecular interaction energy in comparison with those of other crystals, due to the largest overlapping of π-electrons between two quinoid rings among these *p*-quinodimethane derivatives. Moreover, **2c**(Cl)-β and **2d**(Br), leading to substituted [2.2]paracyclophanes formation, have relatively large intermolecular interaction energies, indicative of the effective π-electrons overlapping between two quinoid rings.

Table 1.3 Intermolecular interaction energies of *p*-quinodimethane derivatives dimers and symmetry operations

Compounds	Symmetry operations	Intermolecular interaction energies (kcal/mol)
1a(Me)-α	Translation	−16.16
1a(Me)-β	Translation	−10.58
1b(Et)	Twofold rotation	−31.68
1c(BrEt)	Translation	−19.62
2a(H)	Translation	−9.82
2b(F)	Translation	−18.84
2c(Cl)-α	Translation	−20.31
2c(Cl)-β	Inversion	−24.66
2d(Br)	Inversion	−26.21

1.5 Alternating Copolymerizations in Solid and Solution States

1.5.1 Topochemical Alternating Copolymerization of p-Quinodimethane Derivatives with TCNQ in Solid States

As described previously, both **1a**(Me) and **1b**(Et) form charge-transfer complexes with the strong acceptor monomer TCNQ in acetonitrile, and furthermore, both of them copolymerize with TCNQ in acetonitrile to afford corresponding alternating copolymers [15, 34]. However, the conformational structures of the alternating copolymers of **1a**(Me) and **1b**(Et) with TCNQ were not determined because of their insolubilities in conventional organic solvents.

The **1a**(Me) formed a charge-transfer complex cocrystal with TCNQ by mixing both crystals with relatively high concentrations in acetonitrile and cooling at low temperatures. In the cocrystal, **1a**(Me) and TCNQ are stacked alternatingly with a torsion angle of 58° to form a 1D column as shown in Fig. 1.8 [18]. The distances between the centroids of **1a**(Me) and TCNQ are 4.13 Å and are arranged in a straight line along the *a-b* diagonal direction in the *ab* plane. Along these 1D columns, two different distances are present between the exomethylene carbon atoms: the shorter distance is 3.84 Å and the longer distance is 5.78 Å. And also, intermolecular interactions between **1a**(Me) and TCNQ in the 1D column involve CH_2–H···NC hydrogen bond (2.68 Å), π/π (3.38 Å), and electrostatic interaction of $C{=}O(\delta^-)\cdots(\delta^+)C(CN)_2$ (3.05 Å).

Polymerizations of the cocrystals proceeded through the single-crystal-to-single-crystal transformation, reacting at the exomethylene carbon atoms with the shorter distances without changing the centers of gravity, to yield an alternating copolymer with a *cis*-conformational structure. The molecules have a regular arrangement within

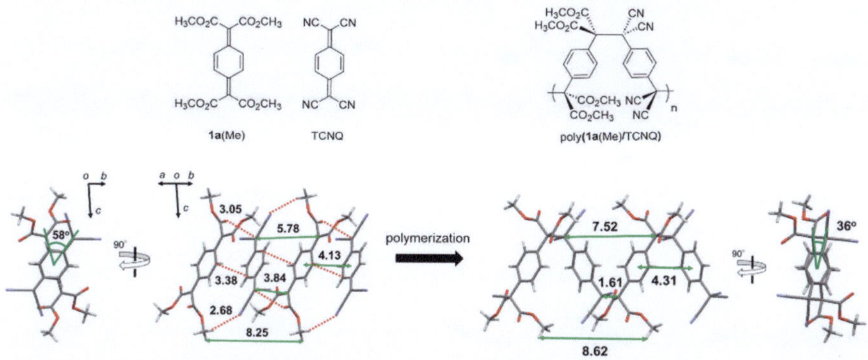

Fig. 1.8 Stacking model for the monomer cocrystal and the alternating copolymer crystal of **1a**(Me) with TCNQ. Intermolecular interactions between **1a**(Me) and TCNQ: CH_2–H···NC hydrogen bond (2.68 Å), π/π (3.38 Å), and electrostatic interaction of C=O(δ^-)···(δ^+)C(CN)$_2$ (3.05 Å), and the shorter distance and the longer distance between the exomethylene carbon atoms and the distance between centroids of **1a**(Me) and TCNQ

the crystals and are in an energetically stable state in comparison with those in solution, where the molecules frequently undergo collisions. Therefore, 1D columnar structures within the crystals are regarded to be molecular arrangements close to a kind of transition state. When the molecules are positioned at suitable distances and angles conductive to the polymerizations, it is considered that the polymerization reactions might proceed.

Previously, efforts have been made to obtain polymers with the linking of benzene rings by vinylene bridges (Fig. 1.9a) and cyclobutane bridges (Fig. 1.9b) through topochemical polymerizations of diacetylenes and diolefins in solid states, where specific phenyl-perfluorophenyl interaction stacking was applied to align molecules for reaction in the crystalline state [35, 36].

Photochemically induced polymerization occurs in both the 1:1 cocrystal composed of diphenylbutadiyne with decafluorodiphenylbutadiyne, and 2,3,4,5,6-pentafluorodiphenyldiacetylene for diacetylenes and in the 1:1 cocrystal composed of *trans,trans*-1,4-bis(2-phenylethenyl)-2,3,5,6-tetrafluorobenzene with *trans,trans*-1,4-bis(2-pentafluorophenylethenyl)benzene for diolefins to afford corresponding polymers, but the conformational structures of these polymers have not been confirmed yet because of the unavailability of single polymer crystals. The alternating copolymer of **1a**(Me) with TCNQ is the first example of a *cis*-conformational structure formation through an ethylene bridge (Fig. 1.9c), instead of cyclobutane bridges and vinylene bridges.

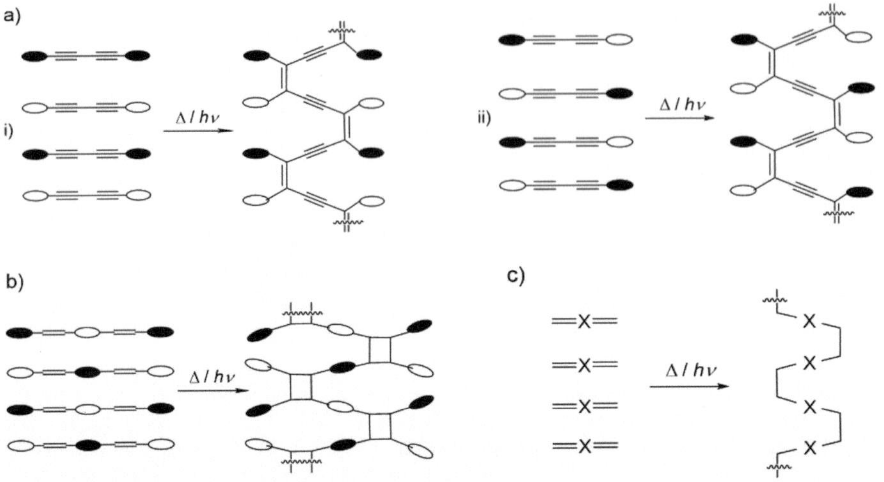

Fig. 1.9 Schematic representation of proposed molecular packing diagrams and polymers with a *cis*-conformation formed by topochemical polymerizations for **a** diacetylene, **b** diolefin, and **c** *p*-quinodimethane

1.5.2 Alternating Copolymerizations of p-Quinodimethane Derivatives with TCNQ in Solution

The next main interest is whether the *cis*-conformational structure of the copolymer crystals would be preserved in solution. As the alternating copolymer composed of TCNQ with **1a**(Me) was insoluble in conventional organic solvents, we attempted cocrystallizations of TCNQ with 7,7,8,8-tetrakis(alkoxycarbonyl)-*p*-quinodimethanes with linear and branched alkoxy groups such as ethoxy (**1b**(Et)), propoxy (**1c**(Pr)), isopropoxy (**1d**(iPr)), butoxy (**1e**(Bu)), isobutoxy (**1f**(iBu)), pentyloxy (**1g**(Pen), hexyloxy (**1h**(Hex)), and dodecyloxy (**1i**(Dod)). **1b**(Et), **1c**(Pr), **1e**(Bu), and **1g**(Pen) formed 1:1 charge-transfer complex cocrystals (**1b**(Et)·TCNQ, **1c**(Pr) TCNQ, **1e**(Bu) TCNQ, and **1g**(Pen) TCNQ) [19]. In contrast, TCNQ did not form charge-transfer complex cocrystals with 7,7,8,8-tetrakis(alkoxycarbonyl)-*p*-quinodimethanes with longer linear alkoxy groups such as hexyloxy (**1h**(Hex)) and dodecyloxy (**1i**(Dod)), and with branched alkoxy groups such as isopropoxy (**1d**(iPr)) and isobutoxy (**1f**(iBu)).

In the cocrystals, four-ester possessing *p*-quinodimethanes and TCNQ were stacked alternatingly to form 1D columns as well as the **1a**(Me) TCNQ cocrystal. For these cocrystals, thermal copolymerizations at 80 °C afforded alternating copolymers, insoluble in conventional organic solvents. But the copolymerization rates are significantly different among cocrystals and increased with an increase in the chain length of the linear alkoxy groups and/or a decrease in the reacting exomethylene carbon distances between 7,7,8,8-tetrakis(alkoxycarbonyl)-*p*-quinodimethanes and TCNQ, resulting in the fastest speed at a copolymerization of **1g**(Pen) and TCNQ

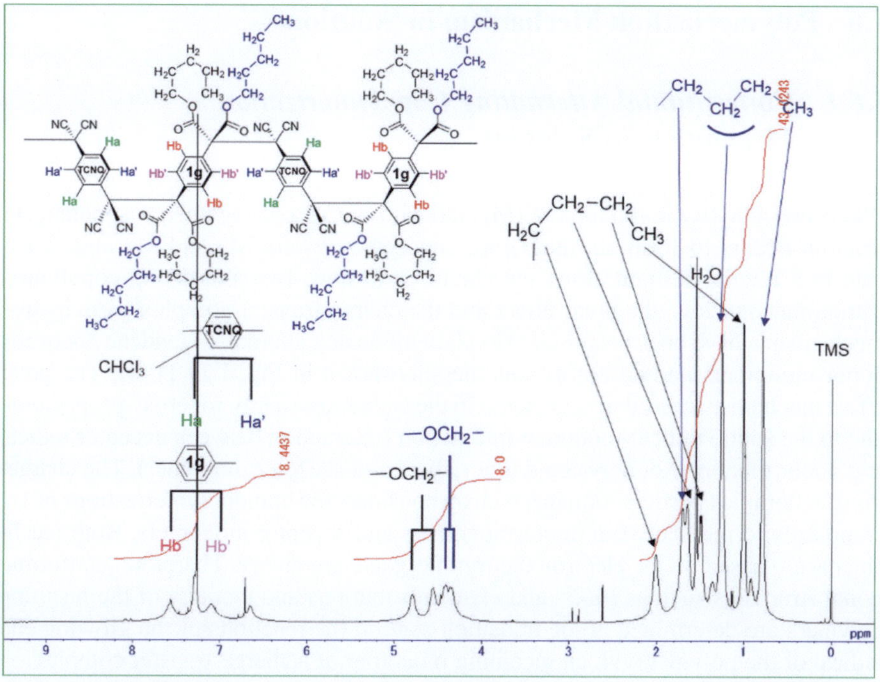

Fig. 1.10 ^{1}H NMR spectrum of an alternating copolymer of **1g**(Pen) with TCNQ in chloroform-*d*. Reprinted with permission from Ref. [19]. Copyright 2022, American Chemical Society

[19]. Interestingly, the solution copolymerization in acetonitrile of **1g**(Pen) with TCNQ afforded a soluble alternating copolymer in chloroform. The ^{1}H NMR spectrum of the alternating copolymer is shown in Fig. 1.10, where each proton can be reasonably assigned to the respective protons of the *cis*-conformational copolymer structure illustrated therein.

^{1}H, ^{13}C, and H–H 2D-COSY NMR spectral analyses of the soluble alternating copolymer and spectroscopic simulations proved that the *cis*-conformational structure is preserved in the solution state as well as in the solid state. The preservation of the *cis*-conformational structure is due to the steric hindrance effect on bulky substituents (alkoxycarbonyl groups) on the carbon–carbon single bonds. From these observations, it is strongly suggested that 1D columnar molecular assemblies, which resemble donor–acceptor alternating stacked arrangements in the crystal, are also formed in solution.

1.6 Polymerization Mechanism in Solution

1.6.1 Conventional Alternating Copolymerization Mechanism in Solution

When two vinyl-type monomers (M_1 and M_2) are mixed together, a spontaneous reaction occurs to form an alternating copolymer, where M_1 and M_2 units alternate in a regular fashion along the chain. Until now, two alternating copolymerization mechanisms, the polar effect and the charge-transfer complex participation mechanisms, have been proposed to explain the strong alternation tendency between donor monomers and acceptor monomers as shown in Fig. 1.11 [1–4]. The polar effect mechanism considers a decrease in the activation energy for cross-propagation due to the subtle radical-monomer interaction (interaction between acceptor radical and donor monomer or between donor radical and acceptor monomer). The charge-transfer complex participation mechanism considers the homopolymerizations of 1:1 complexes formed between donor monomers and acceptor monomers. Both mechanisms are based on the electron theory of organic chemistry. Therefore, conformational structures such as *trans*- and *cis*-conformations and tacticity of the resulting polymers are determined at the transition state in the reactions of the growing end radical of the polymer with an incoming monomer or a charge-transfer complex.

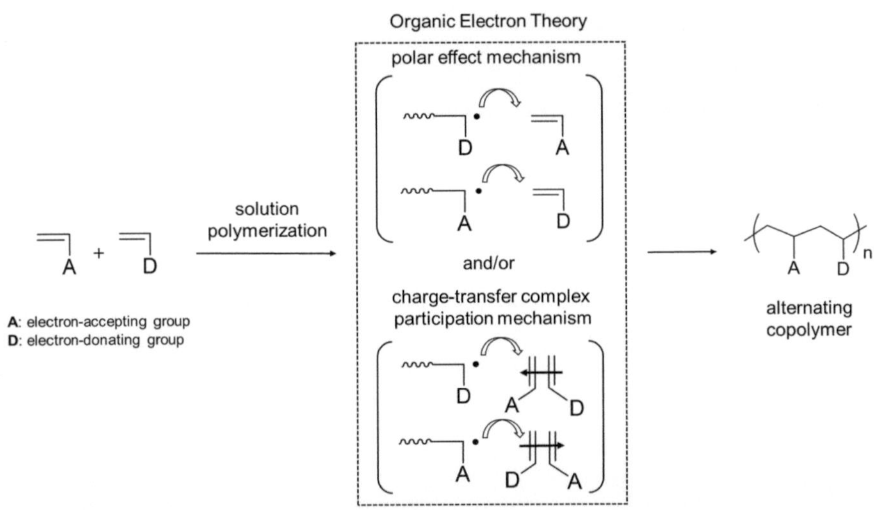

Fig. 1.11 Conventional polymerization mechanisms proposed for the alternating copolymerizations between acceptor monomers and donor monomers in vinyl-type monomer systems

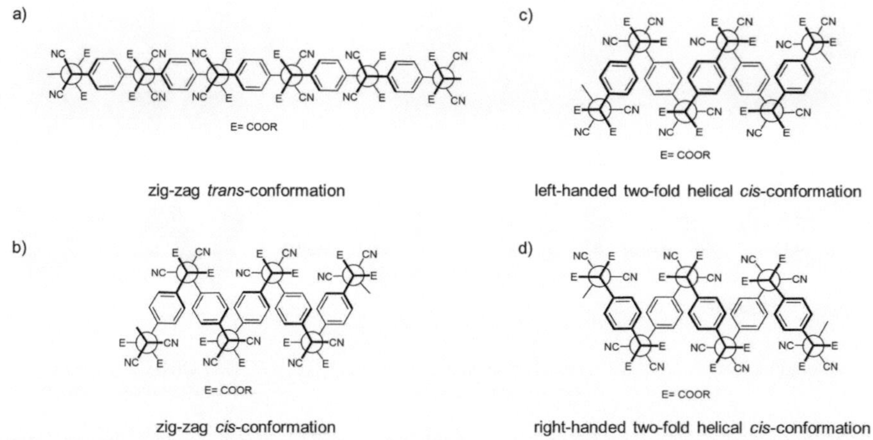

Fig. 1.12 Possible conformational structures for the alternating copolymers between TCNQ and four-ester possessing *p*-quinodimethane derivatives

1.6.2 Novel Polymerization Mechanism in Solution

Alternating copolymers of TCNQ with four-ester possessing *p*-quinodimethane derivatives are possible to exhibit four different conformational structures as follows: (a) zig-zag *trans*-conformation, (b) zig-zag *cis*-conformation, (c) left-handed twofold helical *cis*-conformation, and (d) right-handed twofold helical *cis*-conformation as shown in Fig. 1.12.

The zig-zag *trans*-conformational structure was not found for alternating copolymers at present, but it was observed in homopolymers of **1a**(Me) and **1m**(BrEt), and of typical monomers such as 1,3-diynes, 1,3,5-triynes, 1,3-dienes, and 1,3,5-trienes as shown in Fig. 1.5. The zig-zag *cis*-conformational structure was discovered for alternating copolymers of TCNQ with four-ester possessing *p*-quinodimethane derivatives. Unfortunately, the left- and right-handed twofold helical *cis*-conformational structures have not been found yet. In future, we expect that these conformations would be discovered.

On the basis of the zig-zag *cis*-conformational structure in solid and solution states for alternating copolymers, we propose a novel mechanism for alternating copolymerizations of TCNQ with the *p*-quinodimethane derivatives in solution. Through above-mentioned observations, we demonstrated similarities in the behaviors of alternating copolymerizations in solution and the solid states. The cocrystal structures analyses proved 1D columnar molecular assembly formation through the charge-transfer and π/π intermolecular interactions between the *p*-quinodimethane rings in neighboring molecules. In such 1D columnar molecular assemblies, molecules are positioned in favorable distances and angles for reactions, resulting in spontaneous alternating and stereospecific copolymerizations. In other words, 1D columnar

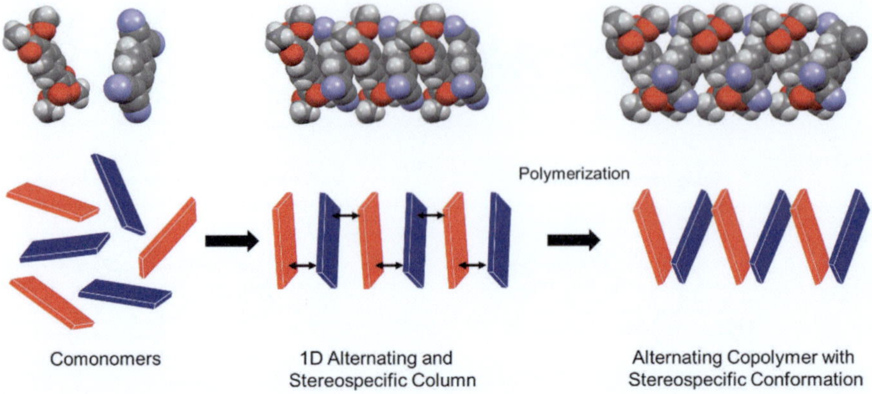

Fig. 1.13 Schematic diagram of the novel mechanism proposed for the formation of the alternating copolymer with a stereospecific conformation in solution

molecular assemblies play an important role in the novel mechanism for alternating copolymerizations in solution.

Recently, supramolecular polymerizations have been developed for the design of functional soft materials [37–40]. In such researches, complex molecular structures are designed to facilitate intermolecular interactions such as hydrogen bonding interaction, resulting in stable linear aggregates (supramolecular polymer) of small molecules like covalent linear polymers in dilute and concentrated solution as well as in the bulk. On the alternating copolymerization between **1a**(Me) and TCNQ, it seems likely that the 1D columnar molecular assembly is formed in solution even though monomer molecules with very simple structures. Here, Fig. 1.13 schematically shows a novel mechanism for alternating polymerizations between *p*-quinodimethane derivatives and TCNQ in solution. We propose the term "Donor–Acceptor Columnar Assembly Mechanism" for this polymerization mechanism.

1.7 Conclusions and Perspectives

We proposed a novel mechanism for alternating copolymerizations between the *p*-quinodimethane derivatives and the preservation of stereospecific conformations in solution. Alternating 1D columnar molecular assemblies were formed through a stereospecific stacking between donor monomers such as four-ester possessing *p*-quinodimethanes and an acceptor monomer such as TCNQ, resulting in very short distances between neighboring reaction centers. Such 1D columns cause the alternating and stererospecific copolymerizations in solution spontaneously or with radical initiators. We suggest that this polymerization mechanism might be applicable to the spontaneous alternating copolymerization between TCNQ and styrene in solution [9]. Furthermore, this concept would be extended to the known vinyl-based alternating copolymerizations in solution [1–4].

References

1. G. Odian, *Principles of Polymerization*, 4th edn., chapter 6 (Wiley-Interscience, Hoboken/New Jersey, 2004), pp. 464–543
2. J.M.G. Cowie, Principles of Alternating Copolymerization, Radical Initiated Alternating Copolymerization, in *Alternating Copolymers*. ed. by J.M.G. Cowie (Plenum Press, New York/N.Y., 1985), pp.1–74
3. M. Coote, T.P. Davis, Copolymerization Kinetics, Chapter 5, in *Handbook of Radical Polymerization*. ed. by K. Matyjaszewski, T.P. Davis (Wiley-Interscience, Hoboken/New Jersey, 2002), pp.263–300
4. G. Moad, D.H. Solomon, *The Chemistry of Radical Polymerization*, chapter 7 (Elsevier, Oxford/UK, 2006), pp. 333–412
5. T.L. Cairns, R.A. Carboni, D.D. Coffman, V.A. Engelhardt, R.E. Heckert, E.L. Little, E.G. McGeer, B.C. McKusick, W.J. Middleton, R.M. Scribner, C.W. Theobald, H.E. Winberg, Cyanocarbon chemistry. I. Preparation and reaction of tetracyanoethylene. J. Am. Chem. Soc. **80**, 2775–2778 (1958)
6. D.S. Acker, W.R. Hertler, Substituted quinodimethans. I. Preparation and chemistry of 7,7,8,8-tetracyanoquinodimethan. J. Am. Chem. Soc. **84**, 3370–3374 (1962)
7. W.R. Hertler, W. Mahler, L.R. Melby, J.S. Miller, R.E. Putscher, O.W. Webster, Cyanocarbons-their history from conducting to magnetic organic charge transfer salts. Mol. Cryst. Liq. Cryst. **171**, 205–216 (1989)
8. L.R. Melby, R.J. Harder, W.R. Hertler, W. Mahler, R.E. Benson, W.E. Mochel, Substituted quinodimethans. II. Anion-radical derivatives and complexes of 7,7,8,8-tetracyanoquinodimethan. J. Am. Chem. Soc. **84**, 3374–3387 (1962)
9. S. Iwatsuki, T. Itoh, K. Horiuchi, Spontaneous alternating copolymerization of tetracyanoquinodimethane with styrene. Macromolecules **11**, 497–500 (1978)
10. S. Iwatsuki, Polymerization of quinodimethane compounds, in *Advances Polymer Science*, vol. 58 (Springer-Verlag, Tokyo/Japan, 1984), pp. 93–120
11. T. Itoh, S. Iwatsuki, Polymerization behavior of electron-accepting substituted quinodimethanes. Macromol. Chem. Phys. **198**, 1997–2016 (1997)
12. T. Itoh, Polymerizations and polymers of quinonoid monomers. Prog. Polym. Sci. **21**, 1019–1059 (2001)
13. T. Itoh, Solution and solid-state polymerizations of substituted *p*-quinodimethanes and *p*-quinone methides. Polymer **46**, 6998–7017 (2005)
14. S. Iwatsuki, T. Itoh, T. Higuchi, K. Enomoto, Equilibrium polymerizations of 7,8-dibenzoyl-, 7,8-diacetyl, and 7,8-dibutoxycarbonyl-7,8-dicyanoquinodimethanes and their copolymerizations with styrene: a new concept on the mechanism of alternating copolymerization. Macromolecules **21**, 1571–1579 (1988)
15. S. Iwatsuki, T. Itoh, Amphoteric behavior of tetrakis(methoxycarbonyl)-quinodimethane in alternating copolymerizations with various comonomers. Macromolecules **13**, 983–989 (1980)
16. T. Itoh, S. Nomura, T. Uno, M. Kubo, K. Sada, M. Miyata, Topochemical polymerization of 7,7,8,8-tetrakis(methoxycarbonyl)quinodimethane. Angew. Chem. Int. Ed. **41**, 4306–4309 (2002)
17. T. Itoh, S. Nomura, H. Nakasho, T. Uno, M. Kubo, N. Tohnai, M. Miyata, Halogen bond effect for single-crystal-to-single-crystal transformation: topochemical polymerization of substituted quinodimrthane. Macromolecules **48**, 5450–5455 (2015)
18. T. Itoh, T. Suzuki, T. Uno, M. Kubo, N. Tohnai, M. Miyata, *Cis*-Specific topochemical alternating copolymerization of 7,7,8,8-tetrakis(methoxycarbonyl)quinodimethane with

7,7,8,8-tetracyanoquinodimethane in the solid state. Angew. Chem. Int. Ed. **50**, 2253–2256 (2011)

19. T. Itoh, T. Suzuki, F. Kondo, T. Uno, M. Kubo, N. Tohnai, F. Sanda, M. Miyata, Preservation of the conformational structures of single-polymer crystals in solution. Macromolecules **55**, 5685–5698 (2022)
20. T. Itoh, F. Kondo, T. Suzuki, K. Inayoshi, T. Uno, M. Kubo, N. Tohnai, M. Miyata, Elucidation of substituent-responsive reactivities via hierarchical and asymmetric assemblies in crystalline *p*-quinodimethane derivatives. Chem. Eur. J. e202301327 (2023)
21. S. Nomura, T. Itoh, H. Nakasho, T. Uno, M. Kubo, K. Sada, K. Inoue, M. Miyata, Crystal structures and topochemical polymerizations of 7,7,8,8-tetrakis(alkoxycarbonyl)quinodimethanes. J. Am. Chem. Soc. **126**, 2035–2041 (2004)
22. S. Iwatsuki, T. Itoh, I. Miyashita, Preparation and a new type of stable quinodimethanes with captodative substituents: 7,8-bis(ethylthio)-, 7,8-bis(phenylthio)-, and 7,8-bis(*tert*-butylthio)-7,8-dicyanoquinodimethanes. Macromolecules **21**, 557–560 (1988)
23. G. Wegner, Topochemishe Reaktionen von Monomeren mit konjugierten Dreifachbindungen/ Topochemical reactions of monomers with conjugated triple bonds. Z. Naturforsch, B. **24b**, 824–832 (1969)
24. G. Wegner, Solid-state polymerization mechanisms. Pure Appl. Chem. **49**, 443–454 (1977)
25. V. Enkelmann, Structural aspects of the topochemical polymerization of diacetylenes. Adv. Polym. Sci. **63**, 91–136 (1984)
26. J. Xiao, M. Yang, J.W. Lauher, F.W. Fowler, Supramolecular solution to a long-standing problem: the 1,6-polymerization of a triacetylene. Angew. Chem. Int. Ed. **39**, 2132–2135 (2000)
27. B. Tieke, Solid state polymerization of butadienes. Polymerization of long chain derivatives of sorbic and muconic acid. Colloid Polym. Sci. **263**, 965–972 (1985)
28. A. Matsumoto, T. Matsumura, S. Aoki, Stereospecific polymerization of diethyl (Z, Z)-hexa-2,4-dienedioate in the crystalline state. J. Chem. Soc. Chem. Commun. **23**, 1389–1390 (1994)
29. A. Matsumoto, K. Sada, K. Tashiro, M. Miyata, T. Tsubouchi, T. Tanaka, T. Odani, S. Nakagawa, K. Inoue, T. Tanaka, S. Saragai, S. Nakamoto, Reaction principles and crystal structure design for the topochemical polymerization of 1,3-dienes. Angew. Chem. Int. Ed. **41**, 2502–2505 (2002)
30. T. Hoang, J.W. Lauher, F.W. Fowler, The topochemical 1,6-polymerization of a triene. J. Am. Chem. Soc. **124**, 10656–10657 (2002)
31. T. Itoh, F. Kondo, T. Uno, M. Kubo, N. Tohnai, M. Miyata, Exclusive formation of bridge-substituted [2.2]paracyclophane by topochemical photocycloaddition reaction of unsymmetrical substituted *p*-quinodimethane. Cryst. Growth Des. **17**, 3606–3610 (2017)
32. S. Nomura, T. Itoh, M. Ohtake, T. Uno, M. Kubo, A. Kajiwara, K. Sada, M. Miyata, Polymerization by insertion of molecular oxygen into crystals of 7,7,8,8-tetrakis(ethoxycarbonyl)quinodimethane. Angew. Chem. Int. Ed. **42**, 5468–5472 (2003)
33. M. Miyata, S. Tsuzuki, in *Supramolecular, hierarchical, and energetical interpretation of organic crystals: generation of supramolecular chirality in assemblies of achiral molecules*, eds. by M. Sakamoto, H. Uekusa Advanced in Organic Crystal Chemistry, chapter 7 (Springer Nature Singapore Pte Ltd., 2020)
34. S. Iwatsuki, T. Itoh. I. Yokotani, Amphoteric polymerization behavior of 7,7,8,8-tetrakis(ethoxycarbonyl)quinodimethane in its alternating copolymerizations. Macromolecules **16**, 1817–1823 (1983)
35. G.W. Coates, A.R. Dunn, L.M. Henling, D.A. Dougherty, R.H. Grubbs, Phenyl-perfluorophenyl stacking interactions: a new strategy for supermolecule construction. Angew. Chem. Int. Ed. Engl. **36**, 248–251 (1997)
36. G.W. Coates, A.R. Dunn, L.M. Henling, J.W. Ziller, E.B. Lobkovsky, R.H. Grubbs, Phenyl-perfluorophenyl stacking interactions: topochemical [2+2] photodimerization and photopolymerization of olefinic compounds. J. Am. Chem. Soc. **120**, 3641–3649 (1998)
37. J.B. Gilory, T. Gädt, G.R. Whittell, L. Chabanne, J.M. Mitcheles, R.M. Richardson, M.A. Winnik, I. Manners, Monodisperse cylindrical micelles by crystallization-driven living self-assembly. Nat. Chem. **2**, 566–570 (2010)

38. W. Zhang, W. Jin, T. Fukushima, A. Saeki, T. Aida, Supramolecular linear heterojunction composed of graphite-like semiconducting nanotubular segments. Science **224**, 340–343 (2011)
39. S. Ogi, K. Sugiyasu, S. Manna, S. Samitsu, M. Takeuchi, Living supramolecular polymerization realized through a biomimetic approach. Nat. Chem. **6**, 188–195 (2014)
40. J. Kang, D. Miyajima, T. Mori, Y. Inoue, Y. Itoh, T. Aida, A rational strategy for the realization of chain-growth supramolecular polymerization. Science **347**, 646–651 (2015)

Chapter 2
Cooperative Photochemical Reactions in Organic Molecular Crystals

Daichi Kitagawa and Seiya Kobatake

Abstract The understanding of the reaction kinetics in organic molecular crystals is very important to design stimuli-responsive organic crystals that exhibit desired functions at arbitrary times. However, there are cooperative structural change processes involving multiple molecules due to intermolecular interactions, which makes the classical kinetic analysis difficult. The Johnson–Mehl–Avrami-Kolmogorov equation quantitatively evaluates cooperative nature of reactions, crucial for understanding kinetics in the crystalline state. However, it lacks insights into elementary reaction processes. Hence, the extended Finke-Watzky model, we have developed, is introduced for more detailed analysis of photoreactions. This model provides insights into elementary reaction processes, offering dynamic quantum yields of photoreactions. Furthermore, the progressive behavior of photoreactions on optical length scale is introduced. While photodimerization in 9-methylanthracene crystals occurs homogeneously on optical length scale, photopolymerization in 2,5-distyrylpyradine crystals exhibits edge-to-center propagation, influenced by cooperative reactions and surface effects. These findings enhance understanding of photochemical reaction kinetics in crystalline materials.

Keywords Molecular crystals · Cooperative reactions · Johnson-Mehl-Avrami-Kolmogorov equation · Finke-Watzky model

2.1 Introduction

Organic molecular crystals have been shed light on as next-generation functional materials because they can show various interesting properties such as elasticity, plasticity, shape-memory, self-healing, and so on [1]. In particular, molecular crystals

D. Kitagawa (✉) · S. Kobatake
Department of Chemistry and Bioengineering, Graduate School of Engineering, Osaka Metropolitan University, 3-3-138 Sugimoto, Sumiyoshi-Ku, Osaka 558-8585, Japan
e-mail: kitagawa@omu.ac.jp

S. Kobatake
e-mail: kobatake@omu.ac.jp

S. Kobatake and H. Uekusa (eds.), *Advances in Organic Crystal Chemistry*,
https://doi.org/10.1007/978-981-96-4479-7_2

whose physicochemical properties change in response to external stimuli (stimuli-responsive molecular crystals) have been widely studied in recent years toward various applications involving photoactuators, electronic elements, optical waveguides, and gas storage/release materials [2–4]. Changes in the physical properties of stimulus-responsive crystalline materials are brought about by changes in the molecular structure or crystalline structure due to external stimuli. Normally, this structural change can be accurately detected by single-crystal X-ray structural analysis, which is the greatest advantage of organic crystalline materials, and it has given a clear answer to the question why the changes in the physical properties take place [5, 6]. However, this method only captures the "static" state before and after a structural change, and cannot capture the "dynamic" property changes during the structural change. In order to design stimuli-responsive organic crystals that exhibit desired functions at arbitrary times, it is important to understand the structural changes based on kinetics. Here, we should notice that in crystals, there are cooperative structural change processes involving multiple molecules due to intermolecular interactions, which are different from those in dilute solution, where molecules exist independently and do not interact each other, as exemplified by structural phase transitions [7, 8]. A proper understanding of kinetics in the crystalline state is a very important issue not only in fundamental research but also in applied research.

Therefore, this chapter focuses on the cooperative reactions in organic molecular crystals, particularly photoreactions. First, we introduce the Johnson–Mehl–Avrami-Kolmogorov (JMAK) equation, which is often used to quantitatively evaluate the cooperative nature of reactions. We then introduce the extended Finke-Watzky (FW) model that we have developed for more detailed analysis in terms of photochemistry. In addition, our recent studies on the progressive behavior of photoreactions in single crystals on the optical length scale (the length scale visible by a conventional optical microscope) are also presented.

2.2 Johnson–Mehl–Avrami-Kolmogorov Equation

The kinetics of photoreactions in dilute solution are well established and follow an exponential-like curve without an induction period, whether unimolecular or bimolecular reactions (Fig. 2.1a). In these cases, the photoreactions can be quantitatively analyzed based on the classical kinetic models. On the other hand, photoreactions in crystals often do not obey the simple exponential-like curve but follow a sigmoidal curve with an induction period (Fig. 2.1b). This is due to the existence of cooperative photoreaction processes, in which the reaction of molecules is facilitated by neighbor reacted molecules (called autocatalytic reaction). This effect makes quantitative analysis of the reaction difficult. To quantitatively evaluate the cooperativity of reactions, many researchers have used the JMAK equation. The JMAK equation (Eq. 2.1) originally appeared around 1940 to describe transformations by nucleation and growth of new phase domains.

$$f = 1 - e^{-kt^n} \tag{2.1}$$

Equation 2.1 can also be written as Eq. 2.2 by setting $k = k'^n$, which is another form commonly used in the analysis.

$$f = 1 - e^{-(k't)^n} \tag{2.2}$$

where f is the conversion ratio expressed by [reacted molecules]/[total molecules], k and k' are the rate constants of reaction, t is the irradiation time, and n is the dimensionality of reaction (also called the Avrami exponent). In addition, $n-1$ indicates the dimension of product phase growth. That is, when $n = 4$, 3, and 2, the product phase growth proceeds three-dimensionally, two-dimensionally, and one-dimensionally, respectively. In these cases, the progress of the reaction follows a sigmoidal curve, with a higher value of n indicating a higher degree of cooperativity in the reaction. When $n = 1$, it indicates that the reaction follows simple first-order kinetics. Furthermore, when $n < 1$, the reaction becomes slower as the conversion ratio increases. This case is called autoinhibition, and the progress of the reaction stops at a certain conversion ratio. Figure 2.2 shows the change in the conversion

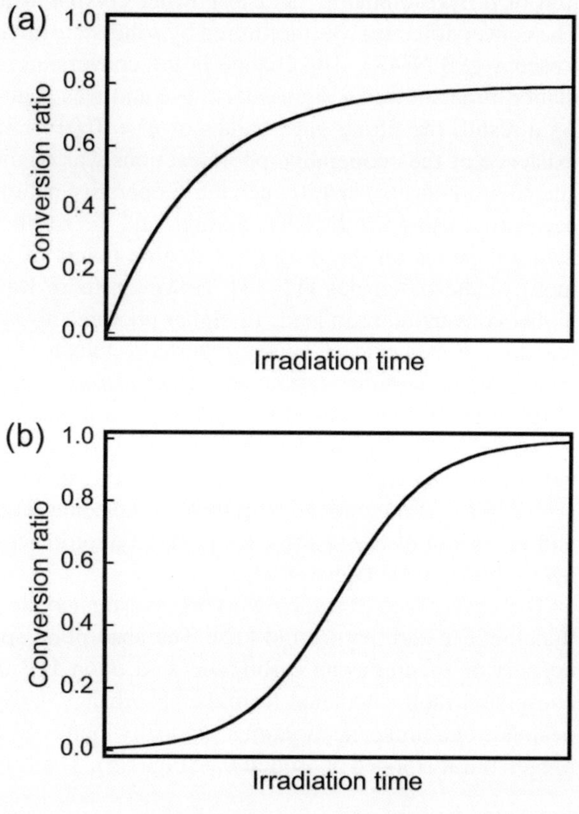

Fig. 2.1 Schematic diagrams of photochemical reactions **a** in solution and **b** in specific crystals. Adapted with permission from Ref. [9]

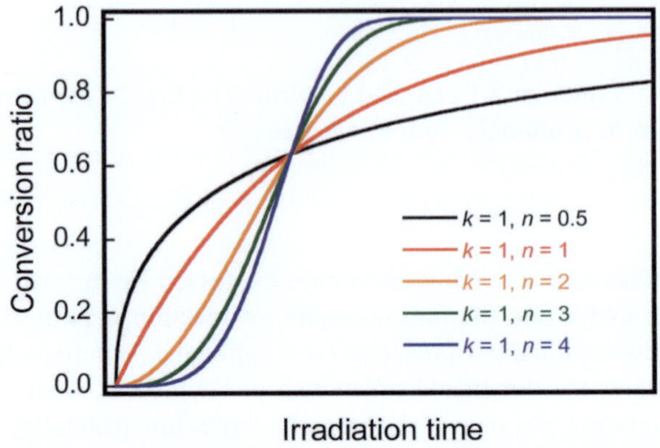

Fig. 2.2 Change in the conversion ratio relative to the irradiation time with various *n* values using Eq. 2.1. Adapted with permission from Ref. [9]

ratio with various *n* values [9]. The shape of the curve strongly depends on the value of *n*, and it becomes more sigmoidal when the value of *n* is larger. Thus, applying the JMAK equation enables the quantitative evaluation of the cooperative nature of the reactions in the crystalline phase.

For instance, Bertmer et al. quantitatively analyzed the photodimerization reaction of *α-trans*-cinnamic acid in powder crystals using the JMAK equation [10]. The conversion ratio was monitored by solid-state nuclear magnetic resonance spectroscopy (SS NMR). The change in the conversion ratio relative to the photoirradiation time showed a sigmoidal curve and was analyzed using Eq. 2.2 (Fig. 2.3). As a result, the fitting gave values of $k' = 0.019\ \text{min}^{-1}$ and $n = 1.66$. Thus, the existence of the cooperative photoreactions was quantitatively confirmed. Bertmer and co-workers also investigated the cooperative nature of other *trans*-cinnamic acid derivatives using the JMAK equation, and the results implied that the cooperative nature depends on the overlap of double bonds of the reactants and the packing motif of the molecules [11–13]. However, there was no clear answer as to why higher conversion ratio leads to higher photoreactivity (the reason for autocatalytic reaction). Benedict and Coppens reported a study to answer this question. They investigated the photodimerization reaction of *α-trans*-cinnamic acid using single-crystal X-ray structural analysis [14]. Interestingly, it was revealed that when the conversion ratio is 0.79, the overlap of the reactive double bonds becomes larger compared with that for the initial state (Fig. 2.4). This is the direct evidence that the molecular arrangement of the remaining unreacted molecules is affected by the photoproducts, and the origin of the positive cooperative photoreaction (autocatalytic reaction) of *trans*-cinnamic acid derivatives.

Moreover, Techert and co-workers investigated the photodimerization reaction of 9-anthracene carboxylic acid [15]. The absorption spectral changes in the powder crystals of 9-anthracene carboxylic acid upon UV irradiation were tracked. The conversion ratio calculated from the absorbance decay corresponding to the disappearance of anthracene monomer gradually increased as the irradiation time became longer, but it stopped at around 0.7 (Fig. 2.5). The fitting using the JMAK equation

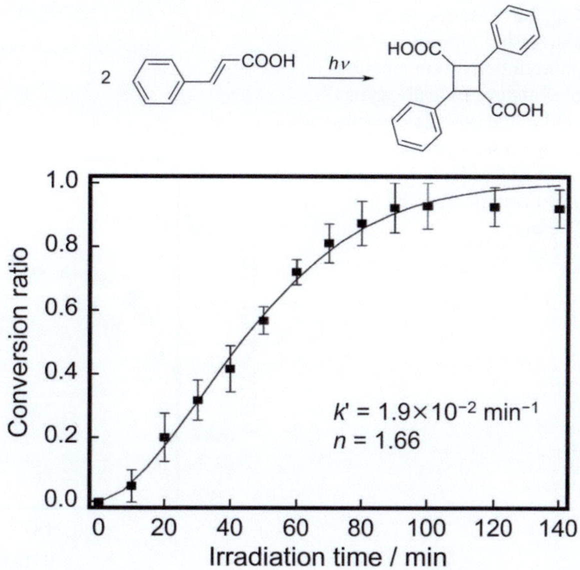

Fig. 2.3 Change in the conversion ratio of the photodimerization reaction of *α-trans*-cinnamic acid in the powder crystals determined by SS NMR spectroscopy. Adapted with permission from Ref. [10]. Copyright 2006 American Chemical Society

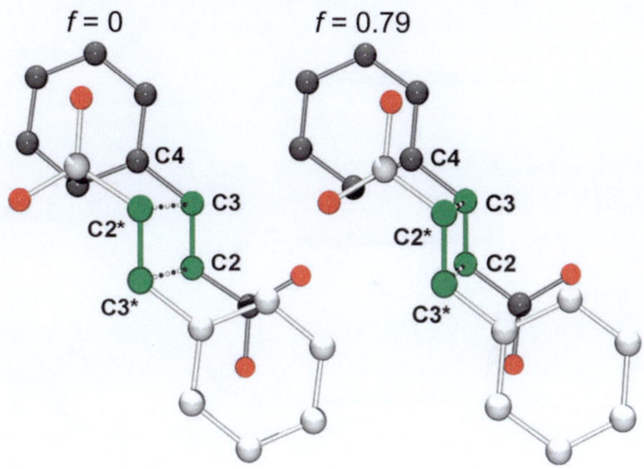

Fig. 2.4 Molecular arrangement of *α-trans*-cinnamic acid in the crystal when the conversion ratio (f) is 0 and 0.79, respectively. Adapted with permission from Ref. [14]. Copyright 2009 American Chemical Society

gave the value $n = 0.55$, indicating the negative cooperative photoreaction (autoinhibition). In other words, the photoreaction proceeds in a self-inhibiting manner. From X-ray structural analysis, it was revealed that the anthracene monomers do not exist in pairs but are stacked and aligned uniaxially as shown in Fig. 2.6. In this case, as the reaction proceeds, some molecules are removed from the dimerization reaction due to statistical considerations. This is one of the plausible mechanisms to explain the negative cooperativity ($n < 1$).

Turowska-Tyrk and co-workers also reported the autoinhibition of the photochemical Norrish-Yang reaction in single crystals of 1,3,5,7-tetraazatricyclo[3.3.1.13,7]decane 4-(2,4,6-triisopropylbenzoyl)benzoate [16].

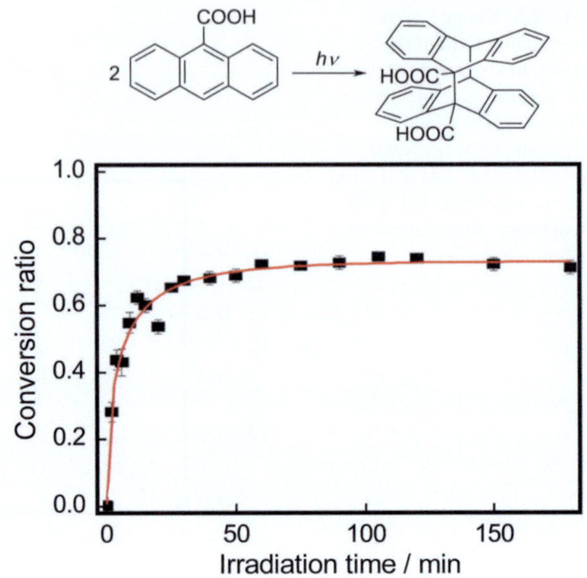

Fig. 2.5 Change in the conversion ratio of the photodimerization reaction of 9-anthracenecalboxylic acid in the powder crystals. Adapted with permission from Ref. [15]. Copyright 2010 American Chemical Society

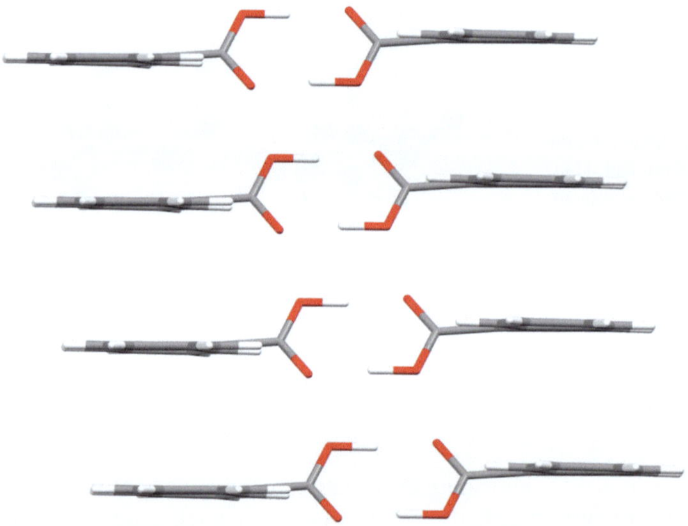

Fig. 2.6 Packing motif of 9-anthracenecalboxylic acid in the crystal (CCDC number: 128359)

In this case, the Norrish-Yang reaction becomes less likely to proceed as the reaction progresses (Fig. 2.7). To analyze the data, they used the following Eq. 2.3 that is the logarithmic form of the JMAK equation derived from Eq. 2.2, known as the Avrami plot.

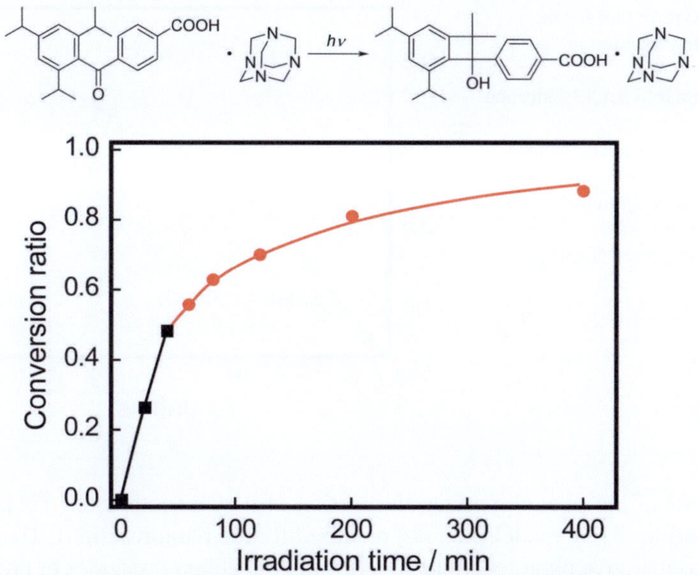

Fig. 2.7 Change in the conversion ratio of the Norrish-Yang reaction of 1,3,5,7-tetraazatricyclo[3.3.1.13,7]decane 4-(2,4,6-triisopropylbenzoyl)benzoate in single crystal. Adapted with permission from Ref. [16]. Copyright 2020 American Chemical Society

$$\ln(-\ln(1-f)) = n\ln(t) + n\ln(k') \tag{2.3}$$

The linear relationship between $\ln(-\ln(1-f))$ and $\ln(t)$ provides information about the value of n. As a result, it was elucidated that the reaction initially proceeds with $n = 1.08$, but after the conversion ratio reaches 50%, the value of n decreases 0.52 (Fig. 2.8). The reason for the autoinhibition is not clear, but the changes in molecular orientation in the crystal as the reaction proceeds would be the most plausible explanation.

2.3 Extended Finke-Watzky Model

As described above, the JMAK equation is very simple and convenient to know the cooperativity of photoreactions in a simple manner. However, in fact, photoreactions involve various processes such as excitation, deactivation, and reaction. The JMAK equation cannot provide useful information on these processes. A new approach from the viewpoint of photochemistry is required. Therefore, we have focused on the FW model that was originally developed to describe the growth of metal nanoparticles [17–20]. As the FW model consists of the rate constants of the elementary processes, it is expected that the detailed information about each reaction process can be obtained from analyses. Venuti and co-workers first used the FW model for

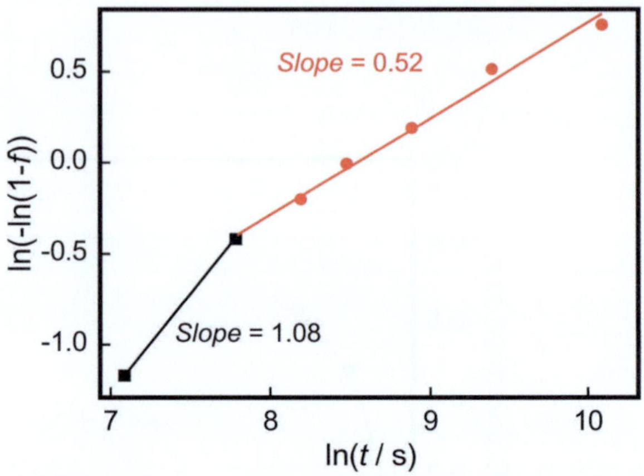

Fig. 2.8 Avrami plot for the Norrish-Yang reaction of 1,3,5,7-tetraazatricyclo[3.3.1.13,7]decane 4-(2,4,6-triisopropylbenzoyl)benzoate in single crystal. Adapted with permission from Ref. [16]. Copyright 2020 American Chemical Society

the analysis of the photodimerization reaction of 9-cyanoanthracene [21], but they used the original FW model (i.e., not modified for the photoreaction). Therefore, to analyze cooperative photoreactions in terms of elementary processes of photochemical reactions, we proposed an extended FW model based on the following reaction scheme (Scheme 2.1) [22].

Where A, A*, and B represent the monomer pair in the ground state, the monomer pair in the excited state, and the photodimer, respectively. σ is the absorption cross section at the excitation wavelength (cm^2 molecule^{-1}), I is the monochromatic light intensity for the photoreaction (photons cm^{-2} s^{-1}), and k_{ex} is the relaxation rate constant of A to the ground state (s^{-1}). k_1 and k_2 are the rate constant (nm^3 molecule^{-1} s^{-1}) of the reaction that is affected by A and B, respectively. Based on these four elementary reaction processes, the following equations can be derived.

$$\frac{\mathrm{d}[\mathrm{A}]}{\mathrm{d}t} = -\sigma I[\mathrm{A}] + k_{\mathrm{ex}}\left[\mathrm{A}^*\right] \tag{2.4}$$

$$\frac{\mathrm{d}\left[\mathrm{A}^*\right]}{\mathrm{d}t} = \sigma I[\mathrm{A}] - k_{\mathrm{ex}}\left[\mathrm{A}^*\right] - k_1\left[\mathrm{A}^*\right][\mathrm{A}] - k_2\left[\mathrm{A}^*\right][\mathrm{B}] \tag{2.5}$$

Scheme 2.1 Elementary reaction processes for the extended FW model

$$\mathrm{A} \xrightarrow{\sigma I} \mathrm{A}^*$$

$$\mathrm{A}^* \xrightarrow{k_{ex}} \mathrm{A}$$

$$\mathrm{A}^* + \mathrm{A} \xrightarrow{k_1} \mathrm{B} + \mathrm{A}$$

$$\mathrm{A}^* + \mathrm{B} \xrightarrow{k_2} \mathrm{B} + \mathrm{B}$$

$$\frac{d[B]}{dt} = k_1[A^*][A] + k_2[A^*][B] \tag{2.6}$$

where [A] and [B] are the concentration (molecule nm^{-3}) of A and B, and [B] = $[A]_0$ − [A] ($[A]_0$ is the initial monomer pair concentration). Using Eqs. 2.4–2.6, Lambert–Beer's law, steady-state approximation, and photodynamic factor, we derived Eq. (2.7) for fitting the absorbance decay of anthracene monomer due to photodimerization reaction.

$$\frac{d\text{Abs}}{dt} = -\frac{\sigma I_0}{\ln(10)}\left(1 - 10^{-\text{Abs}}\right)\left(\frac{\frac{\text{Abs}}{\text{Abs}_0} + \beta\left(1 - \frac{\text{Abs}}{\text{Abs}_0}\right)}{\alpha + \frac{\text{Abs}}{\text{Abs}_0} + \beta\left(1 - \frac{\text{Abs}}{\text{Abs}_0}\right)}\right) \tag{2.7}$$

where $\alpha = k_{ex}/k_1[A]_0$ and $\beta = k_2/k_1$. Fitting the absorbance decay using Eq. 2.7 yields the constants α and β. With the values of α and β, we can know the intrinsic reaction rate constants in Scheme 2.1. This is an advantage of the extended FW model compared with the JMAK equation. Moreover, to make the results interpretable using standard photochemistry concepts, we can define two effective quantum yields for the k_1 and k_2 photochemical reaction pathways as shown in Eqs. 2.8–2.10,

$$\Phi_1 = \frac{k_1[A]}{k_{ex} + k_1[A] + k_2[B]} = \frac{1 - f}{\alpha + 1 - f + \beta f} \tag{2.8}$$

$$\Phi_2 = \frac{k_2[B]}{k_{ex} + k_1[A] + k_2[B]} = \frac{\beta f}{\alpha + 1 - f + \beta f} \tag{2.9}$$

$$\Phi_{1+2} = \Phi_1 + \Phi_2 = \frac{1 - f + \beta f}{\alpha + 1 - f + \beta f} \tag{2.10}$$

where $f = ([A]_0 - [A])/[A]_0 = [B]/[A]_0$ is the fraction of crystal converted to photodimer. Note that whereas the commonly used quantum yield is constant, Φ_1 and Φ_2 are not constant but depend on the dimer fraction f. Φ_{1+2} is the total quantum yield. Moreover, β indicates how the chemical identity of its neighbor affects the reaction rate of A: When $\beta < 1$, it indicates that A is less likely to react if it has a neighboring B (autoinhibition). In contrast, when $\beta > 1$, it means that A is more likely to react (autocatalysis). In the case of $\beta = 1$, Eq. 2.7 yields a standard single exponential decay, while $\beta > 1$ yields a sigmoidal curve. Thus, with the extended FW model, the dynamic reaction quantum yield, which is a physically meaningful value, can be obtained.

To verify the usefulness of the extended FW model, we investigated the photodimerization reaction of 9-methylanthracene (9MA) in the crystalline phase [22]. We prepared both tiny single crystals and polycrystalline films of 9MA and measured the absorbance decay of the monomer absorption signal corresponding to the conversion ratio from monomer to dimer. As can be seen in Fig. 2.9a, b, it was found that the absorbance decay in the single crystal and the polycrystalline film were both sigmoidal curves, but it is much pronounced for polycrystalline film. Fitting these absorbance decay profiles by Eq. 2.7 gave $\alpha = 1.7$ and $\beta = 14$ for the single crystal and $\alpha = 13$ and $\beta = 24$ for the polycrystalline film, respectively. From these constants, we can know the change in the total quantum yield during the reaction as shown in Fig. 2.9c, d. The initial quantum yield in the polycrystalline film is much smaller than that in the single crystal, resulting in the much pronounced sigmoidal behavior in the polycrystalline film than in the single crystal. The lower initial quantum yield in the polycrystalline film can be ascribed to the existence of the deactivation process of the excited state by crystalline defects. Thus, we have succeeded in the quantitative understanding of the photochemical reaction kinetics in the crystalline state involving the cooperative reaction processes by the extended FW model.

Recently, Kearns and D'Alessandro investigated a thermally reversible [2+2] photocyclization reaction in the metal–organic frameworks using Raman spectroscopy and applied both the JMAK and the original FW models for the analysis [23]. They found that significant differences in the goodness-of-fit dependent on the reaction rate are observed between the two models and concluded that the FW model provides a more accurate fit for the data. The debate over which model is more suitable, JMAK or FW models, will continue for a long time. Just as the Arrhenius and

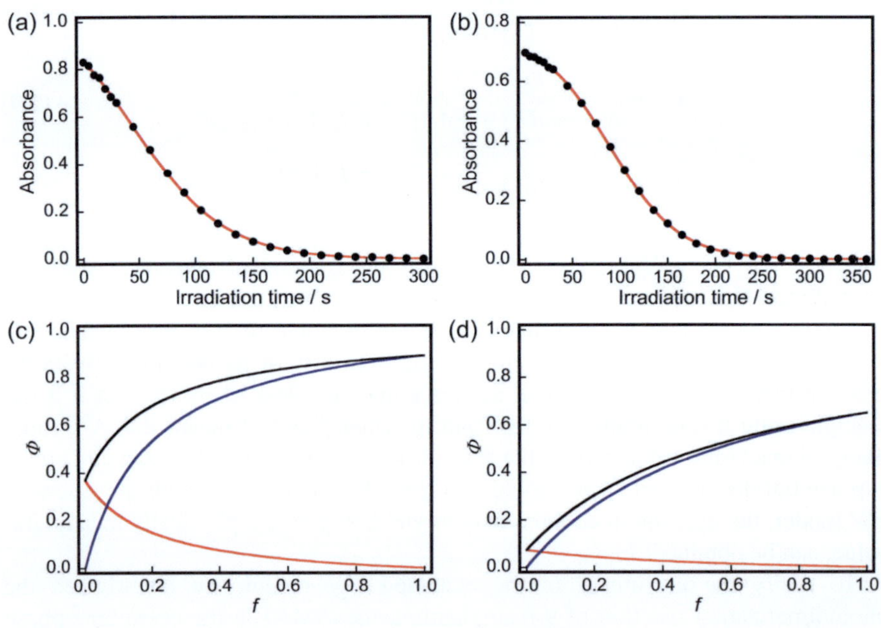

Fig. 2.9 Experimental absorbance decay (black dots) and the fitting (red line) by Eq. 2.7 for **a** the single crystal and **b** the polycrystalline film of 9MA. Change in the quantum yields (Φ_1 (red), Φ_2 (blue), and Φ_{1+2} (black)) relative to the conversion ratio (f) for **c** the single crystal and **d** the polycrystalline film of 9MA. Adapted with permission from Ref. [22]. Copyright 2022 Wiley

Eyring equations are both used when discussing the temperature dependence of reaction rate constants, the JMAK and FW models will both be used as the representative reaction models in solid states.

2.4 Progressive Behavior of Photoreactions on Optical Length Scale

As described before, the cooperative reaction processes are very important when discussing the photochemical reaction kinetics in the crystalline state and originate from the dynamic change in the local environment of individual molecules on the molecular scale. Here, we got interested in how the cooperative reaction processes affect the progressive behavior of photoreactions on optical length scale. In other words, the relationship between microscopic events on the molecular scale and macroscopic events on the optical length scale is not clear. To address this question, we started to directly observe the progress of the photoreactions in high-quality single crystals. Figure 2.10 shows the interference color change of the single crystal of 9MA during the photodimerization reaction, which originates from the change in birefringence associated with photodimerization reaction [22]. The photodimerization reaction of 9MA proceeds with a cooperative reaction process (autocatalytic) on the molecular scale, as described in the previous section. However, on the optical length scale, the interference color changed homogeneously, which means that the photodimerization reaction took place simultaneously over the entire crystal. This result indicates that the scale of the nucleation and growth in the cooperative reaction process of 9MA, in which the photodimerization reaction is facilitated by its neighbor photodimers, is much smaller than optical length scale.

In contrast to the case of 9MA, we found that the photoreaction of 2,5-distyrylpyrazine (DSP) in the single crystal took place heterogeneously on optical length scale [24]. Figure 2.11 shows the change in color during photopolymerization reaction in a single crystal of DSP. The yellow color disappeared from the edge of the crystal, indicating that the photopolymerization reaction proceeded predominantly from the edge of the crystal. We named this unique phenomenon "edge-to-center propagation of photochemical reaction." From the detailed analysis, we concluded that this phenomenon is ascribed to the cooperative reaction process and the surface effect, in which the photoreactivity at the edge is significantly higher than the bulk. Thus, we serendipitously found that novel progressive behavior of photochemical reaction in single crystals on optical length scale. We are now trying to build up the simulation program that reproduces the edge-to-center propagation of photochemical reaction and working to better understand the mechanism.

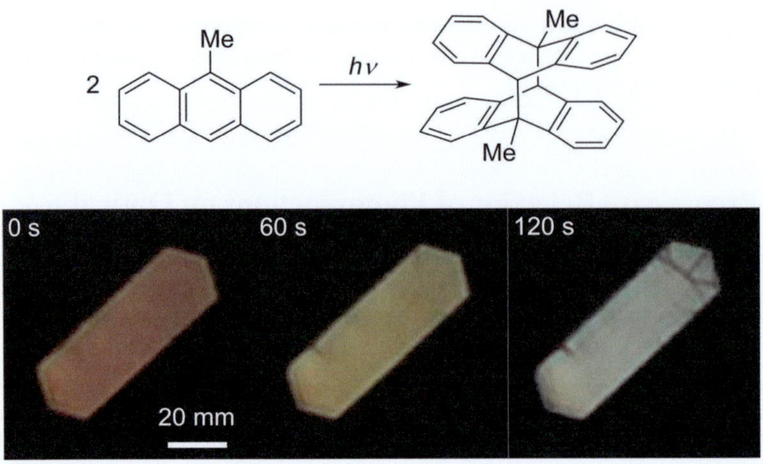

Fig. 2.10 Interference color change of a single crystal of 9MA during the photodimerization reaction. Adapted with permission from Ref. [22]. Copyright 2022 Wiley

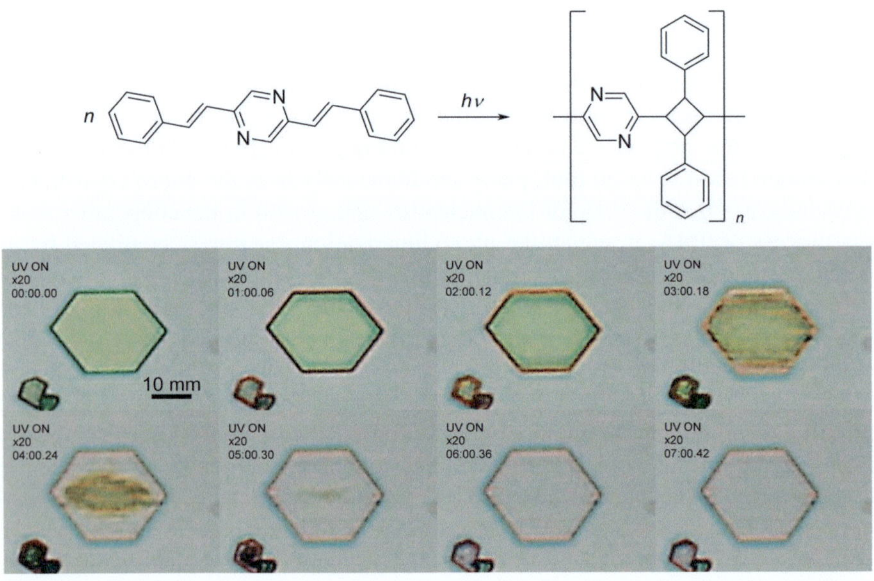

Fig. 2.11 Color change of a single crystal of DSP during the photopolymerization reaction. Adapted with permission from Ref. [24]. Copyright 2022 Wiley

2.5 Summary

This chapter delves into the complex kinetics of photoreactions in organic molecular crystals, focusing particularly on cooperative reactions. While traditional methods like the JMAK equation provide a simplistic understanding of cooperative phenomena, the extended FW model offers a more detailed analysis by considering elementary processes of photochemical reactions. By applying these models, researchers can quantify cooperativity and elucidate underlying mechanisms. For instance, investigations into the photoreactions of various compounds reveal dependencies on molecular structure and packing motif, shedding light on the factors influencing cooperative behavior. Additionally, the progressive behavior of photoreactions on optical length scale is introduced, uncovering intriguing phenomena like "edge-to-center propagation of photochemical reaction." These findings not only enhance our fundamental understanding of crystalline kinetics but also offer insights crucial for designing stimuli-responsive organic crystals with tailored functionalities. Thus, comprehensive analysis and experimental observation would contribute significantly to both fundamental and applied aspects of molecular crystal engineering.

References

1. W.M. Awad, D.W. Davies, D. Kitagawa, J. Mahmoud Halabi, M.B. Al-Handawi, I. Tahir, F. Tong, G. Campillo-Alvarado, A.G. Shtukenberg, T. Alkhidir, Y. Hagiwara, M. Almehairbi, L. Lan, S. Hasebe, D.P. Karothu, S. Mohamed, H. Koshima, S. Kobatake, Y. Diao, R. Chandrasekar, H. Zhang, C.C. Sun, C. Bardeen, R.O. Al-Kaysi, B. Kahr, P. Naumov, Mechanical properties and peculiarities of molecular crystals. Chem. Soc. Rev. **52**, 3098–3169 (2023)
2. P. Naumov, D.P. Karothu, E. Ahmed, L. Catalano, P. Commins, J. Mahmoud Halabi, M.B. Al-Handawi, L. Li, The rise of the dynamic crystals. J. Am. Chem. Soc. **142**, 13256–13272 (2020)
3. D. Yan, Z. Wang, Z. Zhang, Stimuli-responsive crystalline smart materials: from rational design and fabrication to applications. Acc. Chem. Res. **55**, 1047–1058 (2022)
4. W. Wu, K. Chen, T. Wang, N. Wang, X. Huang, L. Zhou, Z. Wang, H. Hao, Stimuli-responsive flexible organic crystals. J. Mater. Chem. C **11**, 2026–2052 (2023)
5. S. Kobatake, S. Takami, H. Muto, T. Ishikawa, M. Irie, Rapid and reversible shape changes of molecular crystals on photoirradiation. Nature **446**, 778–781 (2007)
6. O.S. Bushuyev, A. Tomberg, T. Friscic, C.J. Barrett, Shaping crystals with light: crystal-to-crystal isomerization and photomechanical effect in fluorinated azobenzenes. J. Am. Chem. Soc. **135**, 12556–12559 (2013)
7. K. Nasu, *Photoinduced Phase Transitions* (World Scientific, Singapore, 2004)
8. G. Liu, J. Liu, Y. Liu, X. Tao, Oriented single-crystal-to-single-crystal phase transition with dramatic changes in the dimensions of crystals. J. Am. Chem. Soc. **136**, 590–593 (2014)
9. K. Morimoto, D. Kitagawa, C.J. Bardeen, S. Kobatake, Cooperative photochemical reaction kinetics in organic molecular crystals. Chem. Eur. J. **29**, e202203291 (2023)
10. M. Bertmer, R.C. Nieuwendaal, A.B. Barnes, S.E. Hayes, Solid-state photodimerization kinetics of α-trans-cinnamic acid to α-truxillic acid studied via solid-state NMR. J. Phys. Chem. B **110**, 6270–6273 (2006)

11. I. Fonseca, S.E. Hayes, B. Blumich, M. Bertmer, Temperature stability and photodimerization kinetics of beta-cinnamic acid and comparison to its alpha-polymorph as studied by solid-state NMR spectroscopy techniques and DFT calculations. Phys. Chem. Chem. Phys. **10**, 5898–5907 (2008)
12. I. Fonseca, S.E. Hayes, M. Bertmer, Size effects of aromatic substitution in the ortho position on the photodimerization kinetics of alpha-trans cinnamic acid derivatives. A solid-state NMR study. Phys. Chem. Chem. Phys. **11**, 10211–10218 (2009)
13. I. Fonseca, M. Baias, S.E. Hayes, C.J. Pickard, M. Bertmer, Effects of aromatic substitution on the photodimerization kinetics of β-trans cinnamic acid derivatives studied with 13C solid-state NMR. J. Phys. Chem. C **116**, 12212–12218 (2012)
14. J.B. Benedict, P. Coppens, Kinetics of the single-crystal to single-crystal two-photon photodimerization of α-trans-cinnamic acid to α-truxillic acid. J. Phys. Chem. A **113**, 3116–3120 (2009)
15. R. Moré, G. Busse, J. Hallmann, C. Paulmann, M. Scholz, S. Techert, Photodimerization of crystalline 9-anthracenecarboxylic acid: a nontopotactic autocatalytic transformation. J. Phys. Chem. C **114**, 4142–4148 (2010)
16. K.A. Konieczny, A. Szczurek, J. Bąkowicz, R. Siedlecka, A. Ciesielski, M.K. Cyrański, I. Turowska-Tyrk, The reasons for different kinetics of the Norrish-Yang reaction in crystals. Structural and spectroscopic studies. Cryst. Growth Des. **20**, 5061–5071 (2020)
17. L. Bentea, M.A. Watzky, R.G. Finke, Sigmoidal nucleation and growth curves across nature fit by the Finke-Watzky model of slow continuous nucleation and autocatalytic growth: explicit formulas for the lag and growth times plus other key insights. J. Phys. Chem. C **121**, 5302–5312 (2017)
18. M.A. Watzky, R.G. Finke, Transition metal nanocluster formation kinetic and mechanistic studies. A new mechanism when hydrogen is the reductant: slow, continuous nucleation and fast autocatalytic surface growth. J. Am. Chem. Soc. **119**, 10382–10400 (1997)
19. E.E. Finney, R.G. Finke, Is there a minimal chemical mechanism underlying classical Avrami-Erofe'ev treatments of phase-transformation kinetic data? Chem. Mater. **21**, 4692–4705 (2009)
20. M.A. Watzky, R.G. Finke, Pseudoelementary steps: a key concept and tool for studying the kinetics and mechanisms of complex chemical systems. J. Phys. Chem. A **125**, 10687–10705 (2021)
21. T. Salzillo, S. Zaccheroni, R.G. Della Valle, E. Venuti, A. Brillante, Micro Raman investigation of the photodimerization reaction of 9-cyanoanthracene in the solid state. J. Phys. Chem. C **118**, 9628–9635 (2024)
22. K. Morimoto, D. Kitagawa, F. Tong, K. Chalek, L.J. Mueller, C.J. Bardeen, S. Kobatake, Correlating reaction dynamics and size change during the photomechanical transformation of 9-methylanthracene single crystals. Angew. Chem. Int. Ed. **61**, e202114089 (2022)
23. E.R. Kearns, D.M. D'Alessandro, Variable-temperature photocyclization kinetics in a metal–organic framework (MOF): a comparison of the Johnson–Mehl–Avrami–Kolmogorov and Finke-Watzky models. Cryst. Growth Des. **23**, 6100–6106 (2023)
24. K. Morimoto, D. Kitagawa, H. Sotome, S. Ito, H. Miyasaka, S. Kobatake, Edge-to-center propagation of photochemical reaction during single-crystal-to-single-crystal photomechanical transformation of 2,5-distyrylpyrazine crystals. Angew. Chem. Int. Ed. **61**, e202212290 (2022)

Chapter 3
Versatile and Fast Organic Crystal Actuation by Photothermal Effect and Natural Vibration

Hideko Koshima, Shodai Hasebe, and Yuki Hagiwara

Abstract Mechanically responsive organic crystals that actuate by external stimuli are expected to be applied to actuators and soft robots. Over the past decade, many mechanical crystals have been developed based on photochemical reactions (particularly photoisomerization) and phase transition. However, only a limited number of crystals undergo photoisomerization and phase transition. Furthermore, photoisomerization-induced motion tends to be slow and thick crystals do not actuate. More versatile and faster actuation mechanisms of crystal are required. Recently, we have discovered for the first time very fast crystal actuation (500 Hz) due to the photothermal effect under photoirradiation. Thereafter, for the first time, we have also discovered faster crystal actuation due to the natural vibration induced by the photothermal effect. Furthermore, the natural vibration is greatly amplified by resonance to realize high frequency (700 Hz), large actuation, and high energy conversion efficiency. Both the photothermal effect and natural vibration are universal physical phenomena caused in all crystals that absorb light, and their actuations can be quantitatively simulated. This means that it is now possible to design the operation of crystals. Research on mechanical crystals has reached a new stage where they can be applied as actuation materials to actuators and soft robots.

Keywords Photomechanical crystals · Photothermal effect · Natural vibration · High-speed actuation · Versatile actuation

H. Koshima (✉)
Research Organization for Nano and Life Innovation, Waseda University, 513 Wasedatsurumaki-cho, Shinjuku-ku, Tokyo 162-0041, Japan
e-mail: h.koshima@kurenai.waseda.jp

S. Hasebe · Y. Hagiwara
Department of Advanced Science and Engineering, Graduate School of Advanced Science and Engineering, Waseda University, 2-2 Wakamatsu-cho Shinjuku-ku, Tokyo 162-8480, Japan

S. Kobatake and H. Uekusa (eds.), *Advances in Organic Crystal Chemistry*,
https://doi.org/10.1007/978-981-96-4479-7_3

3.1 Introduction

Conversion of energy into mechanical motion is very important. Living beings have survived and thrived by converting energy such as sunlight and heat into motion. We humans have created a variety of electrically powered machines to improve our daily lives. Building new systems to generate motions is crucial for the development of future society. Recently, mechanically responsive materials that exhibit macroscopic motion by external stimuli such as light, heat, electricity, and chemical reactions have attracted much attention in the fields of chemistry, materials science, and engineering. Such actuation materials are important for applications in sensors, switches, actuators, micromachines, and soft robots [1].

A crystal is the ultimate solid material, with molecules regularly arranged in three dimensions. Historically, research on photomechanically responsive crystals was performed in the 1980s but was not continued [2]. Afterward, in 2007, Irie group reported reversible bending in the rod-like microcrystals of photochromic diarylethene upon ultraviolet (UV) light irradiation, successfully linking the molecular-level shape change to the macroscopic mechanical motion of the crystals [3]. This epoch-making finding broke the preconceived notion that organic crystals are hard and brittle. Since then, numerous mechanically responsive crystals have been developed by several research groups including ours over the last two decades.

Most mechanically responsive crystals developed to date are based on reversible photochemical reactions, especially photoisomerization and photodimerization. The representative crystals are typical photochromic crystals such as diarylethene [3, 4], azobenzene [5], and salicylideneaniline [6], as well as photodimerizable anthracene crystals [7]. The motions include bending [3, 5, 6], twisting [4, 7] and jumping [8]. Phase transition is another underlying actuation mechanism that occurs by heat and light [9–11]. Mechanically responsive crystals have been reviewed in several publications [12, 13]. However, their actuation speeds by photochemical reactions are slow (sub-seconds to minutes) and thick crystals do not bend. Furthermore, photoisomerization and phase transition do not occur in all crystals, but only in a very small fraction of the vast number of crystals. This inevitably limits the number of crystals that may be developed in the future. There is a need for a more versatile actuation mechanism that can be applied to all crystals.

Recently, our group discovered for the first time two new actuation mechanisms, photothermal effect in 2020 [14] and natural vibration in 2023 [15]. Both mechanisms allow crystals to actuate at high speed (500 Hz) by light. Even better, any crystal that absorbs light can actuate, thus solving the issues by photoisomerization and phase transition mentioned above, and expanding the potential of crystals as actuation materials. This chapter introduces crystal actuation by the photothermal effect and natural vibration.

3.2 Actuation Mechanisms

Figure 3.1 summarizes the four actuation mechanisms of organic crystals and their actuation features. Photochemical reaction alone is chemical process, and the other phase transition, photothermal effect, and natural vibration are physical processes. Most mechanically responsive crystals reported to date are based on reversible photochemical reactions, especially photoisomerization and photodimerization. The irradiated light is absorbed near the crystal surface and does not penetrate deep into the crystal. Hence, the thickness of photoproduct layer is very thin (only several μm). The bending mechanism caused by photochemical reaction can be explained by Timochenko's bimetal model [16], consisting of a thin photoproduct layer and an unreacted layer (Fig. 3.1a) [17, 18]. The bending angle is often large due to the large difference in molecular structure between the photoproduct and the reactant. However, thick (> 40 μm) crystals cannot bend because of the thin photoproduct layer. The bending speed is slow, from sub-seconds to minutes due to the chemical process.

Phase transition involves changes in molecular conformation and molecular arrangement in the crystal. Phase transition can occur throughout the crystal, in contrast to photoisomerization which occurs only near the crystal surface. Since the requirements for mechanical crystals are repeatability and durability, a reversible single-crystal-to-single-crystal (SCSC) phase transition is desirable. External stimuli such as heat and light can induce different types of phase transitions. Thermal phase transition and photo-induced phase transition are well-known phase transitions. Photo-triggered phase transition is a new type of phase transition in which

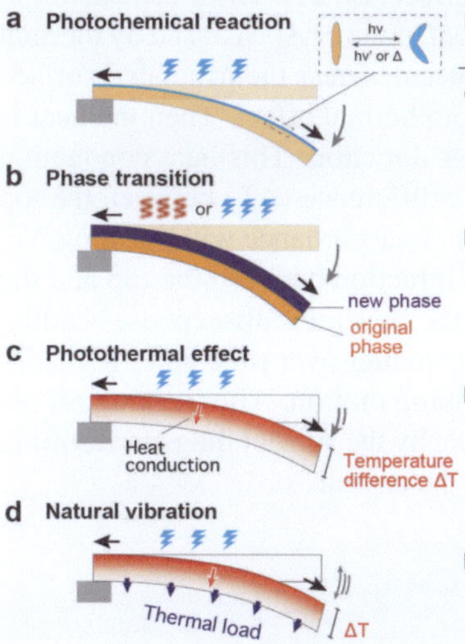

Phenomena	Target crystals	External stimuli	Actuation speed	Actuation design
chemical	limited (thin crystals that undergo photoisomerization)	light	s–min (~5 Hz)	difficult
physical	limited (crystals that undergo a phase transition)	heat, light	ms–min	difficult
physical	any crystals absorbing light	light	ms (~500 Hz)	possible
physical	any crystals	any stimuli (e.g. light)	μs–ms (~1000 Hz)	possible

Fig. 3.1 Actuation mechanisms and characteristics of organic crystals: **a** photochemical reaction, **b** phase transition, **c** photothermal effect, and **d** natural vibration

a photochromic crystal with a thermal phase transition transforms into an identical high-temperature phase in a temperature region lower than the thermal phase transition temperature upon light irradiation, discovered by our group in 2019 [19]. The photo-triggered phase transition is induced by local stress due to photoproduct molecules produced by photoisomerization near the irradiated surface. The mechanism of this phase transition is different from that of well-known photo-induced phase transition, which occurs via an excited state under light irradiation. The requirements for inducing the photo-triggered phase transition are sufficient crystal voids to allow molecular conformation change, smooth photoisomerization, and most likely a chiral molecular arrangement [20]. The stepwise bending and locomotion are realized by the photo-triggered phase transition.

The problem with photochemical reaction and phase transition is that they do not occur in all crystals, but only in a very small portion of the vast number of crystals. For future applications of mechanical crystals, more versatile mechanisms that can move any crystal are required. What follows is a discussion of new versatile actuation mechanisms in which the photothermal effect and natural vibration have been discovered for the first time by our research group.

3.2.1 Photothermal Effect

Photothermal effect is a physical phenomenon by which thermal energy (heat) is very quickly (several picoseconds) generated by nonradiative deactivation of the photoexcited state during an initial photophysical process. When a crystal cantilever is irradiated, it bends very quickly by the photothermal effect [14, 21]. The mechanism can be explained by the non-steady temperature gradient in the crystal caused by thermal conduction (Fig. 3.2). Upon photoirradiation, molecules near the irradiated surface absorb light and generate heat quickly by the photothermal effect. Then the heat is conducted to the back surface along the thickness direction. This heat conduction from the top to back surface creates a temperature difference (ΔT) between the top and the back surfaces along the thickness direction. In accordance with forming ΔT, the expansion difference (ΔL) in the longitudinal direction between the top and the back surfaces changes over time, causing very quick (several milliseconds) bending motion. With continuous irradiation, ΔT changes smaller over time and eventually reaches a constant (steady state), stopping the bending motion. After turning off the light, ΔT decreases quickly over time due to cooling by the surrounding air, resulting in the fast (several milliseconds) straightening of the crystal.

3.2.2 Natural Vibration

Natural vibration is a physical phenomenon during which a material continues to vibrate at a certain frequency (the natural frequency) when subject to an external

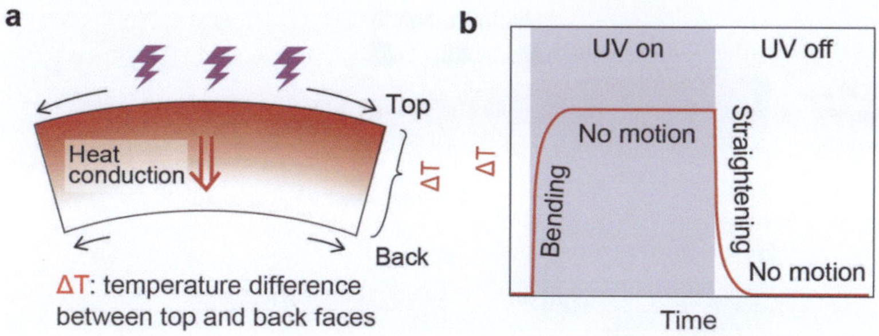

Fig. 3.2 Mechanism of crystal bending by the photothermal effect. **a** Schematic illustration of photothermal bending due to the temperature gradient generated by the heat conduction. **b** The formation of a non-steady temperature gradient in the thickness direction by non-steady thermal conduction creates fast photothermally driven bending. Reproduced from Ref. [21]. Copyright 2021 American Chemical Society

force. Furthermore, when a periodic force is applied near the natural frequency, the vibrational motion is amplified by resonance. For example, during an earthquake, if the natural frequency of a shaking building matches the frequency of the earthquake, the building will resonate and shake violently, which may lead to collapse of the structure [22]. Most musical instruments, such as the guitar and saxophone, create sound (natural vibration) via the movement of strings or reeds and amplify the sound via resonance in a box or tube [23]. Thus, resonated natural vibrations are widely related to our daily life. However, the natural vibration has not been paid attention in terms of organic crystal actuation.

When a crystal cantilever is irradiated, the crystal very quickly causes small, repetitive bending by the natural vibration accompanied by large photothermal bending (Fig. 3.3) [15]. Photothermal bending is thermal deformation due to the non-steady thermal expansion difference between the top and back surface by the temperature gradient along the thickness direction. The natural vibration is also derived from the temperature gradient; however, it is triggered by the thermal load due to the temperature gradient. Thus, the mechanisms of photothermal bending and natural vibration are different while the photothermal bending and the natural vibration appear spontaneously. The natural vibration is greatly amplified by resonance upon pulsed light irradiation at the natural frequency.

3.3 Crystal Actuation by Photothermal Effect

3.3.1 Actuation Behavior by Photothermal Effect

Photothermal bending of a salicylideneaniline crystal with an *o*-amino substituent in enol form (enol-**1**) is addressed (Fig. 3.4) [21]. When a thick (48 μm) platelike crystal cantilever of enol-**1** is irradiated with UV laser light (375 nm, 960 mW cm^{-2}), the

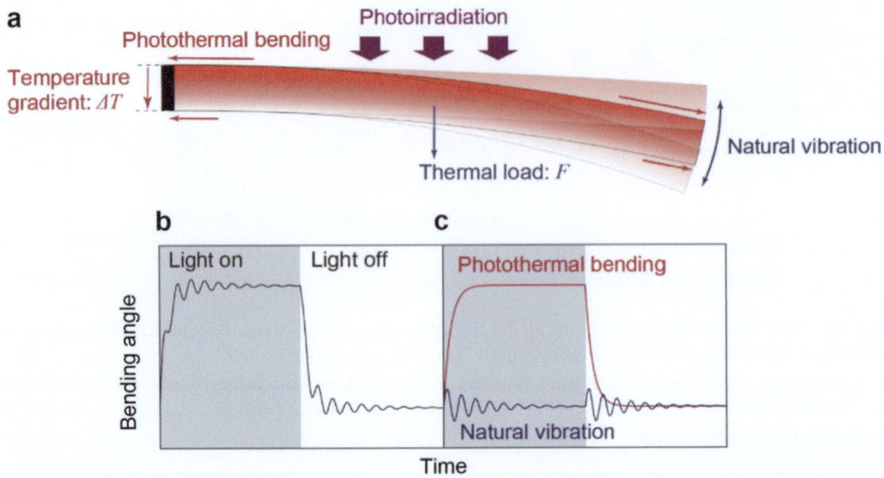

Fig. 3.3 Mechanism of crystal bending by natural vibration. **a** Schematic illustration of natural vibration caused by the thermal load due to the temperature gradient. **b** Small natural vibration accompanied by large photothermal bending. **c** Extracted components of the photothermal bending and the natural vibration from the observed bending

crystal very quickly bends away from the light to bending angle of 0.2° after 21 ms and then maintains the steady state (Fig. 3.4a–c). After stopping irradiation, the crystal returns very quickly to the initial straight shape. The time constants for bending and straightening are 3.3 and 5.9 ms, respectively. The temperature of the irradiated top surface also increased quickly from 27.5 to 38.5 °C by the photothermal effect upon UV irradiation and after light cessation returns quickly to the initial temperature; the time constants are 58 and 51 ms, respectively, which are an order of magnitude slower than those of bending and straightening. This suggests that the photothermal bending is not caused directly by the temperature change of the top surface but, instead, by the temperature gradient extending from the top surface to the back surface in the thickness direction. With an increase in the UV intensity, both the bending angle and surface temperature increase linearly, while both the time constants remain constant (Fig. 3.4d–g). Furthermore, even visible light (488 nm) exposure causes the photothermal bending because the crystal absorbs the visible light. High-frequency bending of 500 Hz is possible by pulsed UV irradiation over 1000 cycles without any deterioration, showing high durability (Fig. 3.4h).

3.3.2 *Two-Step Actuation by Combining Photothermal Effect and Photoisomerization*

The enol-**1** is a photochromic compound that exhibits enol–keto photoisomerization in the crystalline state. When a thin (10 μm) crystal is irradiated with UV

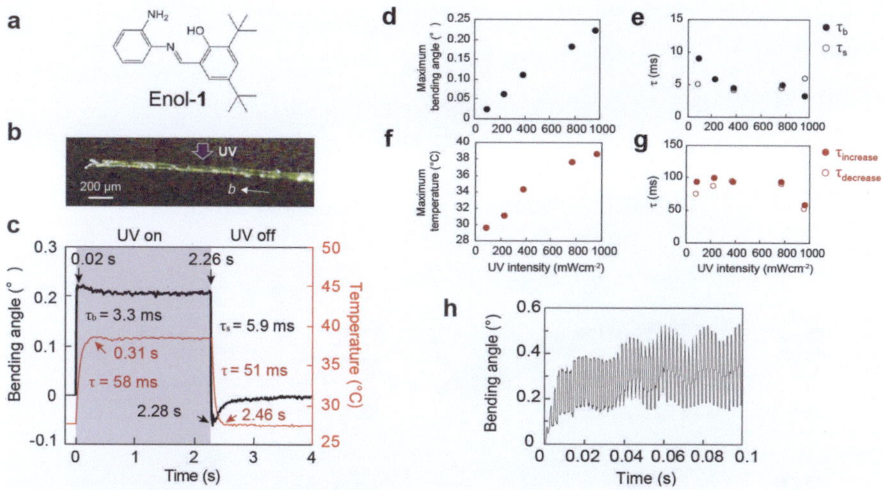

Fig. 3.4 Photothermally driven bending of a thick enol-**1** crystal (1520 × 68.0 × 48.3 μm^3) upon UV laser (375 nm, 960 mW cm^{-2}, spot diameter 2.6 mm) irradiation. **a** Molecular structure of enol-**1**. **b** Photo of the crystal viewed from the side. **c** Time dependence of the bending angle (black) and maximum surface temperature (red). **d–g** UV intensity dependence of the (**d**) maximum bend angle, **e** time constants for bending (black solid circle) and straightening (black open circle), **f** maximum temperature, and (**g**) time constants for temperature increase (red solid circle) and temperature decrease (red open circle). **h** High-frequency bending (500 Hz) of another enol-**1** crystal (1912 × 48.2 × 35.4 μm^3) by pulsed UV exposure (1 ms on, 1 ms off). Reproduced from Ref. [21]. Copyright 2021 American Chemical Society

light, the crystal bends slowly and largely (20°) by photoisomerization from enol to *trans*-keto form, and after stopping irradiation, the bent crystal returns slowly to the initial straight shape by back-isomerization [21]. The time constants for bending and straightening by the photoisomerization and back-isomerization are 0.11 and 1.1 s, respectively which are 20- and 300-fold slower than those by the photothermal effect. With increasing crystal thickness, the degree of bending due to photoisomerization becomes smaller, and instead the fast bending by the photothermal effect appears. In fact, slightly thick (20–40 μm) crystals exhibit a two-step bending, first a fast photothermal bending, then a slow bending by photoisomerization. Thick crystals (>40 μm) exhibit only fast bending due to the photothermal effect, as mentioned in the previous Sect. 3.3.1.

In addition, various actuations by combining photoisomerization and photothermal effect are introduced briefly for some other crystals. A salicylideneaniline derivative with a 4-fluoro substituent in enol form (enol-**2**) forms two polymorphic crystals (**2α** and **2β**, Fig. 3.5) [24]. The **2α** crystal exhibits enol-keto photoisomerization whereas the **2β** crystal does not. The thin **2α** crystal bends away largely and slowly from the light source with some twist because of photoisomerization; in contrast, the thick **2α** crystal shows two-step bending by the combination of photoisomerization and the photothermal effect. Notably, the thin **2β** crystal does

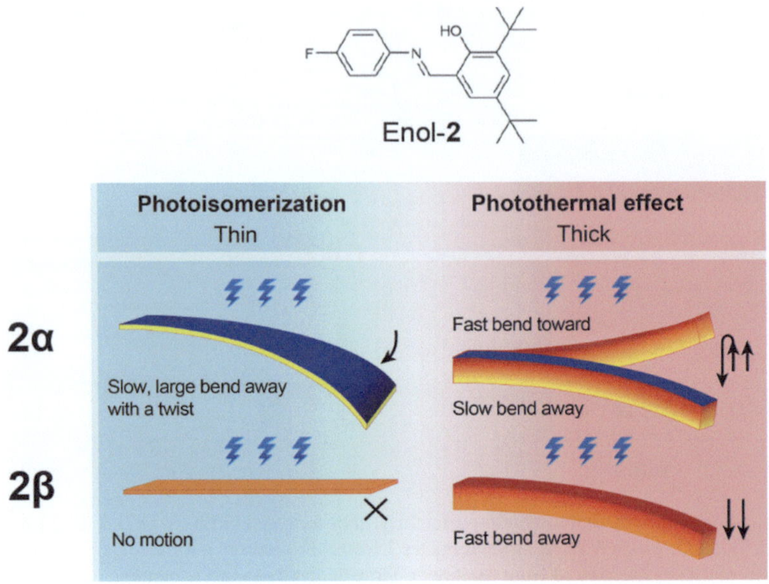

Fig. 3.5 Crystal actuation of **2α** and **2β** polymorphic crystals by photoisomerization and the photothermal effect. Reproduced from Ref. [24]. Copyright 2021 American Chemical Society

not bend due to lack of photoreactivity; however, the thick **2β** crystal exhibits fast bending away from the light because of the photothermal effect, finally achieving 500 Hz high-speed bending. Thus, using the two polymorphic crystals, **2α** and **2β**, the four mechanical motions are induced by photoisomerization and/or photothermal effects.

The actuation switching of the isomorphic crystals is also realized, where bending direction and speed are controlled by crystal thickness and light wavelength [25]. Upon UV light irradiation of the top face, a thin crystal bends toward the light source through photoisomerization; this bending is attributed to local contraction of the irradiated top face. In contrast, a thick crystal bends away rapidly through the photothermal effect. Slightly thick crystals exhibit two-step bending that involves a combination of these two mechanisms, first bending away quickly through the photothermal effect and then bending toward the light source through photoisomerization. The bending motions of thin and slightly thick crystals switch to two-step bending and bending away, respectively, upon UV light irradiation of the side face. Visible light irradiation alters the bending motions of thin and slightly thick crystals to no significant bending and bending away, respectively. These results provide a novel, convenient, and useful approach for switching the crystal bending direction and speed, thereby expanding the versatility of molecular crystals as actuation materials.

3.3.3 Actuation of Hybrid Crystals by Photothermal Effect

Broad-wavelength light-driven high-speed actuation by the hybridization of crystals is achieved [26]. Titanium carbide ($Ti_3C_2T_x$), known as MXene, is an excellent photothermal conversion material that absorbs light over a wide wavelength range, from UV to visible and near-infrared (NIR) light [27]. MXene layer enables the molecular crystals to be actuated rapidly regardless of their thickness over a wide range of wavelengths. The hybrid crystals with MXene are elucidated to be biocompatible. MXene allows for the bending of the hybrid crystals inside agar phantoms mimicking biological soft tissue like brain, successfully demonstrating the possible biological application of crystals. The strategies to hybridize photothermal materials and photomechanical crystals and exploit the photothermal effect over a wide optical wavelength range are expected to pave the way for the development of innovative light-driven mechanical crystals.

3.3.4 Heat Generation by Photothermal Effect

The photothermal effect is a physical phenomenon by which thermal energy (heat) is generated by nonradiative deactivation of the photoexcited state during an initial photophysical process. The generated photothermal energy can be estimated based on the photodynamics revealed by transient absorption spectroscopy measurements (Fig. 3.6a). Just after excitation of the enol-**1** single crystal with femtosecond 400 nm pulsed laser, the sharp positive absorption at 480 nm appears, originating from the absorption of the Franck–Condon (FC) excited-state enol-**1*** (Fig. 3.6b). One picosecond after excitation, this sharp peak begins to decrease with the time constant of 1 ps due to the isomerization to *cis*-keto-**1*** via the excited-state intramolecular proton transfer (ESIPT), and a new positive band at 510 nm and a broad negative band covering of 570–800 nm appear, which are assigned to the absorption and stimulated fluorescence of the *cis*-keto* species, respectively. Both the positive and negative bands decrease with both the time constants of 5 ps, and the broad positive absorption band at around 500 nm remains 1 ns after excitation; this final absorption spectrum is assigned to the photoisomerization product *trans*-keto-**1** with the steady-state absorption spectra.

The fluorescence quantum yield of *cis*-keto-**1*** to *cis*-keto-**1** is 0.006, very small. Most of the photothermal energy (heat) is generated via the nonradiative deactivation process of *cis*-keto-**1*** to *cis*-keto-**1**. In addition, the photothermal energy is generated via the nonradiative deactivation process of FC-enol-**1*** to enol-**1**.

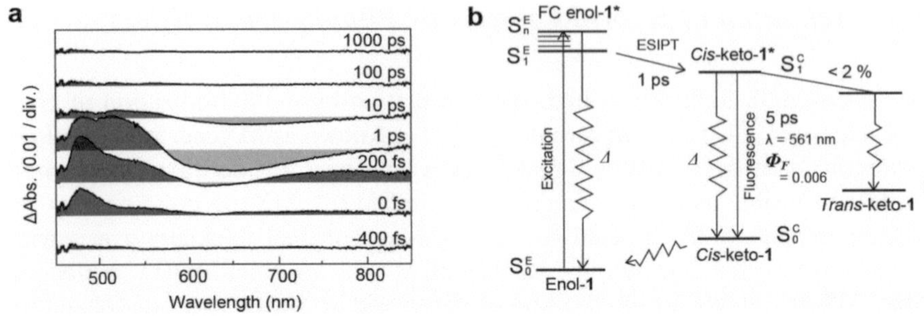

Fig. 3.6 **a** Transient absorption spectra of enol-**1** single crystal after excitation with a femtosecond 400 nm laser pulse. **b** Photodynamics of enol-**1** crystal following photoexcitation. Reproduced from Ref. [21]. Copyright 2021 American Chemical Society

3.3.5 Simulation of Photothermally Driven Crystal Actuation

Crystal bending due to the photothermal effect can be simulated by first calculating the time dependence of the temperature gradient between the irradiated top surface and back surface, and then converting it into a bending angle. However, the direct measurement of the temperature gradient is difficult, especially, for micro-size materials. Here, the photothermal bending of the enol-**1** crystal cantilever shown in Fig. 3.4c is actually simulated. Taking only one-dimensional heat transfer into account, the non-steady thermal conduction equation can be concisely expressed as Eq. 3.1,

$$\frac{\partial T(y,t)}{\partial t} = \alpha \frac{\partial^2 T(y,t)}{\partial y^2} - A(y)(T(y,t) - T_0) \tag{3.1}$$

where y is the displacement along the conduction direction (Fig. 3.7a), t is time, $T(y, t)$ is the temperature at a certain y and t, α is the thermal diffusivity (0.879×10^{-7} m^2 s^{-1} at 20 °C) along the conduction direction which is along the thickness d, A(y) is the coefficient of cooling from ambient air at a certain y ($y = d$ on the back surface where $A(y)$ is constant, and $A(y) = 0$ elsewhere), and T_0 is the ambient temperature. Here assuming that the photothermally generated heat from the whole photoirradiated crystal surface is conducted one-dimensionally from the top to the back surface, Eq. 3.1 can be solved using the finite element analysis (FEA) by approximating the derivatives with finite differences.

Next, it becomes possible to simulate by converting the temperature difference to the bending angle via geometrical change with thermal strain according to Eq. 3.2,

$$\theta(t) = \frac{\Delta L(t)}{2d} = \frac{aL_0(T(0,t)-T(d,t))}{2d} = \frac{aL_0\Delta T(t)}{2d} \tag{3.2}$$

where $\theta(t)$ is the bending angle at a certain time t, d is the crystal thickness (4.83×10^{-5} m), $\Delta L(t)$ is the difference in length between the two sides, a is the coefficient of thermal expansion along the length direction (1.29×10^{-4} °C^{-1}), L_0 is the initial

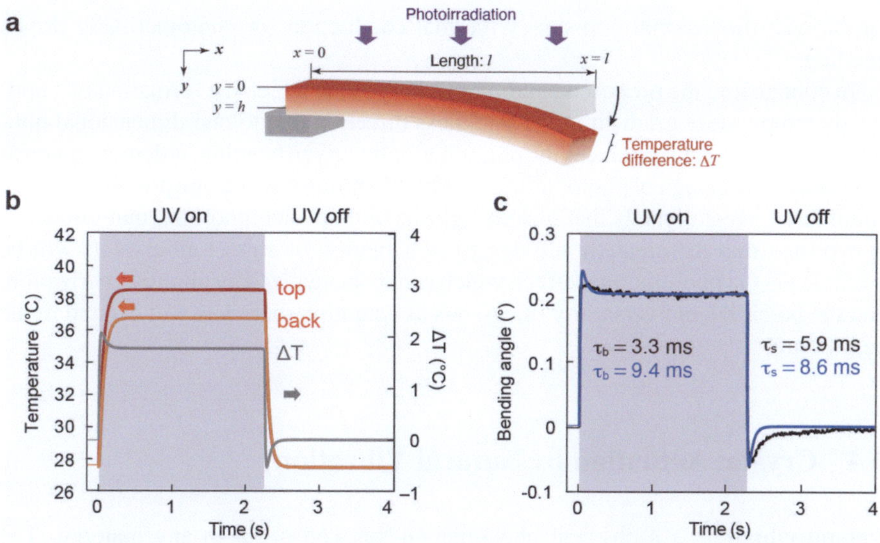

Fig. 3.7 Simulation of photothermally driven bending of a thick enol-**1** crystal upon UV laser irradiation (the same crystal and same irradiation condition in Fig. 3.4). **a** Schematic illustration of crystal bending. **b** Time dependence of top surface (red, measured) and back surface (orange, calculated) temperature and temperature difference between two surfaces (ΔT, gray, calculated). **c** Measured (black) and simulated (blue) bending angle. The numbers indicate the time constants for bending and straightening. Reproduced from Ref. [21]. Copyright 2021 American Chemical Society

length of the crystal (1.524×10^{-3} m), $T(0, t)$ and $T(d, t)$ are the temperatures of the top and back surfaces, respectively, and $\Delta T(t)$ is the temperature difference between the two surfaces.

Figure 3.7b shows the calculated time dependence (orange) of the back surface temperature and the temperature difference (ΔT, gray) between both surfaces when the time dependence of the top surface temperature measured by an infrared (IR) camera (red) is used. Under UV laser irradiation, the back surface temperature increases from 27.5 to 36.7 °C in 428 ms and returns to its initial temperature (27.5 °C) 398 ms after turning off the UV light. ΔT increases up to 2.20 °C in 38 ms and then reaches a steady state, with a temperature difference of 1.81 °C after 300 ms. After turning off the laser, ΔT decreases to − 0.532 °C in 32 ms and then returns to almost its initial state (−0.001 °C) in 340 ms. Figure 3.7c shows a comparison between the measured (black) and simulated (blue) bending angles. Under UV laser irradiation, the simulated bending angle increases to 0.243° after 36 ms and then reaches a steady state of 0.207° after 270 ms, which is almost the same as that of the measured bending angle (0.206°). The time constants τ_b and τ_s of simulated bending and straightening are 9.4 and 8.6 ms, respectively, which are nearly coincident with the measured time constants ($\tau_b = 3.3$ ms, $\tau_s = 5.9$ ms), confirming the validity of the simulation based on the one-dimensional non-steady thermal conduction for photothermally driven bending.

In conclusion, the proposed bending mechanism, in which the formation of a non-steady temperature gradient in the thickness direction due to one-dimensional non-steady heat conduction induces the photothermally driven bending, is demonstrated to be correct. The design of photothermally driven bending has become possible through simulation. Most crystals that absorb light have their own photothermal effects. It is expected that the creation and design of actuation of almost all crystals will be possible via the photothermal effect, which cannot be realized by photoisomerization, and the potential and versatility of crystals as actuation materials will expand in the near future.

3.4 Crystal Actuation by Natural Vibration

Natural vibration is a physical phenomenon that can occur in any material. The natural vibration is greatly amplified by resonance. The resonated natural vibrations are widely related to our daily life such as musical instruments and collapse of skyscrapers due to earthquake resonance. However, the natural vibration has not been paid attention in terms of organic crystal actuation. Recently, rapid photothermally driven bending of crystals has been reported, but the bend angle is small, 0.5° [14, 21, 24, 25]. One strategy to enhance photothermal bending is to use crystals with large thermal expansion coefficients, as suggested by Eq. 3.2. A β-phase crystal (**3β**) [28] of 2,4-dinitroanisole (**3**) (Fig. 3.8a) has a relatively large thermal expansion coefficient. While observing the bending of **3β** crystal due to the photothermal effect under UV light irradiation, we unexpectedly discovered for the first time that small, fast natural vibration (0.2°) occurs in conjunction with large photothermal bending (1°) [15]. The natural vibration is considerably amplified by resonance upon pulsed light irradiation at the natural frequency to realize large bending (4°), high frequency (700 Hz), and high energy conversion efficiency from light to kinetic energy. The bending behavior is successfully simulated. Any light absorbing crystal can be actuated by photothermally induced natural vibration. The resonated natural vibration can expand the potential and versatility of organic crystals as actuation materials.

3.4.1 Actuation Behavior by Natural Vibration Induced by Photothermal Effect

When a hexagonal rod-shaped **3β** crystal cantilever is uniformly irradiated from the top with UV laser (375 nm) (Fig. 3.8b), the photothermally driven bending away from the light source quickly reaches a bending angle of 0.85° in 9.4 ms, then gradually increases to 1.22° in 100 ms (Fig. 3.8c). This bending angle is greater than the

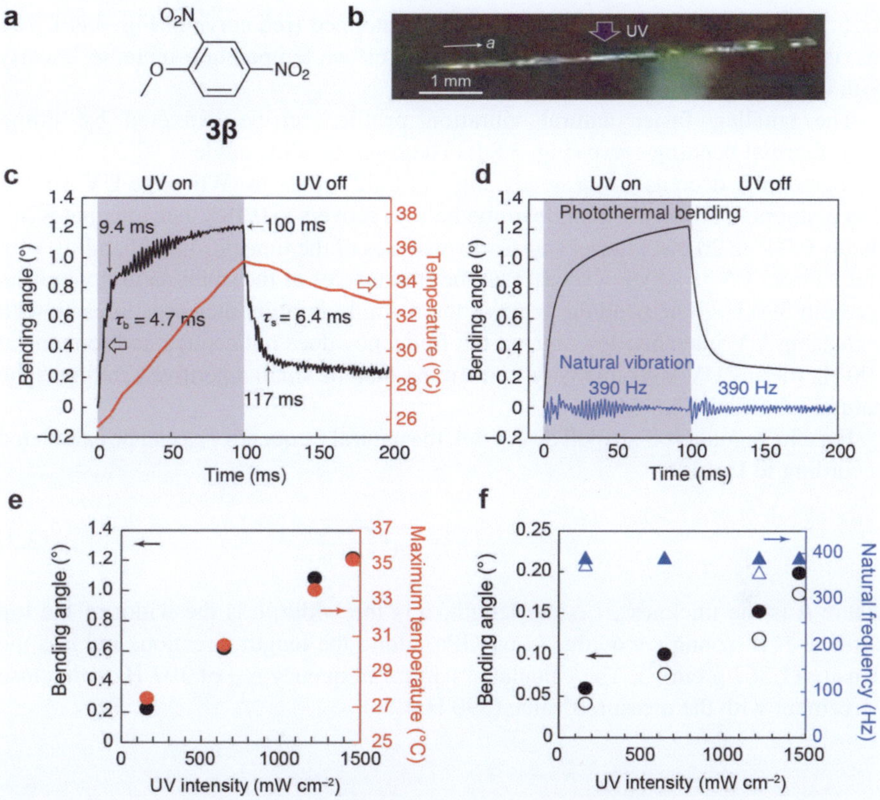

Fig. 3.8 Bending behavior of **3β** crystal (length 6075 μm, width 151 μm, width of the top face 40 μm, thickness 105 μm) by the photothermal effect and the natural vibration (390 Hz) by irradiation with UV laser light (375 nm, 1456 mW cm^{-2}). **a** Molecular structure of **3β**. **b** Photo of the crystal viewed from the side. **c** Time dependence of the bending angle (black) and the maximum temperature (red) of the irradiated surface with and without UV irradiation for 100 ms. **d** Time profiles of the extracted components of the photothermal bending and the natural vibration. **e** UV intensity dependence of the photothermally driven bending angle (black circle) and the surface temperature (red circle) upon UV irradiation for 100 ms. **f** UV intensity dependence of the bending angle (black circle) of the natural vibration and natural frequency (blue triangle). Solid and open plots indicate the results upon UV irradiation and after UV cessation, respectively. Reproduced with permission of Hagiwara et al. [15], licensed under Creative Commons Attribution 4.0 International License

0.2°–0.5° previously reported for other crystals [14, 21, 24, 25] due to the larger thermal expansion coefficient (2.47 × 10^{-4} °C^{-1}) of the **3β** crystal compared with those (0.80–1.30 × 10^{-4} °C^{-1}) reported for other crystals. After stopping the UV exposure at 100 ms, the crystal bends up quickly to 0.32° in 17 ms, then slowly returns to 0.23° in 100 ms, with the time constants (τ_{on}, τ_{off}) of 4.7 and 6.4 ms for UV light on and off conditions, respectively.

The surface temperature, which is simultaneously monitored by an IR camera, increases from 25.6 to 35.3 °C upon 100 ms of UV irradiation and decreases to

33.1 °C in 100 ms after the UV irradiation is stopped (red curve in Fig. 3.8c). The maximum bending angle and the maximum surface temperature increase linearly with increasing UV intensity (Fig. 3.8e).

The smaller, faster natural vibration profile can be extracted by fitting photothermal bending curve (Fig. 3.8d). The small bending angle is ~0.20° initially, then gradually decreases to a steady angle of 0.02° at 80 ms. When the UV irradiation is stopped, the bending angle immediately recovers to 0.18°, then attenuates to a steady 0.04° in 20 ms. Fourier transform analyses of the time profile of bending with and without UV irradiation reveal that the frequencies of the small natural vibration are both 390 Hz. The bending angle of the natural vibration increases linearly with increasing UV intensity. In contrast, the frequency does not change and remains at 390 Hz for any UV laser intensity, confirming that the small repetitive bending is the natural vibration (Fig. 3.8f).

For a hexagonal rod-shaped **3β** crystal, the natural frequency f_{cal} can be calculated according to Eq. 3.3,

$$f_{\mathrm{cal}} = \frac{h}{4\pi}\left(\frac{1.875}{l}\right)^2\sqrt{\frac{E(b+3c)}{6\rho(b+c)}} \tag{3.3}$$

where h is the thickness, l is the length, b is the width, c is the width of the top surface, E is Young's modulus (1.65 GPa) along the length direction, and ρ is the density (1.567 g cm^{-3}). The calculated natural frequency f_{cal} of 397 Hz is in close agreement with the measured value (390 Hz).

3.4.2 Amplification of Natural Vibration by Resonance

The bending angle of the natural vibration is one order of magnitude smaller than that of the photothermally driven bending. However, when the crystal is irradiated with pulsed UV light at the natural frequency of 390 Hz, the natural vibration is dramatically amplified by resonance (Fig. 3.9a). The bending angle begins at 0.20° in the first cycle and increases with irradiation time to the maximum of 3.4° after 150 ms; the difference in angle between the maximum of 2.4° and the minimum of −1.0° represents a 17-fold resonance amplification (Fig. 3.9b). When the surface temperature reaches a constant 40.8 °C after 850 ms under pulsed UV light, the resonated bending angle reaches a steady value of 2.3° and maintains the angle during 850–1190 ms (133 cycles). Once the UV irradiation is stopped at 1190 ms, the bending angle rapidly decreases, and the natural vibration almost disappears within 200 ms. This resonated bending is observed for at least 460 cycles without any fatigue of the crystal under pulsed UV irradiation.

To determine the exact natural frequency of the crystal, the bending behavior while irradiating it with UV pulses of a wide range of frequencies, to give the natural frequency (the most amplified frequency) of 390 Hz (black open circle, Fig. 3.9c). The fitted natural frequency f_{fit} determined by the forced vibration model is also

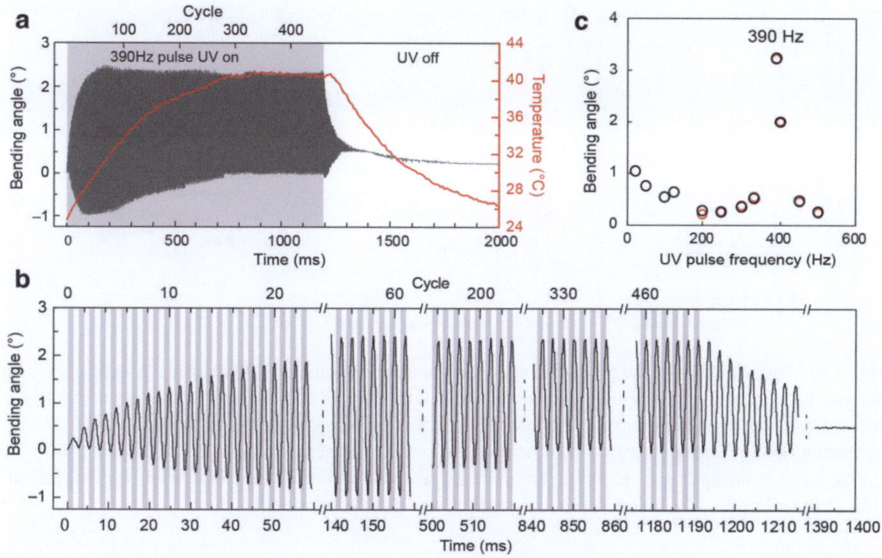

Fig. 3.9 Amplification of natural vibration by resonance of the same **3β** crystal cantilever as in Fig. 3.8 with a natural frequency of 390 Hz. **a** Time profiles of the resonated natural vibration and surface temperature upon exposure to 390 Hz pulsed UV irradiation. **b** Partially enlarged view of (**a**). **c** UV pulse frequency dependence of the measured (black) and fitted (red) amplified bending angles. Reproduced with permission of Hagiwara et al. [15], licensed under Creative Commons Attribution 4.0 International License

390 Hz (red open circle, Fig. 3.9c). In addition, bending amplification by resonance is observed not only at the natural frequency f, but also at its odd fractions of $f/3$, $f/5$, $f/7$, $f/9$, $f/11$, and $f/13$, and various bending patterns can be created. This amplification of the odd fractions is caused by the start of down-bending and up-bending due to the natural vibration at the same time when the UV light is turned on and off, respectively.

3.4.3 Relationship Between Crystal Shape and Natural Vibration

To clarify the relationship between crystal shape and natural vibration as well as photothermally driven bending, the bending behavior of five **3β** crystals (I–V) of different shapes is examined, with different geometries ranging from 5 to 8 mm in length and 50 to 220 μm in thickness. When the crystals are continuously irradiated with UV light for 100 ms, they exhibit photothermally driven bending of 1.1°–1.7°, which increases slightly in proportion to the thickness. The natural vibrations of 0.06°–0.3° are extracted and are greatly amplified to 1.9°–4.0° due to resonance upon irradiation with pulsed UV light at each natural frequency; these are nearly

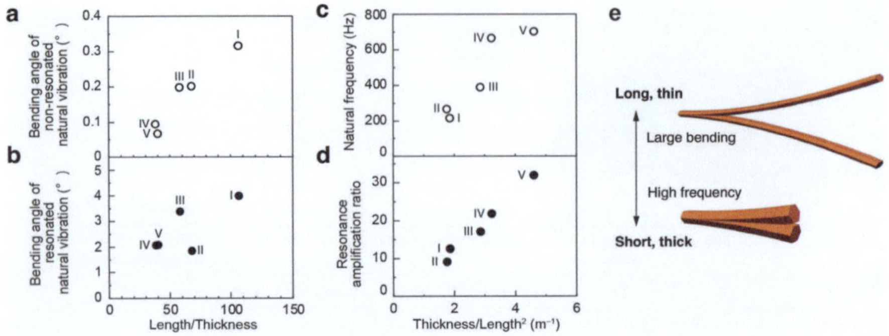

Fig. 3.10 Relationship between crystal shape and natural vibration of **3β** crystals. **a**, **b** Aspect ratio (length/thickness) dependence of the maximum bending angle of the non-resonated natural vibration (**a**) and the resonated natural vibration (**b**). **c**, **d** Thickness/length^2 dependence of the natural frequency (**c**) and the resonance amplification ratio (**d**). **e** Schematic diagram of crystal shapes to realize large bending angles and high frequencies. Reproduced with permission of Hagiwara et al. [15], licensed under Creative Commons Attribution 4.0 International License

proportional to the aspect ratio (length/thickness) of the crystals (Fig. 3.10a, b). The natural frequencies are in the range of 200–700 Hz and are nearly proportional to thickness/length^2, in accordance with Eq. 3.3 (Fig. 3.10c). Moreover, the resonance amplification ratio ranging from 9 to 32 also increases in proportion to the thickness/length^2 (Fig. 3.10d). Summarizing, the frequency and the bending angle can be changed and tuned by changing the crystal shape; larger bending can be realized by resonating thinner and longer crystals, and conversely, higher natural frequency can be realized by using thicker and shorter crystals (Fig. 3.10e). However, precise control over uniformity, dimensions, and shape of crystals is still challenging; it is an essential requirement toward the development of crystal actuators with the desired output.

3.4.4 Simulation of Crystal Actuation by Natural Vibration

Photothermal bending is caused by the non-steady thermal elongation difference between the irradiated and back surfaces, which is due to the non-steady temperature gradient formed by the heat conduction of photothermal energy [21]. On the other hand, the natural vibration is caused by the thermal load in the crystal generated by the photothermal effect. Despite these different bending mechanisms, simulation of actual bending behavior is attempted by combining both photothermally driven bending and natural vibration.

Figure 3.11a presents a schematic illustration of the geometrical parameters of a hexagonal crystal cantilever for photothermally driven bending and natural vibration. Upon photoirradiation, heat is generated at the irradiated surface by the photothermal effect, conducting into the thickness direction according to the non-steady heat

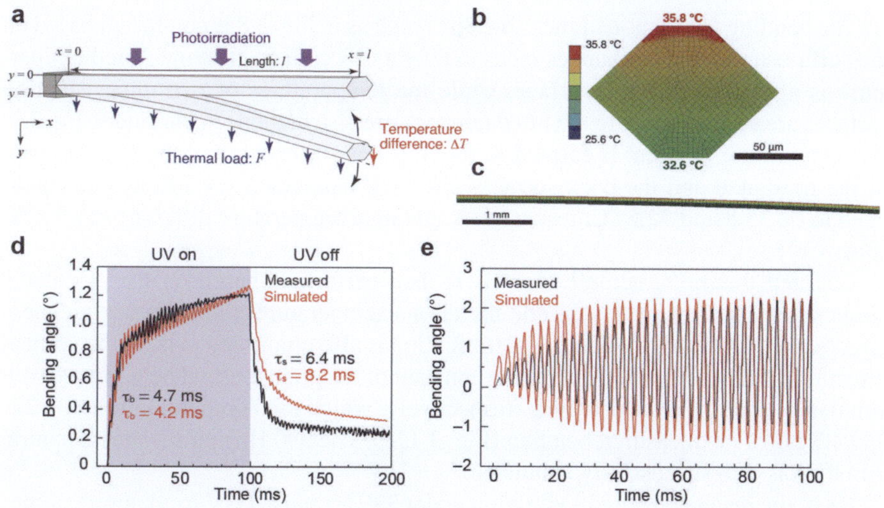

Fig. 3.11 Simulation of **3β** crystal bending by the combination of the photothermal effect and the natural vibration. **a** Schematic illustration of crystal bending. **b, c** Simulated temperature distribution after UV laser irradiation for 100 ms: the (100) cross section (**b**) and the side view of the bent crystal (**c**). **d** Time profiles of measured (black) and simulated (red) bending angles with and without UV light irradiation for 100 ms. The inset values of τ_{on} and τ_{off} indicate the time constants for bending and straightening. **e** Time profiles of measured (black) and simulated (red) bending angles, amplified under 390 Hz pulsed UV irradiation. Reproduced with permission of Hagiwara et al. [15], licensed under Creative Commons Attribution 4.0 International License

conduction equation (Eq. 3.1), where α is the thermal diffusivity (1.34×10^{-7} m^2 s^{-1} for **3β** crystals) along the thickness direction.

In the elastic cantilever system, the photothermal bending and the natural vibration are explicitly included in the elastic curve equation (Eq. 3.4) [29],

$$F(x, t) = \rho A \frac{\partial^2 \delta(x,t)}{\partial t^2} + c\frac{\partial x}{\partial t} + EI\frac{\partial^4 \delta(x,t)}{\partial x^4} \tag{3.4}$$

where A is the cross-sectional area, x is the displacement along the length direction, $\delta(x, t)$ is the deflection at a certain x and t, c is damping coefficient, and $F(x, t)$ is the applied thermal load at a certain x and t derived from the temperature gradient. Using the tip deflection $\delta(l, t)$ calculated according to Eq. 3.4, the combined bending angle θ of the photothermal bending and the natural vibration can be calculated using Eq. 3.5.

$$\theta(t) = \frac{\delta(l,t)}{l} \tag{3.5}$$

FEA using the ANSYS software package is applied to simulate the bending angle [30]. The temperature gradient is calculated by non-steady heat conduction analysis, and then the actual bending is simulated by coupled thermo-structural analysis.

The bending simulation of the same **3β** crystal as in Fig. 3.8 is conducted based on the surface temperature changes measured by an IR camera. The measured temperature is applied to the top surface, while the temperatures of two upper slanting surfaces are set according to the fitted temperature distribution. According to Eq. 3.1, the temperature gradient is formed in the thickness direction and the temperatures of the irradiated and the back surfaces after UV irradiation for 100 ms are calculated to be 35.8 and 32.6 °C, respectively, giving a temperature difference of 3.2 °C (Fig. 3.11b, c).

Coupled thermo-structural analysis is then performed based on the calculated temperature gradient. Due to the one-directional temperature gradient in the thickness direction (Fig. 3.11b), the bending deformation is also confirmed to be in the vertical direction. The bending angle and time constant of the photothermal bending are well simulated to be 1.26° and 4.2 ms, respectively, which are comparable to the 1.22° and 4.7 ms of the measured bending (Fig. 3.11d). The 390 Hz non-resonated natural vibration is also successfully simulated.

Next, the resonated natural vibration under 390 Hz pulsed UV irradiation is simulated. First, the time dependence of the temperature difference between the irradiated and back surfaces is calculated. Then, coupled thermo-structural analysis is performed; the calculated time profile of amplified bending angle is nearly coincident with the measured one (Fig. 3.11e). This successful simulation elucidates the mechanism of high-speed bending based on FEA models. Such FEA models are expected to enable a deeper understanding of mechanical motions of crystals from the viewpoint of not only materials chemistry but also materials engineering.

3.5 Evaluation of Actuation Performance

Finally, the actuation performance of mechanically responsive crystals is evaluated with different actuation mechanisms (Fig. 3.12). The tip deflection speed v_l of the resonated natural vibration of **3β** crystal is in the range of 0.2–0.6 m s^{-1} (black solid circle), which is more than ten times faster than the 0.01–0.03 m s^{-1} (black open circle) of the non-resonated natural vibration. The tip deflection speed (0.001–0.05 m s^{-1}) by the photothermal effect itself of the **3β** crystal (red circle) and other crystals (red triangle [21], square [24], and rhombus [25]) is slightly slower to comparable to that of the non-resonant vibration, but still fast because the photothermal effect takes place at the picosecond level [21, 24] in the primary photophysical process. On the other hand, the tip deflection speed by photoisomerization (blue triangle [21], square [24], and rhombus [25]) is slow (0.0001–0.001 m s^{-1}) because photoisomerization takes place at the nanosecond level [21, 24] in the secondary photochemical process. To the best of our knowledge, the bending speed of the resonated natural vibration reported herein is the fastest of reported mechanically responsive crystals. The higher Young's modulus (1.65 GPa) of **3β** crystals than those (0.01–1 GPa) of polymers realized the high natural frequency (200–700 Hz, Fig. 3.10c), contributing to the fast tip deflection speed.

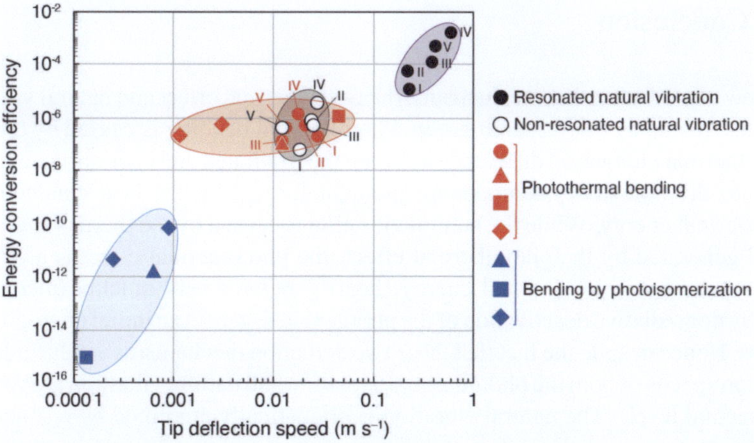

Fig. 3.12 Actuation performance of mechanical crystals with different actuation mechanisms. Relationship between the energy conversion efficiency and the tip deflection speed: the resonated natural vibration (black solid circle in purple ellipse), the non-resonated natural vibration (black open circle in gray ellipse), and the photothermally driven bending (red circle in red ellipse) of **3β** crystals; the photothermally driven bending (red triangle [21], red square [24], and red rhombus [25] in red ellipse) and the photoisomerization-driven bending (blue triangle [21], blue square [24], and blue rhombus [25] in blue ellipse) of other crystals. Reproduced with permission of Hagiwara et al. [15], licensed under Creative Commons Attribution 4.0 International License

The energy conversion efficiency η from incident light energy U_L to kinetic energy U_K is evaluated according to Eq. 3.6 [31],

$$\eta = \frac{U_K}{U_L} = \frac{\frac{1}{2} m_e v_l^2}{I_L l b t_I} \tag{3.6}$$

where m_e is the effective weight of the crystal ($m_e = \frac{104}{405} m$ where m is the crystal weight), I_L is the UV irradiation intensity per area, and t_I is the irradiation time. The energy conversion efficiency of the resonated natural vibration is 10^{-5}–10^{-3} (black solid circle), which is two to three orders higher than that (10^{-8}–10^{-6}) of the non-resonated natural vibration (black open circle) and the 10^{-7}–10^{-6} of the photothermally driven bending (red circle). Compared with other crystals, this efficiency is two and three orders of magnitude higher than the 10^{-7}–10^{-6} of bending by the photothermal effect (red triangle [21], square [24], and rhombus [25]) and ten orders of magnitude higher than that the 10^{-15}–10^{-10} of photoisomerization (blue triangle [21], square [24], and rhombus [24]). Thus, the resonated natural vibration induces actuations with the fastest speed and the highest energy conversion efficiency.

3.6 Conclusion

Two new crystal actuation mechanisms, the photothermal effect and natural vibration were discovered by our research group. Photothermal bending is caused by the non-steady thermal elongation difference between the irradiated and back surfaces, which is due to the non-steady temperature gradient formed by the heat conduction of photothermal energy. While the natural vibration is caused by the thermal load in the crystal generated by the photothermal effect, the photothermal effect is a physical phenomenon by which thermal energy (heat) generates very quickly (picosecond level) by nonradiative deactivation of the photoexcited state in an initial photophysical process. Hence despite the fact that these two actuation mechanisms are different, the actuation speeds of both the photothermal actuation and natural vibration are very fast (millisecond level). The natural vibration is dramatically amplified by resonance to induce actuations with the fastest speed, large actuation, and high energy conversion efficiency. Both the photothermal effect and natural vibration are general physical phenomena caused in all crystals occurring in all crystals as long as light absorption occurs. Actuation behavior can be simulated, making possible to design actuation of crystals; this is important for application of mechanical crystals to actuators and soft robots. These versatile and fast crystal actuation mechanisms, especially resonated natural vibration, are expected to contribute to the development of actuators and soft robotics with high-speed and high-efficient actuation capabilities.

References

1. H. Koshima (ed.), *Mechanically Responsive Materials for Soft Robotics* (Wiley-VCH, Weinheim, 2020)
2. G.A. Abakumov, V.I. Nevodchikov, Thermo- and photomechanical effects on crystals of a free-radical complex. Dokl. Phys. Chem. **266**, 1407–1410 (1982)
3. S. Kobatake, S. Takami, H. Muto, T. Ishikawa, M. Irie, Rapid and reversible shape changes of molecular crystals on photoirradiation. Nature **446**, 778–781 (2007)
4. D. Kitagawa, H. Nishi, S. Kobatake, Photoinduced twisting of a photochromic diarylethene crystal. Angew. Chem. Int. Ed. **52**, 9320–9322 (2013)
5. H. Koshima, N. Ojima, H. Uchimoto, Mechanical motion of azobenzene crystals upon photoirradiation. J. Am. Chem. Soc. **131**, 6890–6891 (2009)
6. H. Koshima, K. Takechi, H. Uchimoto, M. Shiro, D. Hashizume, Photomechanical motion of salicylideneaniline microcrystals. Chem. Commun. **47**, 11423–11425 (2011)
7. L. Zhu, R.O. Al-Kaysi, C.J. Bardeen, Reversible photoinduced twisting of molecular crystal microribbons. J. Am. Chem. Soc. **133**, 12569–12575 (2011)
8. P. Naumov, S.C. Sahoo, B.A. Zakharov, E.V. Boldyreva, Dynamic single crystals: kinematic analysis of photoinduced crystal jumping (The photosalient effect). Angew. Chem. Int. Ed. **52**, 9990–9995 (2013)
9. T. Shima, T. Muraoka, N. Hoshino, T. Akutagawa, Y. Kobayashi, K. Kinbara, Thermally driven polymorphic transition prompting a naked-eye-detectable bending and straightening motion of single crystals. Angew. Chem. Int. Ed. **53**, 7173–7178 (2014)
10. T. Taniguchi, H. Sugiyama, H. Uekusa, M. Shiro, T. Asahi, H. Koshima, Walking and rolling of crystals induced thermally by phase transition. Nat. Commun. **9**, 538 (2018)

11. T. Seki, K. Sakurada, M. Muromoto, H. Ito, Photoinduced single-crystal-to-single-crystal phase transition and photosalient effect of a gold(I) isocyanide complex with shortening of intermolecular aurophilic bonds. Chem. Sci. **6**, 1491–1497 (2015)
12. H. Koshima, S. Hasebe, Y. Hagiwara, T. Asahi, Mechanically responsive organic crystals by light. Isr. J. Chem. **61**, 683–696 (2021)
13. W.M. Awad, D.W. Davies, D. Kitagawa, J.M. Halabi, M.B. Al-Handawi, I. Tahir, F. Tong, G. Campillo-Alvarado, A.G. Shtukenberg, T. Alkhidir, Y. Hagiwara, M. Almehairbi, L. Lan, S. Hasebe, D.P. Karothu, S. Mohamed, H. Koshima, S. Kobatake, Y. Diao, R. Chandrasekar, H. Zhang, C.C. Sun, C. Bardeen, R.O. Al-Kaysi, B. Kahr, P. Naumov, Mechanical properties and peculiarities of molecular crystals. Chem. Soc. Rev. **52**, 3098–3169 (2023)
14. Y. Hagiwara, T. Taniguchi, T. Asahi, H. Koshima, Crystal actuator based on a thermal phase transition and photothermal effect. J. Mater. Chem. C **8**, 4876–4884 (2020)
15. Y. Hagiwara, S. Hasebe, H. Fujisawa, J. Morikawa, T. Asahi, H. Koshima, Photothermally induced natural vibration for versatile and high-speed actuation of crystals. Nat. Commun. **14**, 1354 (2023)
16. S. Timoshenko, Analysis of bi-metal thermostats. J. Opt. Soc. Am. **11**, 233–255 (1925)
17. D. Kitagawa, S. Kobatake, Crystal thickness dependence of photoinduced crystal bending of 1,2-bis(2-methyl-5-(4-(1-naphthoyloxymethyl)phenyl)-3-thienyl)perfluorocyclopentene. J. Phys. Chem. C **117**, 20887–20892 (2013)
18. D. Kitagawa, C. Iwaihara, H. Nishi, S. Kobatake, Quantitative evaluation of photoinduced bending speed of diarylethene crystals. Crystals **5**, 551–561 (2015)
19. T. Taniguchi, H. Sato, Y. Hagiwara, T. Asahi, H. Koshima, Photo-triggered phase transition of a crystal. Commun. Chem. **2**, 19 (2019)
20. Y. Hagiwara, A. Takanabe, T. Asahi, H. Koshima, Photo-triggered phase transition of crystals and photoactuation. Chem. Eur. J. **30**, e202401590 (2024)
21. S. Hasebe, Y. Hagiwara, J. Komiya, M. Ryu, H. Fujisawa, J. Morikawa, T. Katayama, D. Yamanaka, A. Furube, H. Sato, T. Asahi, H. Koshima, Photothermally driven high-speed crystal actuation and its simulation. J. Am. Chem. Soc. **143**, 8866–8877 (2021)
22. R.K. Mobley, *Vibration Fundamentals* (Elsevier, Burlington, USA, VT, 1999)
23. V. Bucur, *Handbook of Materials for String Musical Instruments* (Springer, New York, 2016)
24. S. Hasebe, Y. Hagiwara, K. Takechi, T. Katayama, A. Furube, T. Asahi, H. Koshima, Polymorph-derived diversification of crystal actuation by photoisomerization and the photothermal effect. Chem. Mater. **34**, 1315–1324 (2022)
25. S. Hasebe, Y. Hagiwara, K. Hirata, T. Asahi, H. Koshima, Crystal actuation switching by crystal thickness and light wavelength. Mater. Adv. **3**, 7098–7106 (2022)
26. D.W. Kim, Y. Hagiwara, S. Hasebe, N.O. Dogan, M. Zhang, T. Asahi, H. Koshima, M. Sitti, Broad-wavelength light-driven high-speed hybrid crystal actuators actuated inside tissue-like phantoms. Adv. Funct. Mater. **33**, 2305916 (2023)
27. D. Xu, Z. Li, L. Li, J. Wang, Insights into the photothermal conversion of 2d mxene nanomaterials: synthesis, mechanism, and applications. Adv. Funct. Mater. **30**, 2000712 (2020)
28. H. Takahashi, R. Tamura, Low temperature phase transition induced biaxial negative thermal expansion of 2,4-dinitroanisole. CrystEngComm **17**, 8888–8896 (2015)
29. S.S. Rao, *Vibration of Continuous Systems* (Wiley, Weinheim, 2007)
30. ANSYS® Academic Research Mechanical, @@2021 R1 (ANSYS Inc., Cannonsburg, USA, PA, 2021)
31. S. Serak, N. Tabiryan, R. Vergara, T.J. White, R.A. Vaia, T.J. Bunning, Liquid crystalline polymer cantilever oscillators fueled by light. Soft Matter **6**, 779–783 (2010)

Chapter 4
Photoresponsive Crystals with Bio-mimetic Functions

Ryo Nishimura, Satoshi Yokojima, and Kingo Uchida

Abstract Reversible topographical changes due to the crystal growth of photogenerated isomers are observed on a photochromic diarylethene microcrystalline film surface by alternate irradiation of UV and visible light. These phenomena are attributed to the self-aggregation of photogenerated isomers (open- or closed-ring isomers) of the diarylethenes from the eutectic mixtures of both isomers by UV or visible light irradiations. The shapes of the grown crystals are dependent on the crystal habits of both isomers. By leveraging these phenomena, we prepared surfaces mimicking the microstructures of the lotus leaf with its double roughness structures and the termite wing with its dual wettability. Then, we designed an object transportation system made from vertically arranged diarylethene crystals mimicking the Cilia of *Paramecium aurelia*. Focusing on photosalient phenomena, specifically the photoinduced scattering of the crystal into small pieces, we prepared a crystalline system mimicking *Impatiens*. Thus prepared crystals with hollow or capsule structures break into small pieces with the inclusions scattering upon UV irradiation.

Keywords Biomimetic function · Lotus effect · Double roughness structures · Photosalient phenomenon · Hollow structure · Crystalline capsule · Object transportation system

---

R. Nishimura
Department of Chemistry and Research Center for Smart Molecules, Rikkyo University, Tokyo, Japan

S. Yokojima
School of Pharmacy, Tokyo University of Pharmacy and Life Sciences, Tokyo, Japan

K. Uchida (✉)
Department of Materials Chemistry, Faculty of Science and Technology, Ryukoku University, Otsu, Japan
e-mail: uchida@rins.ryukoku.ac.jp

S. Kobatake and H. Uekusa (eds.), *Advances in Organic Crystal Chemistry*,
https://doi.org/10.1007/978-981-96-4479-7_4

4.1 Introduction

Diarylethenes are well-known photochromic compounds that undergo a cyclization reaction to form colored closed-ring isomers by UV irradiation and revert to colorless open-ring isomers by visible light irradiation. Their photochromism has excellent properties: thermal stability of both isomers, high fatigue resistance, rapid response, high quantum yield, and reactivity even in the crystalline state [1, 2].

The crystals of diarylethenes, whose reactive carbon atoms have a distance of less than 4.2 Å, undergo photochromism upon alternate irradiation with UV and visible light [2]. The photochromism in the crystalline state sometimes induces topographical changes on their surfaces. Irie et al. reported the reversible formation of the steps and valleys on the crystalline surface [3]. On the other hand, we found the photoinduced crystal growth of a photogenerated closed-ring isomer via eutectic mixtures of the closed- and open-ring isomers on the surface [4]. These differences among the phenomena are due to the ambient temperature after photoirradiation to the crystals [5].

On the other hand, plants and insects have specific structures on their surfaces, and these structures have specific functions [6]. For example, a lotus leaf has a double roughness structure (DRS) where the surface of the leaf is covered with protrusions of about 10 μm in diameter and 14 μm in height [7, 8]. All such protrusions are covered with wax tubes with a diameter of about 0.2 μm and a height of about 0.6 μm [8]. Thanks to these structures, the surface of a lotus leaf shows both superhydrophobicity and self-cleaning properties. The surface of a moth eye has a well-known anti-reflective property, where it is covered in a regular array of conical protuberances, typically of about 200 nm in height and spacing [9]. If one could mimic such structures, the surface properties would be reproducible. In 2006, the reversible crystal growth of a diarylethene derivative was discovered, and it was accompanied with changes in the reversible wetting property [4]. This finding became a breakthrough into a new surface science. Consequently, this led to the fabrication of bio-mimicking surfaces by applying these photoinduced crystal growth phenomena.

In recent years, the photoresponse of organic crystals, i.e., photoinduced bending and photosalient phenomena, has attracted much attention. Photoinduced bending was discovered by Irie et al. [10, 11] and then extensively studied by Kobatake's group [12, 13]. Photosalient phenomena were also closely studied by Naumov et al. [14, 15]. By applying these phenomena, the biomimetic photoresponse crystalline systems have also been developed [16].

4.2 Photoinduced Reversible Topographical Changes on Diarylethene Surface

Topographical changes in diarylethene crystals were first reported by Irie et al. in 2001 [3]. Upon UV irradiation of a single crystalline surface, micrometer-sized steps and valleys were formed, and these disappeared upon visible light irradiation. These phenomena were explained by the size difference between the open- and closed-ring isomers. Here, changes in wettability were not studied in this system.

New types of photoinduced changes appear in the morphology of the photochromic diarylethene crystalline surface of 1,2-bis(2-methoxy-5-trimethylsilylthien-3-yl)perfluorocyclopentene (**1o**) (Fig. 4.1), and these are accompanied by wettability changes [4]. Upon UV irradiation (254 nm), the color becomes deep blue within 5 min, yet at room temperature, no change in surface morphology is observed at this moment. After 10 min irradiation, the crystal is stored in the dark, when small needle-shaped crystals immediately start to grow. After storage for 24 h, the surface is covered with many needle-shaped crystals (Fig. 4.1b). Then upon irradiation of the surface with visible light ($\lambda > 500$ nm), the blue color and the needle-shaped crystals disappear (Fig. 4.1c). Topographical changes have also been observed on the surface of the microcrystalline film prepared by coating a chloroform solution of **1o** on a glass substrate [4]. From X-ray analysis, the photogenerated needle-shaped crystals are determined to be the crystals of closed-ring isomer **1c**.

Measurement of the contact angle (CA) of a water droplet is a simple and direct method for evaluating wettability on a surface. The CA of the film surface before UV irradiation is 120°. Upon irradiation of the surface with UV light (254 nm), the

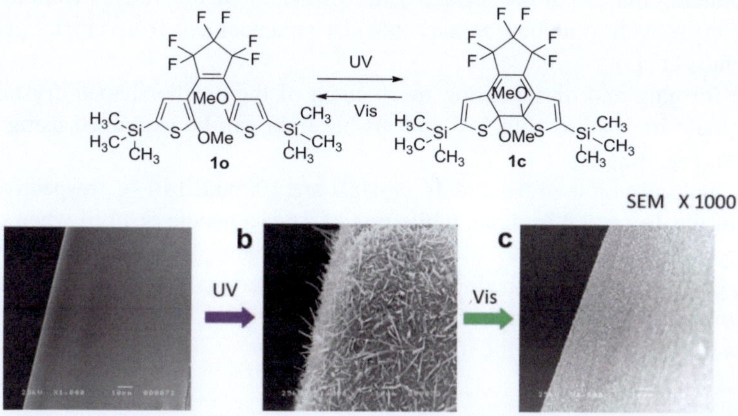

Fig. 4.1 Photoinduced topographical changes on a single crystal of **1o**. **a** SEM image of a crystal surface of **1o** from a side view before UV irradiation. **b** After UV (254 nm, 12 W) irradiation for 10 min followed by storing for 24 h at 30 °C in the dark. **c** SEM image from a side view of the surface after irradiation with visible light ($\lambda > 500$ nm, 500 W, 20 min) and storage in the dark for 24 h. Reprinted with permission from Ref. [4]. Copyright 2006, John Wiley and Sons

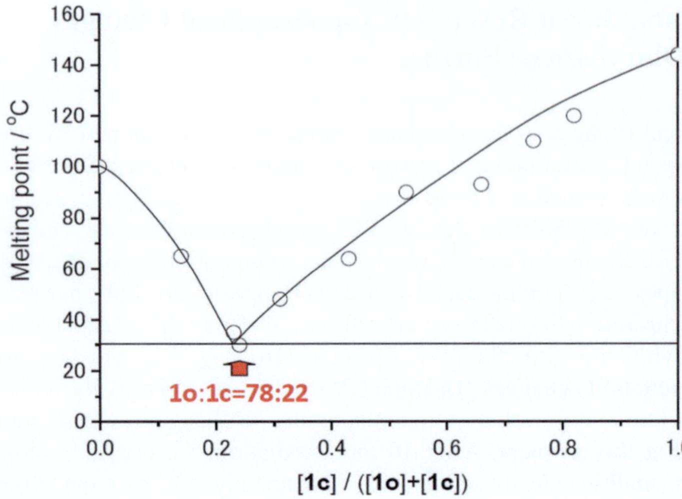

Fig. 4.2 Phase diagram of mixtures of open- (**1o**) and closed-ring (**1c**) isomers of diarylethene crystals formed by irradiation with UV light (313 nm). Reprinted with permission from Ref. [4]. Copyright 2006, John Wiley and Sons

color becomes blue, and the angle decreases to around 90° after a short irradiation time (1–2 min). After 24 h, the angle gradually increases to 163°, showing superhydrophobicity. The colored surface is irradiated with visible light, and then the blue color disappears after 20 min as the CA gradually decreases and finally recovers to the initial 120° as the fibrils melt [4]. Furthermore, the needle shape is important for enhancing the CA of the surface, since diarylethene derivatives without needle-shaped crystals do not show remarkable CA enhancement from crystal growth on the surface [17].

The forming and disappearing mechanism of the needle-shaped crystals of **1c** by alternate irradiation with UV and visible light can be explained using a phase diagram (Fig. 4.2) [4].

The melting points of **1o** and **1c** crystals are 100 and 140 °C, respectively. The shape of the **1o** crystal is cubic, while that of **1c** is a needle or fibril when obtained by recrystallization from the solutions of these isomers. The needle-shaped crystals tend to grow in one direction along the a-axis of the crystal [4]. Upon UV irradiation, **1o** is converted to **1c**. When the conversion exceeds 22%, crystal growth of **1c** starts to form fibril structures on the film surface.

4.3 Photoinduced Fabrication of Double Roughness Structures (DRS) Mimicking the Surface of a Lotus Leaf

In nature, many plants and insects have specifically structured surfaces with specific surface properties. These structures have received much attention, and artificial systems have been fabricated by mimicking such natural surfaces. Historically, the most important finding is the "lotus effect." The super-water-repellent and self-cleaning effects of the lotus leaf have been attributed to the DRS of surfaces with micro- and nanostructures.

A hydrophobic surface whose CA of water droplets exceeds 150° is classified as a superhydrophobic surface. The superhydrophobicity of a lotus leaf was initially reported by W. Barthlott and N. Ehler in 1977 [18], when they described such self-cleaning and superhydrophobic properties for the first time as the "lotus effect." Such surfaces have been given much attention for their many practical uses, such as self-cleaning materials.

We prepared and reported a photoresponsive superhydrophobic surface with DRS mimicking the lotus leaf using diarylethene **1o** (Fig. 4.3). Initially, the microcrystalline surface of **1o** prepared by solution casting was irradiated with UV light. Due to the cyclization reaction, the colorless **1o** near the surface converted to blue-colored **1c** (Fig. 4.3a 1 → 2). Then, the needle-shaped crystals of **1c** grew on the surface, which was kept at 50 °C in the dark (Fig. 4.3a 2 → 3). After storage at 50 °C for 24 h, the surface was covered with rod-shaped crystals of **1c** (Fig. 4.3a 3 → 4). The topographical change on the surface is due to Ostwald ripening [17]. In order to fabricate the DRS, initially thicker rod-shaped crystals, which do not melt upon visible light irradiation, were prepared, and then these played the role of a larger-sized structure, which is the substrate for the fine needle-shaped crystals of the DRS.

In order to make thick rod-shaped crystals of **1c**, Ostwald ripening was repeatedly applied. Then, the surface in Fig. 4.3a-4 was again irradiated with UV light and stored at 70 °C for 3 h in the dark (Fig. 4.3a-5): the small rod-shaped crystals of **1c** grew from the newly generated **1c** on the surface (Fig. 4.3a-6), and they were incorporated in the larger rod-shaped crystals (Fig. 4.3a-7). The average width and length of these rod-shaped crystals at the state illustrated in Fig. 4.3a-7 are 10 and 21 μm, respectively. Upon visible light irradiation to the surface of this 4.3a-7 state, a cycloreversion reaction from **1c** to **1o** proceeds, and the surface is covered with cubic-shaped crystals of **1o** (Fig. 4.3a-8) on the 10-μm rough structures. The size of each bumpy rod-shaped crystal is almost the same as that of a projection on a lotus leaf that is covered with plant wax tubes. The enlarged crystals do not melt upon visible right irradiation. Upon UV irradiation to the surface, each rod-shaped crystal is covered with needle-shaped crystals of **1c** of 0.2−0.5 μm width and of 3−5 μm length, thus preparing DRSs (Fig. 4.3a-9) [19].

After the formation of DRSs, CA and sliding angle (SA) on the surface changed to 161.9 ± 2.0° and 2°, respectively. The advanced (CA_{ad}) and receding (CA_{rec}) CAs of a water droplet were 162.0 ± 1.6° and 158.3 ± 1.4°, respectively. Consequently, a

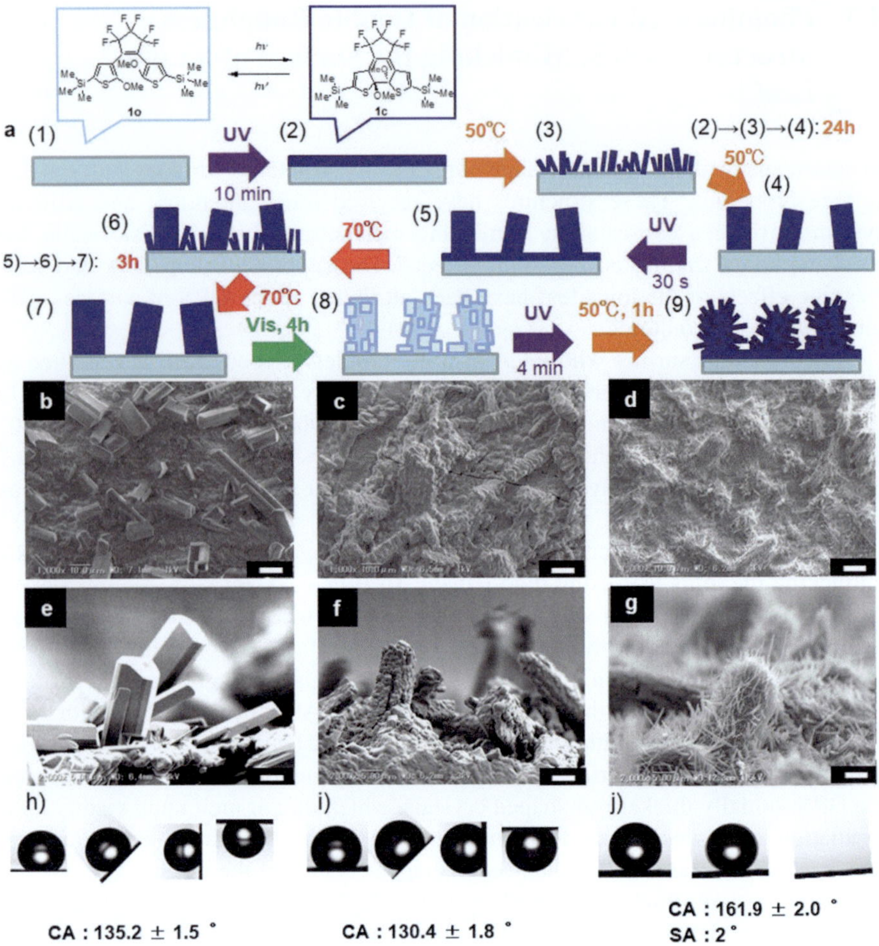

Fig. 4.3 **a**-1–9 Preparation of lotus surface with DRSs: pale blue and dark blue parts correspond to crystals of **1o** and **1c**, respectively. SEM images of each surface: **b** surface with rod-shaped crystals of **1c**; **c** after visible light irradiation to surface of b; **d** after second UV irradiation to surface of c, followed by storing at 50 °C for 1 h in the dark; **e**–**g** side view of the surfaces of b, c, d, respectively; **h**–**j** water droplet (1.5 μL) on surfaces of a-7, a-8, and a-9, respectively. Scale bars: 10 μm for images b–d, which are magnified 1000 times; 5 μm for images e–g, which are magnified 2000 times. Reprinted with permission from Ref. [19]. Copyright 2016 American Chemical Society

remarkable lotus effect was observed. Additionally, the surface shows the bouncing ability of a water droplet, while the microcrystalline surfaces covered with crystals do not show this ability, as shown in Figs. 4.1b and 4.3a-3. The DRS mimicking a natural lotus leads to observation of the bouncing ability, reflecting the importance of this structure's role.

Lotus plants that grow at the waterside need to keep their leaves clean and free of dirt to photosynthesize during the summertime. To wash away dirt such as spider webs

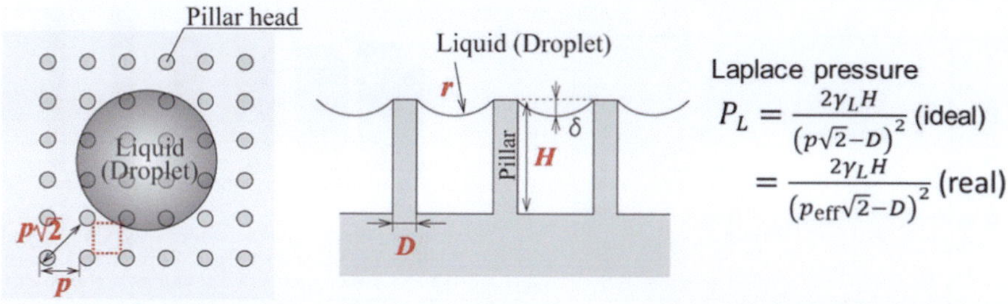

Fig. 4.4 Left: Schematic representations of Laplace pressure and multi-pillar (rod) surface as ideal surface, Center/Right: γ_L: surface tension of liquid, H: height of surface structure, p_{eff}: surface structure spacing, D: diameter of surface structure. Reprinted with permission from Ref. [19]. Copyright 2016 American Chemical Society

and bird droppings, lotus leaves are thought to use the kinetic energy of raindrops to remove such stubborn debris. Therefore, the bouncing ability of the leaf surface would be indispensable. A water droplet released from a 15-cm height corresponds to rainfall. The DRS shown in Fig. 4.3a-9, d and g can bear a water droplet released from a 10-cm height but not from a 15-cm height.

The Laplace pressure of the surface of a natural lotus leaf and our DRS by using a pillar model (Fig. 4.4) was calculated [20]. From the equation in the figure, Laplace pressure is strongly dependent on the spacing between surface structures. The average spacings of a natural lotus leaf and a double roughness crystalline surface were measured. The former average spacing is 18.6 μm, while the latter one is 27.3 μm. From these values, the calculated Laplace pressures for lotus leaf and the crystalline surface are 2827 and 2344 Pa, respectively. Therefore, the crystalline surface with DRS cannot bounce a water droplet from a 15-cm height. To enhance the Laplace pressure, it is necessary to make a crystalline surface with smaller spacing of the structures.

In order to make a surface with smaller spacing structures, a detailed study of the process in Fig. 4.3 was carried out. Consequently, the temperature of the a-3 to a-4 process in Fig. 4.3 was found to be the most important condition. The average spacing of the rod-shaped crystals changed from 37.1 μm (at 45 °C) to 23.5 μm (at 50 °C), 19.1 μm (at 55 °C), 15.2 μm (at 60 °C), 14.7 μm (at 65 °C), and 12.2 μm (at 70 °C) as shown in Fig. 4.5. However, the rod-shaped crystals generated above 60 °C became narrower, and then they melted following the visible light irradiation process; therefore, suitable crystals were those obtained at 55 °C. The average interval is almost the same as that of a natural lotus leaf. Accordingly, the Laplace pressure is almost the same as that of a lotus leaf, and the same bouncing property from a 15-cm height was observed on the microcrystalline surface [20]. These results indicate that the surface structures on a lotus leaf are designed to bounce rainfall as a way to clean the surface. This process could be used to maintain a clean surface area for photosynthesis in the summertime.

However, the DRS using **1o** and **1c** has two problems. In the fabrication process of the DRS, it is necessary to irradiate the surface with UV and visible light repetitively

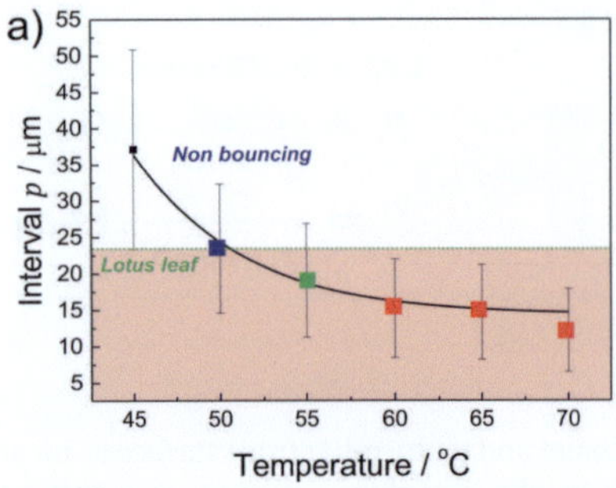

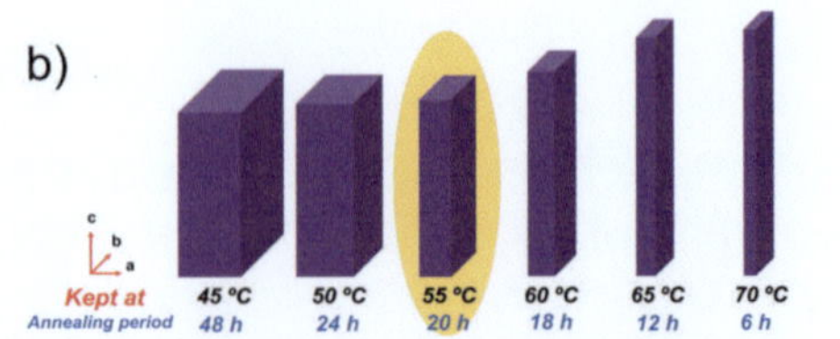

	45 (°C)	50 (°C)	55 (°C)	60 (°C)	65 (°C)	70 (°C)
a axis (µm)	7.10 ± 1.3	7.11 ± 1.6	6.12 ± 1.7	4.83 ± 1.6	3.28 ± 1.2	2.09 ± 1.0
b axis (µm)	11.4 ± 2.5	9.05 ± 2.1	8.83 ± 1.3	8.21 ± 2.0	7.59 ± 2.3	4.61 ± 1.4
c axis (µm)	18.7 ± 3.3	19.2 ± 3.5	19.6 ± 1.9	22.6 ± 3.8	26.8 ± 4.1	27.7 ± 6.7
Volume (μm^3)	1518.1	1236.7	1058.6	894.6	667.2	266.8

Fig. 4.5 Average intervals (a) and sizes (b) of photogenerated rod-shaped crystals of **1c** depending on the temperature of crystal growth as in Fig. 4.3a-3–4. Reprinted with permission from Ref. [20]. Copyright 2019 American Chemical Society

while controlling temperature (Fig. 4.3a) for a long period (32 h). In addition, the formed structure can be easily broken by the impact of falling water droplets.

Therefore, the analog **2o** (Fig. 4.6) was prepared to study the photoinduced crystal growth process, since DAEs having a six-membered perfluorocyclohexene ring at the ethene moiety show a remarkable difference in crystal growth and photoresponse compared to a derivative having a five-membered perfluorocyclopentene ring, as found in our previous studies [21, 22].

A microcrystalline surface of **2o** was prepared by solution casting at 30 °C for the fabrication of DRSs. A schematic illustration of the procedure used to fabricate the DRS is shown in Fig. 4.6 [23].

After casting a chloroform solution of **2o** on a glass substrate, a microcrystalline surface was generated immediately after evaporation of the solvent in 20 min under a normal-pressure atmosphere. During the process, this surface was covered with cube-like crystals, which were stacked platelet crystals of **2o** (Fig. 4.6c). Such a phenomenon has not been observed for **1o**.

After UV irradiation for 5 min to the microcrystalline surface of **2o** at 24 °C (room temperature), each crystal of **2o** was covered with needle-shaped crystals of **2c** (Fig. 4.6d, e). The sizes of the needle-shaped crystals on average were 0.2−0.35 µm in width and 0.7−1.2 µm in length. These needle-crystals grow during the 5 min of UV irradiation at room temperature. Then, the CA increased to 164°. The fast crystal growth of **2c** was attributed to the rough crystalline surface of **2o**. The photogenerated DRS of **2c** reverted to the original surface of **2o** upon irradiation with visible light for 20 min at the eutectic temperature (104.2 °C).

The durability of superhydrophobicity of the surfaces of **1** and **2** was then compared (Fig. 4.7). After 50 water droplets (5.5 µL) were released from a height of 15 cm, the CA of the DRS of **1** decreased from 161° to 126°, while that of the DRS of **2** changed from 163° to 161° while maintaining superhydrophobicity. The remarkable decrease in CA for the DRS of **1** was attributed to breaking the needle-shaped crystals. The durability of the needle-shaped crystals of **2c** was due to the aspect ratio of the **2c** crystals. The length and width of the needle-shaped

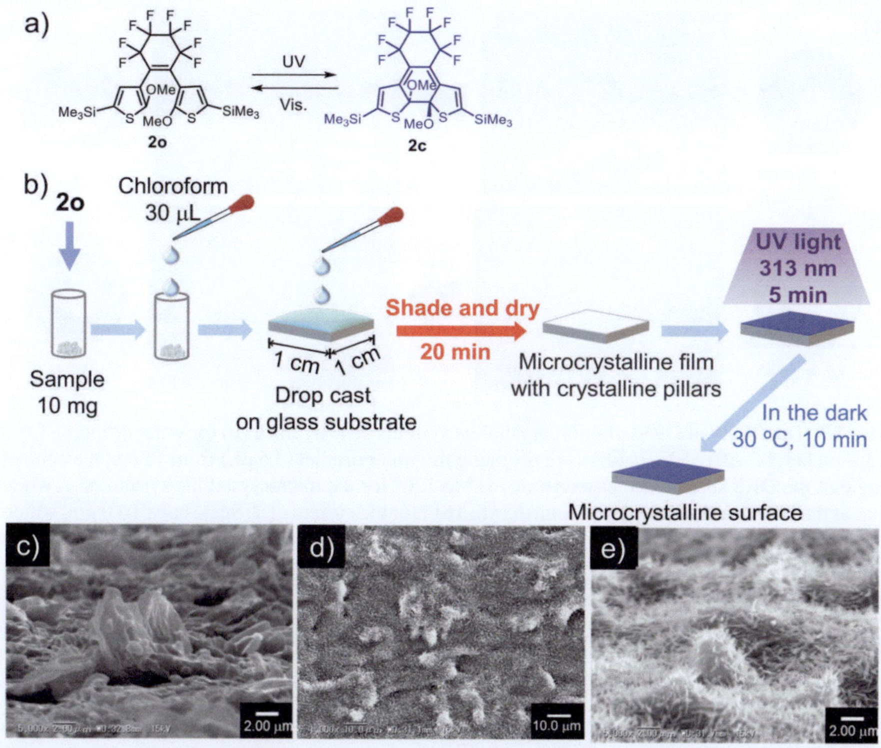

Fig. 4.6 **a** Molecular structures of a DAE **2**, **b** Schematic illustration of one-step formation of DRSs of **2**, **c** SEM image of the crystalline surface of **2o** before UV irradiation (slightly high-angle view), **d** DRS after 5 min UV irradiation followed by storage in the dark for 10 min (top view), and **e** magnified SEM image of (**d**) (slightly high-angle view). Scale bars: 2.0 μm for (**c**) and (**e**), which are magnified 5000 times, and 10 μm for (**d**), which is magnified 1000 times. Reprinted with permission from Ref. [23]. Copyright 2024 American Chemical Society

crystals of **1c** were 15–25 μm and 1.0–1.1 μm, respectively, while those of **2c** were 2.5–5.5 μm and 0.4–0.5 μm, respectively. This is the key point for the durability of the superhydrophobicity of the microcrystalline surface of **2** [23].

The importance of the submicron-sized structure of DRSs was also demonstrated using a natural lotus leaf. The surface of the lotus leaf is covered with projections having a diameter of ca. 10 μm, and these are covered with nanostructures consisting of 0.1-μm ϕ-sized plant wax tubes. After removal of the wax tubes by washing with ethanol, the lotus effect was lost, also indicating the importance of the submicrometer-sized structure to the lotus effect [24].

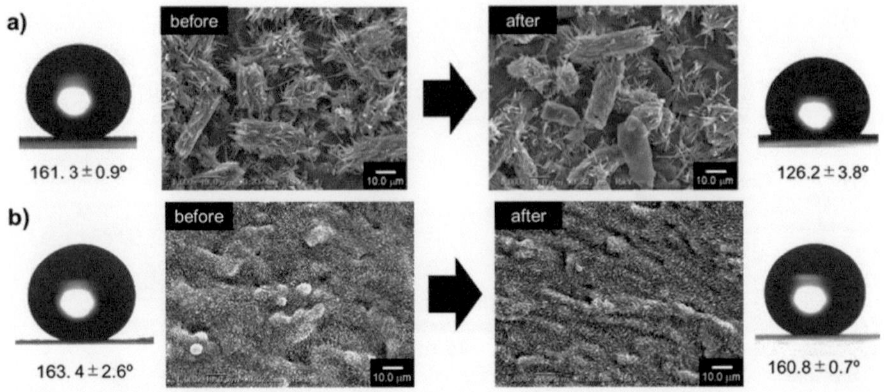

Fig. 4.7 Durability of DRSs of microcrystalline systems of **1** (**a**) and **2** (**b**) for water dropping from a 15-cm height. After 50 repetitions of dropping the water droplet (5.5 μL) from 15 cm, the contact angle of the DRS surface decreased from 161° to 126° for the microcrystalline system of **1**, while that of the DRS surface decreased from 163° to 161° for the system of **2**. Scale bars: 10.0 μm, which are magnified 1000 times. Reprinted with permission from Ref. [23]. Copyright 2024 American Chemical Society

4.4 Photoinduced Fabrication of Microcrystalline Surface Mimicking Termite Wing

The unique micro- and nanostructures on such surfaces were also reported for the termite wing by Watson et al. [25]. They described the surface of a wing of the termites *Nasutitermes* sp. and *Microcerotermes* sp. These termites fly during the rain period to avoid attacks by predators. Watson et al. reported the delicate structures, or "dual wettability," on their wings, and these specific structures work to minimize interaction with water. The termite's wing surface is covered with two types of projections of different sizes. One is a hair-like projection (macrotrichia) that is around 50 μm long and 1 μm wide, and the other is a star-shaped projection (micrasters) of 5–6 μm in height and width. These hair-like macrotrichia protect the wing from contact with large water droplets, whereas the small star-shaped micrasters work to keep the small droplets on the wing surface. Then the small droplets gather to build up large ones by surface tension and are finally removed via the hair arrays.

We focused on the collection ability of small water droplets and reproduced the surface structure using two diarylethene derivatives **1o** and **3o**. Diarylethene **3o** (Fig. 4.8) also shows the photoinduced crystal growth phenomenon upon UV irradiation [26, 27]. The size of photogenerated needle-shaped crystals of **3c** is much smaller than those of **1c**. Therefore, a mixed microcrystalline surface was produced by casting the chloroform solution containing **1o** and **3o** with equimolar amounts on a glass substrate followed by evaporation and storage for 9 days at 30 °C in the dark after UV irradiation. The larger needle-shaped crystals that grew on the mixed microcrystalline surface were crystals of **1c**, and their lengths and widths were about 16 and 1.5 μm, respectively, whereas the smaller ones were crystals of **3c** with lengths and widths of about 1.9 and 0.2 μm, respectively (Fig. 4.9) [28].

As can be seen in Fig. 4.9, needle-shaped crystals of **1c** and **3c** were independently grown on the surface, even in the mixed condition, and their sizes (Fig. 4.9e, h) were almost the same as their pure microcrystalline films (Fig. 4.9d, f, g, i).

Then a water spraying test was carried out to examine the difference in dynamic wetting properties on these microcrystalline surfaces. $\mathbf{Surf}_{\mathbf{1c+3c}}$ should have the dual wettability of capturing and repelling water owing to the presence of two different sizes of needle-shaped crystals. This kind of wettability is the same as that of a termite wing. On the surface of the termite wing, by spraying the microdroplet (20–150 μm), small-sized droplets (< 100 μm) were maintained on the wing surface

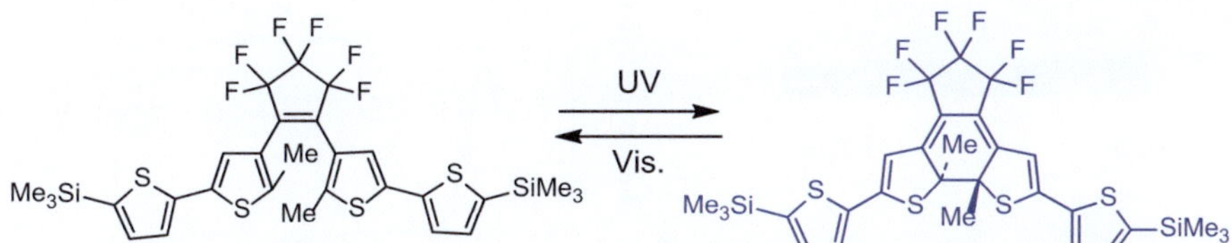

3o **3c**

Fig. 4.8 Photochromism of diarylethene **3** and the molecular structures

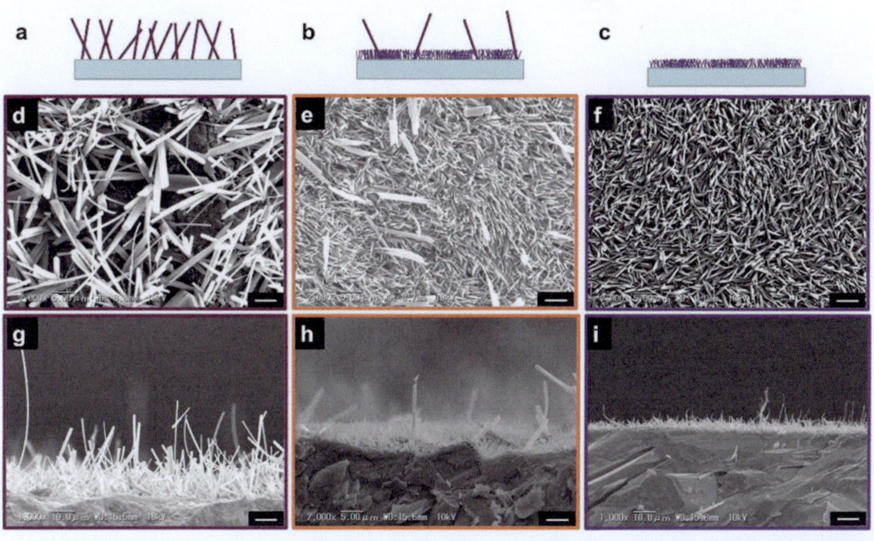

Fig. 4.9 Comparison of the topologies on photogenerated microcrystalline surfaces. **a** $\mathbf{Surf}_{\mathbf{1c}}$, **b** $\mathbf{Surf}_{\mathbf{1c+3c}}$, and **c** $\mathbf{Surf}_{\mathbf{3c}}$: pale blue substrate and dark purple crystals correspond to crystals of open- and closed-ring isomers, respectively. SEM images of each surface; **d**, **g** surfaces with needle-shaped crystals of **1c**, **e**, **h** surface with needle-shaped crystals of **1c** and **3c**, **f**, **i** surface with needle-shaped crystals of **3c**. Scale bars: 5 μm for (**d**–**f**) and (**h**) images, which are magnified 2000 times; 10 μm for (**g**) and (**i**) images, which are magnified 1000 times

[25]. To compare the dual wettability with that of the termite wing, a spray test using microdroplet (20–640 μm) was carried out on **Surf$_{1c}$**, **Surf$_{3c}$**, and **Surf$_{1c+3c}$**. Two cases were observed: microdroplets showing bouncing and those showing non-bouncing phenomena (Fig. 4.10).

The sizes of all droplets contacting the surfaces were measured and summarized to show the distribution of the droplet sizes. From these results, the bouncing and non-bouncing ranges were separated by mixing the two diarylethenes **1c** and **3c**. The

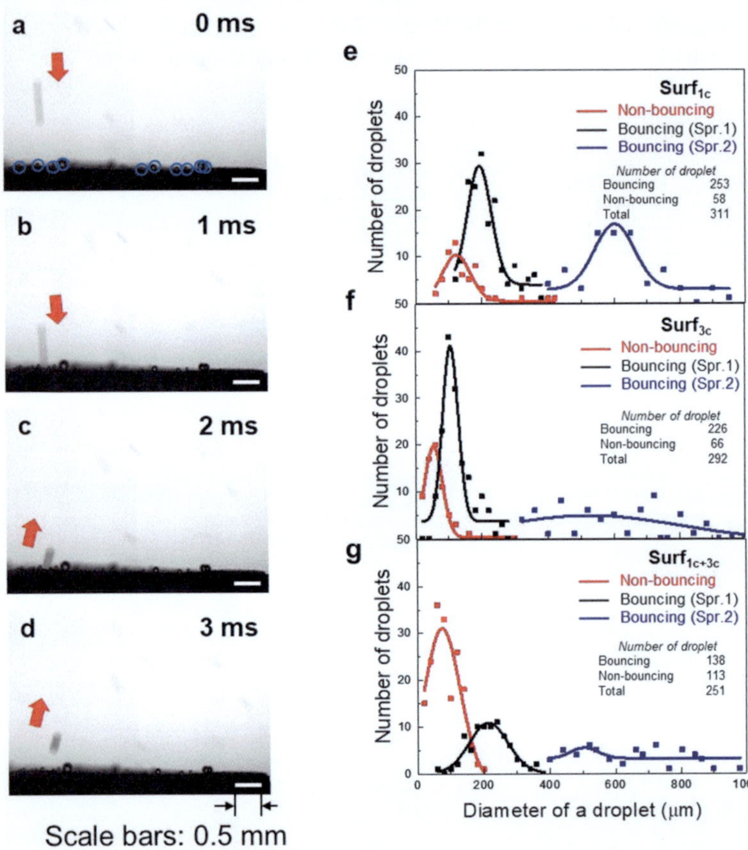

Fig. 4.10 Optical images of a microdroplet and size distributions. **a–d** Optical afterimages of the movement of a water droplet during bouncing on **Surf1c+3c** for 3 ms. Red arrows indicate the direction of the droplet's movement. Blue circles are droplets adsorbing on the surface. **e–g** Distributions of diameters of water droplets contacting **Surf$_{1c}$**, **Surf$_{3c}$**, and **Surf$_{1c+3c}$**. **e** Distribution of diameters of water droplets contacting **Surf$_{1c}$**, **f Surf$_{3c}$**, and **g Surf$_{1c+3c}$**. In graphs (**e–g**), red dots show diameter distribution of non-bouncing droplets and black dots show diameter distribution of bouncing droplets generated by sprayer 1 (Spr. 1: Generating a microdroplet of 40–400 μm in diameter). Blue dots show diameter distribution of bouncing droplets generated by sprayer 2 (Spr. 2: generating microdroplets of 400–1000 μm in diameter). Number of droplets used for preparation of distribution is indicated inside graphs e–g). Scale bars for all optical images are 0.5 mm

surface successfully mimicked that of a termite wing. By using a soft lithography technique and leveraging the properties of a polymer, the surfaces were stabilized against daylight and impact from water droplets while keeping their dual wettability [29].

4.5 Photoinduced Generation of Superhydrophilic Microcrystalline Surface

The previous two sections described photoinduced formation of superhydrophobic surfaces by crystal growth. Next, we explain the preparation of a photoresponsive superhydrophilic surface. Snail shells show superhydrophilicity and self-cleaning properties. Water droplets immediately expand over the surface when they are attached to it. To construct a superhydrophilic surface, it is necessary to prepare a derivative that has a contact angle of less than 65° when the surface is flat and molecular switches should be polar but still not dissolve in water. To achieve this, we prepared a derivative **4o** having an ionic liquid structure at both ends of the molecule [30]. Figure 4.11 shows the molecular structures of the open- (**4o**) and closed-ring (**4c**) isomers, the phase diagram of **4o** and **4c**, and their crystal shapes estimated by the Bravais, Friedel, Donnay, and Harker (BFDH) crystal morphology. The melting points of **4o** and **4c** are 203.2 and 190.2 °C, respectively. The eutectic temperature is 161.1 °C. At eutectic points, the ratio of **4o** to **4c** is 66:34. The T_g of **4o** is 64.4 °C.

The microcrystalline surface of **4o** was also prepared on a glass substrate by the solution casting method. The acetonitrile solution was coated on the substrate. After the solvent was evaporated, the microcrystalline surface of **4o** was formed. Then the film was stored at 70 °C for 24 h in the dark and a crystalline surface of **4o** was generated. The CA of a water droplet was 0° on the surface, showing superhydrophobicity. Upon UV irradiation for 5 min, the surface melted, and a flat surface was formed.

On the surface, the ratio of **4o** and **4c** was around 6:4, which is almost the same ratio at the eutectic point. Thus, the surface was flat due to the eutectic mixture. Furthermore, the CA of a water droplet on the surface was 60°; it has been reported that a contact angle smaller than 65° on a flat polymer surface decreases when the surface roughens, while a contact angle larger than 65° on other surfaces increases when the surface roughens [31]. On the flat surface, visible light (λ > 430 nm) was irradiated for 5 min followed by storage at 70 °C for 24 h in the dark, a rough surface covered with bumpy crystals of **4o** formed, and the CA recovered to 0° (Fig. 4.12). The flat and superhydrophilic rough surfaces reversibly formed by alternating irradiation with UV and visible light at least three times, indicating the surface was not dissolved in water.

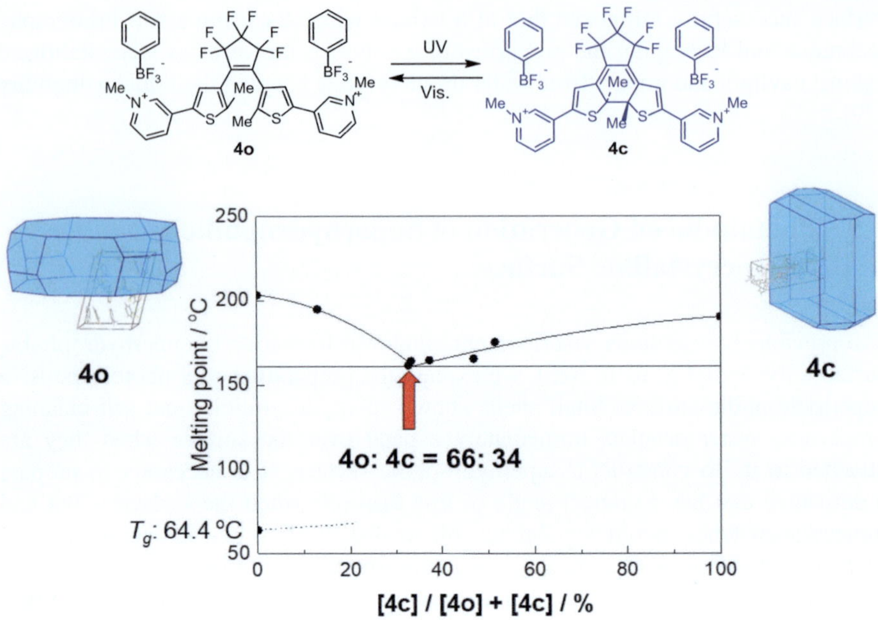

Fig. 4.11 Photochromism of diarylethene **4,** estimated crystal shapes by the crystal units of **4o** and **4c**, and phase diagram of mixtures of open- (**4o**) and closed-ring (**4c**) isomers. Reprinted with permission from Ref. [30]. Copyright 2016 The Royal Society of Chemistry

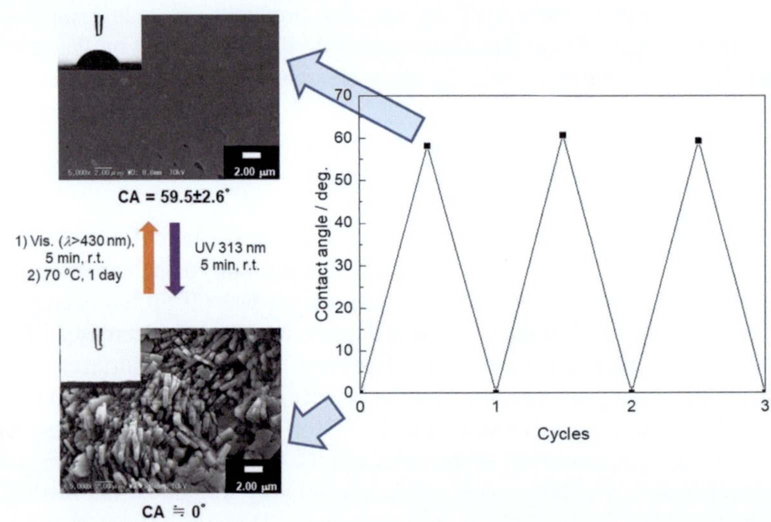

Fig. 4.12 Reversible formation of a superhydrophilic rough surface of **4o** upon alternate irradiation with UV and visible light. Scale bars: 2.00 μm, which is magnified 5000 times. Reprinted with permission from Ref. [30]. Copyright 2016 The Royal Society of Chemistry

4.6 Object Transportation System Mimicking the Cilia of *Paramecium aurelia*

In the above, we described biomimetic surfaces by applying photoinduced crystal growth phenomena. This section describes a biomimetic function using the photoinduced bending phenomena of photochromic diarylethene crystals. Photoinduced bending of diarylethene crystals was initially reported by Irie et al. [10]. They also reported the lifting of a metal ball attached to a needle-shaped crystal upon UV irradiation [11]. We designed an object transportation system using the bending behavior of surface-assembled diarylethene crystals [32]. The molecular structure of the diarylethene derivative used for the crystals and the concept of the system are shown in Fig. 4.13. The needle-shaped crystals show bending phenomena upon UV irradiation. Therefore, the object on the crystals was expected to move due to the bending of these crystals.

The crystals were prepared by sublimation of the powders in a glass cell (thickness = 1 mm). Initially, sublimation of **5o** onto the glass substrate under an ambient atmosphere made dot-shaped amorphous microstructures consisting of **5o** generated densely at 250 °C for 20 min instead of the standing platelet crystals of **5o** (Fig. 4.14b, c). The average diameter and height of the dotted structures were 0.30 and 0.43 μm, respectively. Consequently, the rough surface provided nucleation sites for crystal growth. The use of this dotted surface with the Au alloy as a target substrate resulted in the accelerated growth of thin platelet crystals of **5o** on each of the dots on the glass substrate (Fig. 4.14e, f). The average length and width of the platelet crystals of **5o** on the glass substrate were 384.5 and 10.5 μm, respectively.

Next, the object transportation on the crystal arrays was examined by photoirradiation. UV light was irradiated from various directions. First, the object transportation surface was irradiated with UV light ($\lambda = 254$ nm) from the bottom of the screen and all of the crystals bent away from the light. Accompanied with this movement,

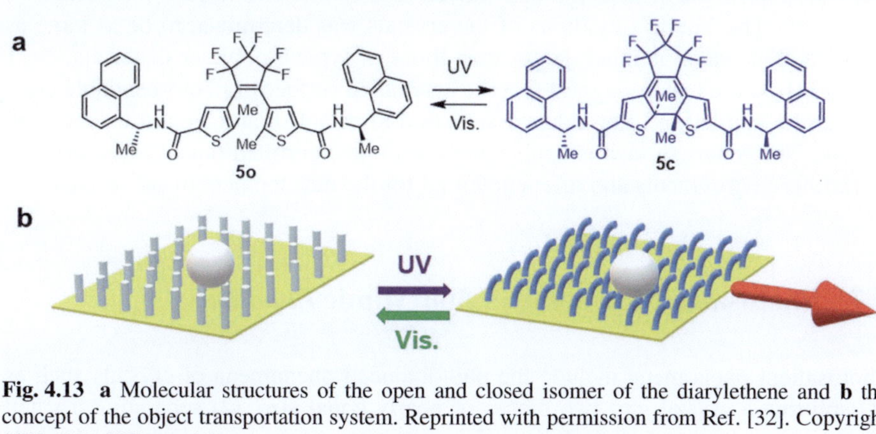

Fig. 4.13 **a** Molecular structures of the open and closed isomer of the diarylethene and **b** the concept of the object transportation system. Reprinted with permission from Ref. [32]. Copyright 2019, John Wiley and Sons

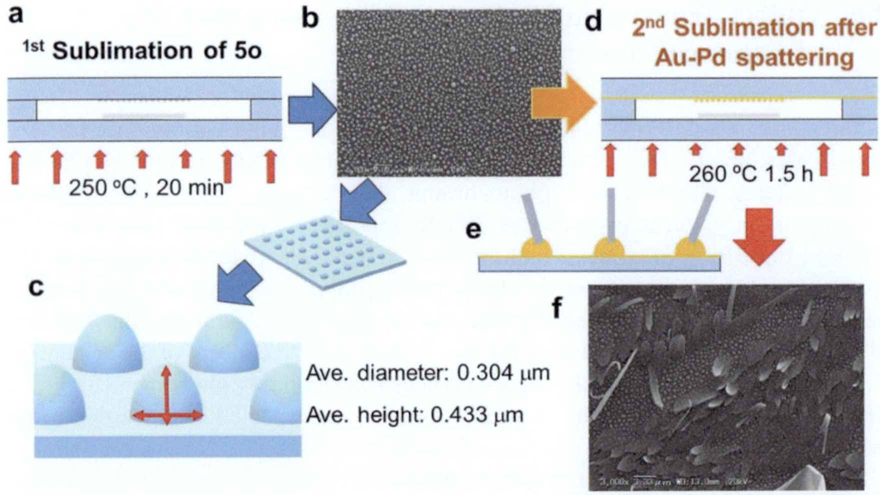

Fig. 4.14 Schematic illustration of the generation of amorphous dotted microstructures and sublimation onto the dotted substrate with Au sputtering (**a**–**d**). SEM image of the dotted substrate (**b**) after sublimation of the crystalline film of **5o** onto the dotted substrate with Au sputtering. Scale bar: 2.00 μm, which is magnified 5000 times. Next sublimation after Au-spattered dotted structure formed bending crystals (**e**), with SEM image (**f**). Scale bar: 3.33 μm, which is magnified 3000 times. Reprinted with permission from Ref. [32]. Copyright 2019, John Wiley and Sons

the polystyrene microsphere (PB) (average diameter 1 mm) on the crystals moved upward (Fig. 4.15a). This was followed by UV irradiation from the right side that induced the sphere to roll to the left side (Fig. 4.15b), corresponding to the direction of UV incidence. Next, the surface was irradiated with visible light applied from the top orthogonal to the surface, while the crystals reverted to straight and the sphere reverted to its original position (Fig. 4.15c). Furthermore, upon UV light irradiation from the top of the screen, the ball moved in the direction toward the bottom of the screen. The Young's modulus of the crystals was determined to be as large as 5.5 GPa. This value is much larger than those of typical polymer materials (ca. 1 GPa) [33]. These results suggest that the assembled surface of photo-bendable crystals can be applied in light-induced actuation to establish an object transportation system. This smart surface is applicable to remote-controlled object transportation in various environments and offers potential for the development of soft robots.

4.7 Photosalient Phenomena that Mimic *Impatiens*

Photosalient phenomena include the photoinduced phenomena of crystals such as (1) splitting a crystal into two pieces of nearly equal size, (2) separating a small piece of a crystal that propels the remaining part into a spinning or linear motion, (3) crystal explosion, (4) minor repositioning without lifting off the base and causing

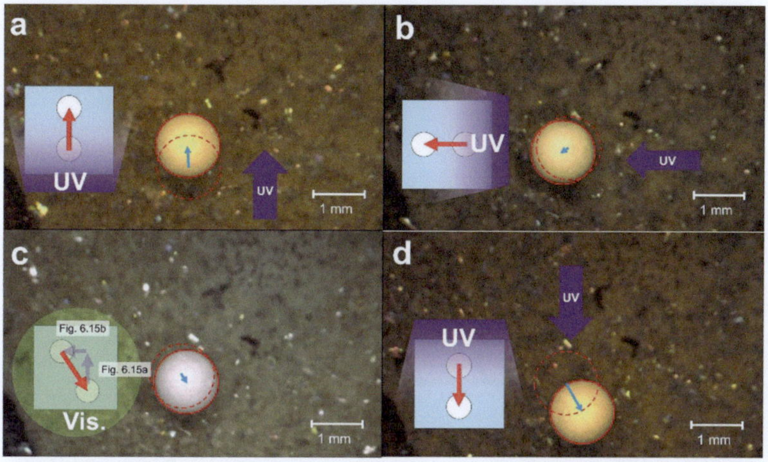

Fig. 4.15 Photographs showing transportation of PB upon UV and visible light irradiation. **a** After UV light irradiation from the bottom of the screen, **b** after UV light irradiation from the right side of the screen, **c** after visible light irradiation orthogonal to the surface, **d** after UV light irradiation from the top of the screen. Scale bars: 1 mm. Reprinted with permission from Ref. [33]. Copyright 2019, John Wiley and Sons. Reprinted with permission from Ref. [32]. Copyright 2019, John Wiley and Sons

apparent splitting or separation of debris, and (5) rolling or flipping the crystal [14]. Such remarkable phenomena of "burst" crystals have received much attention due to their rapid response and fast speed. Naumov et al. intensively studied crystal jumping and breaking using a light stimulus and named this the photosalient effect (salient is derived from the Latin *salire,* which means leaping) [14].

Compared to the systems behind the photoinduced crystal growth and the photoinduced bending of diarylethene crystals described in the above sections, the response of photosalient phenomena is much faster and more remarkable. As its photosalient system, the photoinduced crystal jump of diarylethene microcrystals upon UV irradiation has been reported [34].

In the following, we describe a photoresponsive crystalline system that mimics the *Impatiens*, which disperse their seeds at a distance as the seedpod explodes open. In other words, this is a photoreactive crystalline system that exhibits photoinduced scattering of inclusions by light stimuli. The crystals of diarylethene **6o** are reported to form a crystal that undergoes photoinduced bending [10]. Here, a derivative **7o** whose perfluorocyclopentene moiety of **6o** is replaced by a six-membered perfluorocyclohexene ring was prepared [22]. In addition, a derivative **8o**, which has two *m*-trimethylsilyl phenyl groups at both ends of the molecule, was also prepared (Fig. 4.16) [35].

On sublimation, **7o** formed hollow crystals (Fig. 4.17a, b) in 4.8% yield as explained by the Iwanaga model [36] (Fig. 4.17c), while **6o**, having a five-membered perfluorocyclopentene ring, formed no hollow crystals [22].

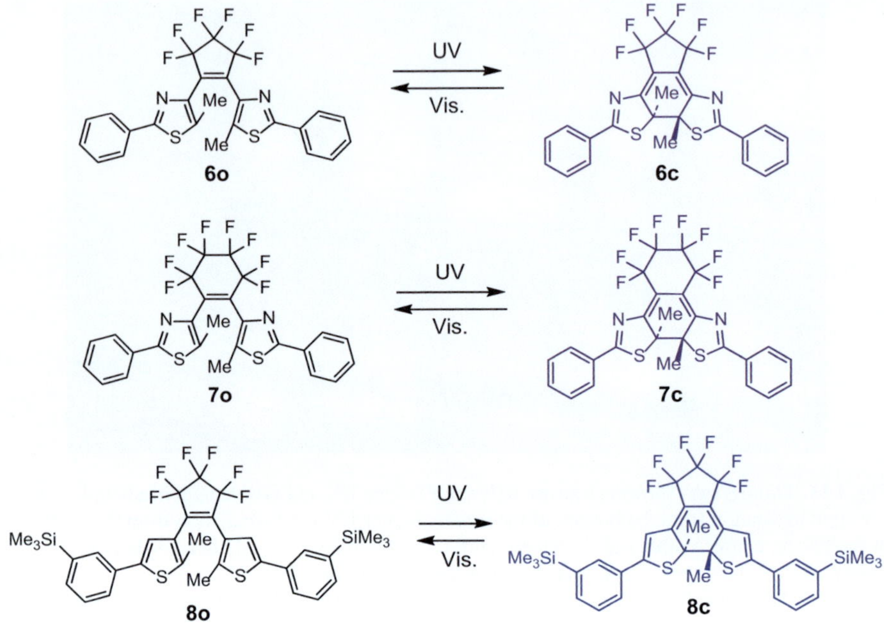

Fig. 4.16 Molecular structures of diarylethenes. A derivative whose crystal shows photoinduced bending phenomenon (**6**) and derivatives (**7** and **8**) that form crystalline systems mimicking *Impatiens* with photosalient phenomena

The needle-shaped crystal of **6o** shows photoinduced bending, while that of **7o** shows not only bending but also photosalient phenomena, i.e., photoinduced breaking into small pieces. This is due to the relatively large changes in size of the molecules from the open- to closed-ring state [21].

Into a hollow crystal of **7o**, fluorescent polymer beads of 1 μm diameter were filled by the capillary effect (Fig. 4.17d). Upon UV irradiation, the hollow crystal turned to purple, which is the color of **7c**, and started breaking (Fig. 4.17e). Finally, the crystal broke into small pieces as the fluorescent beads scattered (Fig. 4.17f). This resembles how *Impatiens* spray their seeds [22].

The **7o** crystals prepared by sublimation are cone-shaped, so one side of the crystals is always opened. Therefore, it is impossible for liquids or solutions to be included in the hollow, which makes photoreactive crystalline capsules desirable.

Crystalline capsules were prepared by recrystallization from a solution of **8o** and fluorescence dye **5(6)-FAM** for inclusions (Fig. 4.18). The capsule is shown in Fig. 4.18c under white light.

When the crystalline capsule is irradiated with 450 nm light in the dark, green fluorescence from the encapsulated **5(6)-FAM** solution can be observed without any photoresponse of the crystal. Upon UV (365 nm) irradiation to the capsule, it is immediately broken, and the release of green fluorescence is observed (Fig. 4.19). Such photoresponse is observable upon pulsed laser light of near-infrared (802 nm)

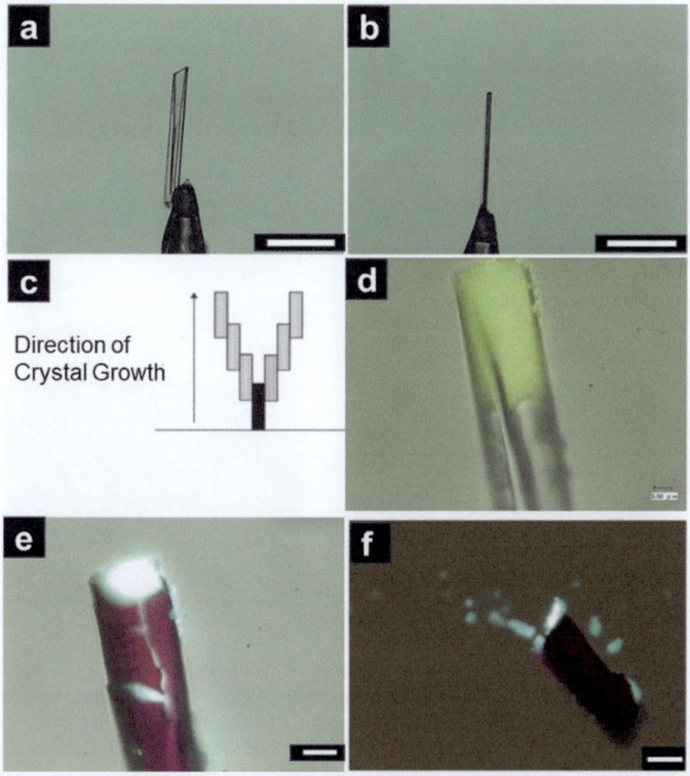

Fig. 4.17 Hollow crystal obtained by sublimation. **a** Front and **b** side view of the crystal. **c** Crystal growth by Iwanaga model. **d** Hollow crystal filled with fluorescent microbeads having diameter of 1 μm. **e** Upon UV irradiation to the crystal, its color changes to red–purple, and finally scattering occurs with microbeads (**f**). Scale bars: **a**, **b** 100 μm, which are magnified 100 times; **d** 5 μm, which is magnified 2000 times; **e** 10 μm, which is magnified 1000 times; **f** 20 μm, which is magnified 500 times. Reprinted with permission from Ref. [22]. Copyright 2017, John Wiley and Sons

light corresponding to the biological window (650–1350 nm) in which biological tissue is penetrated. This is an important factor for applications, such as drug delivery systems, that operate inside living bodies.

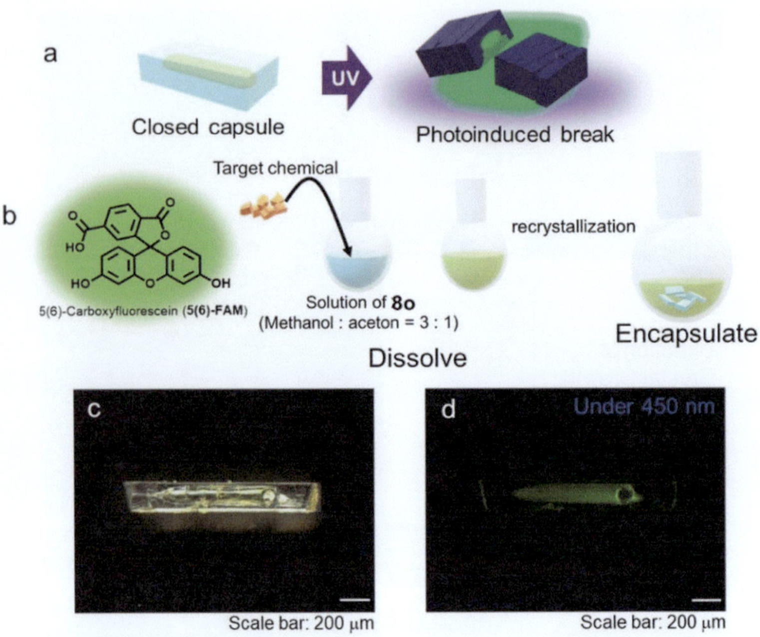

Fig. 4.18 **a** Image of photo-release from a crystalline capsule. **b** Schematic illustration of making crystalline capsules of **8o** in the existing green fluorescent dye 5(6)-carboxyfluorescein (**5(6)-FAM**). **c** Crystalline capsule of **8o** including a **5(6)-FAM** solution under white light. **d** Same capsule irradiated with 450 nm light in the dark, which does not induce the photoreaction of **8o**. Green fluorescence from **5(6)-FAM** solution is observed. Reprinted with permission from Ref. [35]. Copyright 2021 The Royal Society of Chemistry

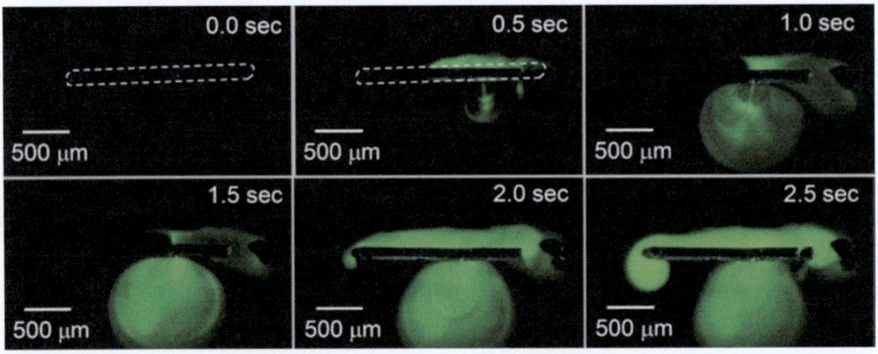

Fig. 4.19 Crystalline capsule irradiated with 365 nm light, which induces the photoreaction of **8o** to release the fluorescent solution of **5(6)-FAM** by a photosalient phenomenon. The fluorescence area spreads on the water surface over time. Reprinted with permission from Ref. [35]. Copyright 2021 The Royal Society of Chemistry

4.8 Conclusion

Diarylethenes are thermally irreversible photochromic compounds, so they can be applied to photoinduced crystal growth phenomena even at elevated temperatures (below 100 °C). The shapes of photogenerated crystals are dependent on their crystal habits. However, needle-shaped crystals are most applicable to control the wetting properties on surfaces. By controlling the irradiation periods and temperatures or by mixing the diarylethene derivatives, it is possible to effectively reproduce complicated surface structures such as a lotus leaf having DRSs or a termite wing having dual wettability. Moreover, this research attempts to understand the natural surface structures of plants and insects. A superhydrophilic surface was also reproduced by selecting an ionic-structured diarylethene. As a result, a superhydrophilic surface was successfully fabricated as well as superhydrophobic surfaces. In addition, by controlling crystal shapes to mimic natural surfaces, we can expect to generate additional outstanding surface functions by photoirradiation.

The photosalient phenomenon characterized by breaking crystals at very high speed was applied to hollow-shaped crystals. These systems mimicking *Impatiens* allow photoinduced release of the inclusions in the crystals in very short periods. Such a photoresponsive crystalline system will give rise to photo-actuators and new types of robots in the near future.

References

1. M. Irie, T. Fukaminato, K. Matsuda, S. Kobatake, Photochromism of diarylethene molecules and crystals: memories, switches, and actuators. Chem. Rev. **114**, 12174–12277 (2014)
2. M. Irie, *Diarylethene Molecular Photoswithes; Concepts and Functionalities* (Wiley-VCH, Weinheim, 2021)
3. M. Irie, S. Kobatake, M. Horichi, Reversible surface morphology changes of a photochromic diarylethene single crystal by photoirradiation. Science **291**, 1769–1772 (2001)
4. K. Uchida, N. Izumi, S. Sukata, Y. Kojima, S. Nakamura, M. Irie, Photoinduced reversible formation of microfibrils on a photochromic diarylethene microcrystalline surface. Angew. Chem. Int. Ed. **45**, 6470–6473 (2006)
5. N. Fujinaga, N. Nishikawa, R. Nishimura, K. Hyodo, S. Yamazoe, Y. Kojima, K. Yamamoto, T. Tsujioka, M. Morimoto, S. Yokojima, S. Nakamura, K. Uchida, Photoinduced topographical changes on microcrystalline surfaces of diarylethenes. CrystEngComm **18**, 7229–7235 (2016)
6. F. Xia, L. Jiang, Bio-inspired, smart, multiscale interfacial materials. Adv. Mater. **20**, 2842–2858 (2008)
7. W. Barthlott, C. Neinhuis, Purity of the sacred lotus, or escape from contamination in biological surfaces. Planta **202**, 1–8 (1997)
8. C. Neinhuis, W. Barthlott, Characterization and distribution of water-repellent, self-cleaning plant surfaces. Ann. Bot. **79**, 667–677 (1997)
9. P.B. Clapham, M.C. Hutley, Reduction of lens reflexion by the "Moth eye" principle. Nature **244**, 281–282 (1973)
10. S. Kobatake, S. Takami, H. Muto, T. Ishikawa, M. Irie, Rapid and reversible shape changes of molecular crystals on photoirradiation. Nature **446**, 778–781 (2007)
11. M. Morimoto, M. Irie, A diarylethene cocrystal that converts light into mechanical work. J. Am. Chem. Soc. **132**, 14172–14178 (2010)

12. D. Kitagawa, K. Kawasaki, R. Tanaka, S. Kobatake, Mechanical behavior of molecular crystals induced by combination of photochromic reaction and reversible single-crystal-to-single-crystal phase transition. Chem. Mater. **29**, 7524–7532 (2017)
13. D. Kitagawa, H. Tsujioka, F. Tong, X. Dong, C.J. Bardeen, S. Kobatake, Control of photomechanical crystal twisting by illumination direction. J. Am. Chem. Soc. **140**, 4208–4212 (2018)
14. P. Naumov, S.C. Sahoo, B.A. Zakharov, E.V. Boldyreva, Dynamic single crystals: kinematic analysis of photoinduced crystal jumping (The photosalient effect). Angew. Chem. Int. Ed. **52**, 9990–9995 (2013)
15. P. Naumov, S. Chizhik, M.K. Panda, N.K. Nath, E. Boldyreva, Mechanically responsive molecular crystals. Chem. Rev. **115**, 12440–12490 (2015)
16. K. Uchida, R. Nishimura, E. Hatano, H. Mayama, S. Yokojima, Photochromic crystalline systems mimicking bio-functions. Chem. Eur. J. **24**, 8491–8506 (2018)
17. K. Uchida, N. Nishikawa, N. Izumi, S. Yamazoe, H. Mayama, Y. Kojima, S. Yokojima, S. Nakamura, K. Tsujii, M. Irie, Phototunable diarylethene microcrystalline surfaces: lotus and petal effects upon wetting. Angew. Chem. Int. Ed. **49**, 5942–5944 (2010)
18. W. Barthlott, N. Ehler, Raster-Elektronen-mikroskopie der epidermis-oberflaechen von spermatophyten. Trop. Subtrop. Pflanzenwelt **19**, 1–105 (1977)
19. R. Nishimura, K. Hyodo, H. Sawaguchi, Y. Yamamoto, Y. Nonomura, H. Mayama, S. Yokojima, S. Nakamura, K. Uchida, Fractal surfaces of molecular crystals mimicking lotus leaf with phototunable double roughness structures. J. Am. Chem. Soc. **138**, 10299–10303 (2016)
20. R. Nishimura, H. Mayama, Y. Nonomura, S. Yokojima, S. Nakamura, K. Uchida, Crystal growth technique for formation of double roughness structures mimicking lotus leaf. Langmuir **35**, 14124–14132 (2019)
21. E. Hatano, M. Morimoto, K. Hyodo, N. Yasuda, S. Yokojima, S. Nakamura, K. Uchida, Photosalient effect of a diarylethene with a perfluorocyclohexene ring. Chem. Eur. J. **22**, 12680–12683 (2016)
22. E. Hatano, M. Morimoto, T. Imai, K. Hyodo, A. Fujimoto, R. Nishimura, A. Sekine, N. Yasuda, S. Yokojima, S. Nakamura, K. Uchida, Photosalient phenomena that mimic *Impatiens* are observed in hollow crystals of diarylethene with a perfluorocyclohexene ring. Angew. Chem. Int. Ed. **56**, 12576–12580 (2017)
23. Y. Hashimoto, A. Hase, R. Shiromae, R. Nishimura, M. Morimoto, Y. Hattori, H. Mayama, S. Yokojima, S. Nakamura, K. Uchida, Straightforward fabrication of double roughness structures on microcrystalline film of a diarylethene derivative. Langmuir **40**, 7661–7668 (2024)
24. M. Yamamoto, N. Nishikawa, H. Mayama, Y. Nonomura, S. Yokojima, S. Nakamura, K. Uchida, Theoretical explanation of the lotus effect: superhydrophobic property changes by removal of nanostructures from the surface of a lotus leaf. Langmuir **31**, 7355–7363 (2015)
25. G.S. Watson, B.W. Cribb, J.A. Watson, How micro/nanoarchitecture facilitates anti-wetting: an elegant hierarchical design on the termite wing. ACS Nano **4**, 129–136 (2010)
26. N. Nishikawa, H. Kiyohara, S. Sakiyama, S. Yamazoe, H. Mayama, T. Tsujioka, Y. Kojima, S. Yokojima, S. Nakamura, K. Uchida, Photoinduced formation of superhydrophobic surface on which contact angle of a water droplet exceeds 170° by reversible topographical changes on a diarylethene microcrystalline surface. Langmuir **28**, 17817–17824 (2012)
27. N. Nishikawa, S. Sakiyama, S. Yamazoe, Y. Kojima, E. Nishihara, T. Tsujioka, H. Mayama, S. Yokojima, S. Nakamura, K. Uchida, Photoinduced self-epitaxial crystal growth of a diarylethene derivative with antireflection moth-eye and superhydrophobic lotus effects. Langmuir **29**, 8164–8169 (2013)
28. R. Nishimura, K. Hyodo, H. Mayama, S. Yokojima, S. Nakamura, K. Uchida, Dual wettability on diarylethene microcrystalline surface mimicking a termite wing. Commun. Chem. **2**, 90 (2019)
29. Y. Hashimoto, A. Hase, A. Tani, R. Nishimura, Y. Hattori, H. Mayama, S. Yokojima, S. Nakamura, K. Uchida, Polymer replica on microcrystalline surface with dual wettability, mimicking a termite wing. Polym. J. **56**, 847–853 (2024)

30. K. Takase, K. Hyodo, M. Morimoto, Y. Kojima, H. Mayama, S. Yokojima, S. Nakamura, K. Uchida, Photoinduced reversible formation of a superhydrophilic surface by crystal growth of diarylethene. Chem. Commun. **52**, 6885–6887 (2016)
31. X. Feng, L. Jiang, Design and creation of superwetting/antiwetting surfaces. Adv. Mater. **18**, 3063–3078 (2006)
32. R. Nishimura, A. Fujimoto, N. Yasuda, M. Morimoto, T. Nagasaka, H. Sotome, S. Ito, H. Miyasaka, S. Yokojima, S. Nakamura, B.L. Feringa, K. Uchida, Object transportation system mimicking the Cilia of *Paramecium aurelia* making use of the light-controllable crystal bending behavior of a photochromic diarylethene. Angew. Chem. Int. Ed. **58**, 13308–13312 (2019)
33. C.L. van Oosten, K.D. Harris, C.W.M. Bastiaansen, D.J. Broer, Glassy photomechanical liquid-crystal network actuators for microscale devices. Eur. Phys. J. E **23**, 329–336 (2007)
34. I. Colombier, S. Spagnoli, A. Corval, P.L. Baldeck, M. Giraud, A. Leaustic, P. Yu, M. Irie, Diarylethene microcrystals make directional jumps upon ultraviolet irradiation. J. Chem. Phys. **126**, 011101 (2007)
35. A. Nagai, R. Nishimura, Y. Hattori, E. Hatano, A. Fujimoto, M. Morimoto, N. Yasuda, K. Kamada, H. Sotome, H. Miyasaka, S. Yokojima, S. Nakamura, K. Uchida, Molecular crystalline capsules that release their contents by light. Chem. Sci. **12**, 11585–11592 (2021)
36. H. Iwanaga, T. Yoshiie, T. Yamaguchi, N. Shibata, Growth mechanism of hollow CdSe and CdS crystals. J. Cryst. Growth **51**, 438–442 (1981)

Chapter 5
Mechanical Properties in Organic Molecular Crystals Toward Photonic Applications

Shotaro Hayashi and Takumi Matsuo

Abstract Elastic molecular crystals are crystalline molecular substances with mechanical deformability and robustness. Over the years, deformable organic crystals have received extensive attention, but these lack robustness, thus; elasticity in molecular crystals offers reversibility in crystal deformation. The properties of flexibility with optoelectronic characteristics are essential for various applications. It is preferable to develop the elastic molecular crystals of π-conjugated molecular systems to afford controlled HOMO–LUMO bandgap for optical and electric applications. Recent advances in the development of elastic molecular crystals and their analytical challenges open the way for successful photonic applications such as optical waveguides, optical resonators, and amplified spontaneous emission. However, in addition to the lack of guidelines for crystal design, photonics measurements lack uniformity and reproducibility. Therefore, a thorough appropriate characterization method of molecular crystals is vital for understanding the photonics of mechanically flexible molecular crystals. This chapter highlights recent elastic molecular crystals and emphasizes their potential photonic applications and appropriate characterization methods. Basic organic crystal photonics, an important advancement point in flexible molecular crystals, optical waveguides, optical resonators, and analytical techniques, are also covered.

Keywords Flexibility · π-Conjugated molecules · Elastic molecular crystals · Optical waveguides · Optical resonators · Lasing

S. Hayashi (✉) · T. Matsuo
School of Engineering Science, Center of FOREST, Research Institute, Kochi University of Technology, Kochi, Japan
e-mail: hayashi.shotaro@kochi-tech.ac.jp

S. Kobatake and H. Uekusa (eds.), *Advances in Organic Crystal Chemistry*,
https://doi.org/10.1007/978-981-96-4479-7_5

5.1 Introduction

5.1.1 Scope of the Book Chapter

π-Conjugated molecules exhibit a narrow bandgap due to the delocalization of π-electrons; thus, the molecular crystals display impressive optical functions depending on their arrangement and orientation (Fig. 5.1a) [1, 2]. In particular, the light-emitting properties open the door of possibilities for various photonics applications (optical switches, waveguides, cavities, and lasers) [3–6]. Such crystal photonics is attracting attention because of its chemical tunability. However, because the measurement method depends on the system and methodology, the results of many studies vary. Therefore, conducting the evaluation correctly is an important issue for the future. The optical properties of organic crystals without robustness are fatal in such photonics. Many crystals are brittle and will cleave or plastically deform in response to stress. On the other hand, flexible (elastic) molecular crystals are considered potentially flexible optical materials [7–10]. Elastic molecular crystals are materials with mechanical strength and deformability (Fig. 5.1b). Therefore, these are formally shown as elastic properties. In other words, these show that the stress-deformed crystal relaxes and macroscopically recovers to its original shape. The semiconducting property in such crystals offers potential as flexible and wearable crystal devices [3–6, 11, 12]. Therefore, creating elastic molecular crystals consisting of π-conjugated molecules is important for realizing flexible crystal photonics. This chapter introduces elastic molecular crystals and some examples of measurement methods in molecular crystal photonics.

Like many flexible materials, the deformation of crystals is based on the Poisson effect (Fig. 5.1c). This interpretation is important in understanding flexibility and has a particularly significant impact on optical resonators, as discussed later. First, when many materials are expanded in one direction, they tend to compress in the other two directions perpendicular to the direction of expansion. A similar effect is observed in elastic molecular crystals. However, because the deformation of the crystal originates from an anisotropic supramolecular structure, the compression rate in the other two directions perpendicular to the direction of expansion tends to be different. As in many optical fiber studies, understanding Poisson's ratio should be considered important in the flexible photonics of molecular crystals (Fig. 5.1c) [13].

5.1.2 Photonic Applications of Organic Single Crystals

5.1.2.1 Organic Crystal Photonics

Organic photonics uses organic molecules that are luminescent in the solid state. In photonics using organic molecular crystals, the high refractive index allows light to be confined inside the crystal, and it is expected that excellent functions will be

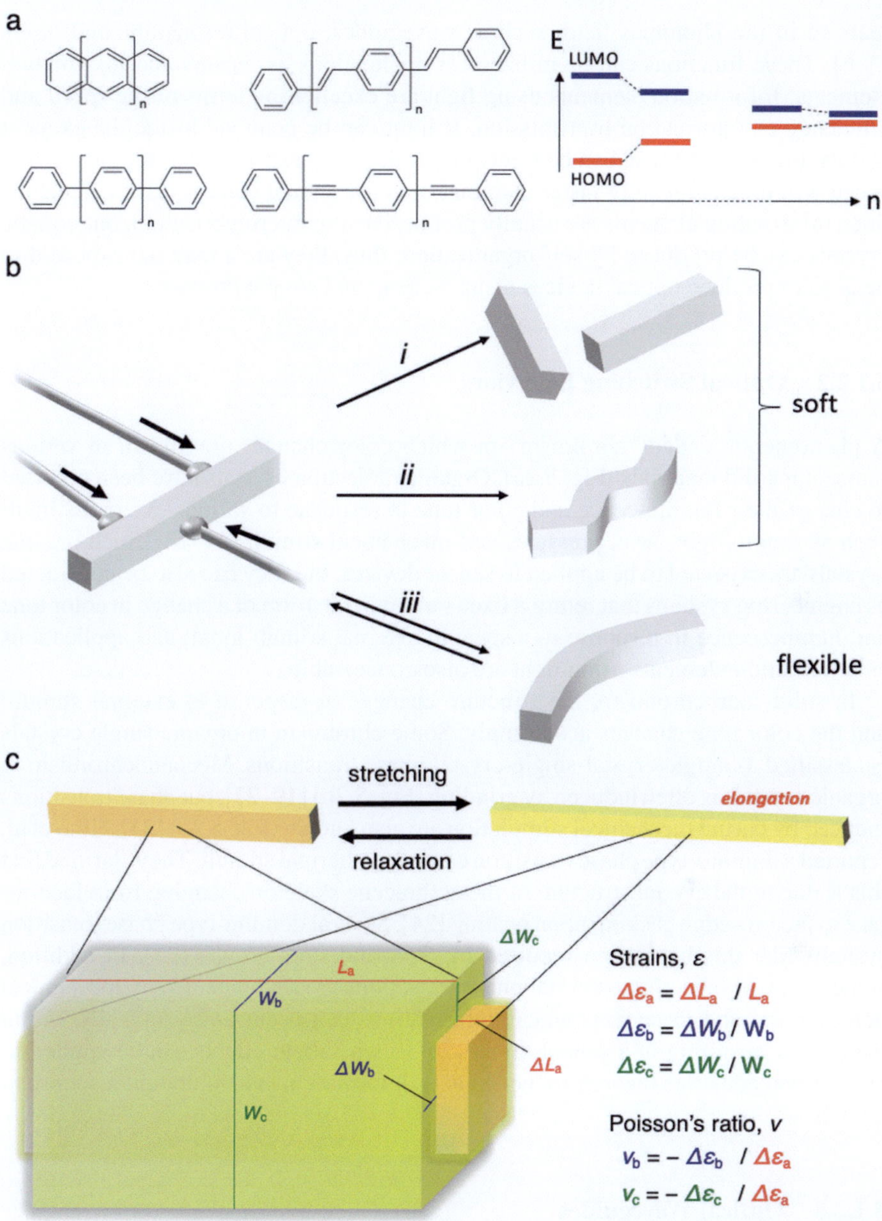

Fig. 5.1 **a** Structures of polyaromatic hydrocarbon as an example of π-conjugated molecules and their band gap. **b** Schematic illustration of bending stress for organic crystals. **c** Schematic illustration of stretching stress for the elastic organic crystals

realized in the photonics field, such as waveguides, optical resonators, and lasers [3–6]. These functions can be utilized for applications as sensors and information elements. Information elements using light are excellent in terms of the speed and efficiency of information transmission. If light can be confined inside the element and its direction of travel can be freely controlled, it is expected that the calculation speed will be significantly faster than that of conventional current-driven elements. Such information elements are usually produced using microfabrication, but organic crystals can be produced by self-organization; thus, they are a material expected to be used to produce optical devices using a cheap and simple process.

5.1.2.2 Optical Switching Behaviors

A phenomenon called "chromism" in which color changes are known in various molecular solid materials (Fig. 5.2a). Organic molecular crystals have been reported to change their luminescence and color tone in response to various external stimuli such as steam, light, heat, pressure, and mechanical stimuli [14–18]. Such organic crystals are expected to be applied to sensor devices, but they can also be considered as engineering systems that return a fixed output in the form of a change in color tone and luminescence in response to a specific external stimuli input, and applications as information elements using light are also conceivable.

In solid-state chromism, the structure changes in response to external stimuli, and the color tone changes accordingly. Some chromism in organic single crystals is classified as single-crystal-single-crystal phase transitions. Mechanochromism in organic crystals is often induced by grinding (Fig. 5.2b) [19–22], but phase transitions induced by partial mechanical stimulation are also known (Fig. 5.2c) [23]. Hino et al. reported a domino-type phase transition caused by thermal stimuli. They clarified that this is due to the crystal structure of the anthracene skeleton changing from face-to-face to face-to-edge packing upon heating [24]. Several domino-type phase transition systems like this have been reported by Seki and Ito's groups [25]. In addition, some single-crystal-single-crystal phase transitions are accompanied by mechanical deformation, and these mechanical deformations can occur anisotropically within the crystal, resulting in a salient effect [26]. Such salient effects can be applied to sensors and actuators that return mechanical deformation as an output in response to stimuli.

5.1.2.3 Optical Waveguides

One of the applications of molecular crystals has been studied in recent years as optical waveguides (OWGs). Common OWGs can be broadly divided into passive and active processes, and their schematic diagrams are shown in Fig. 5.3 [27]. When light meets one side of the waveguide, the light comes out from the other terminal after some loss. Materials widely used as waveguides include quartz, silicon, and polymers. The characteristic of these materials is based on their flexibility, which

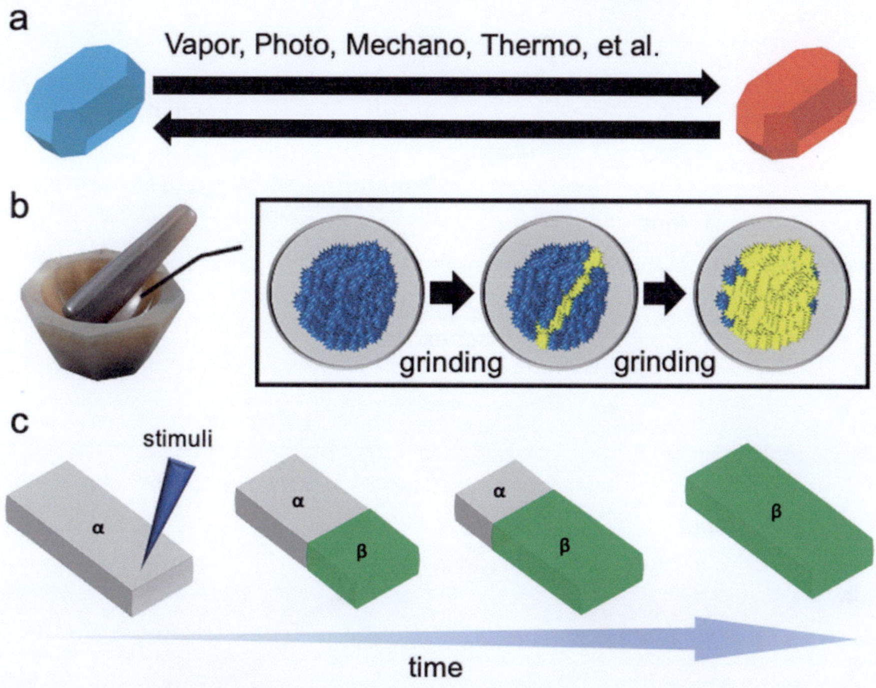

Fig. 5.2 **a** Stimuli-induced luminescence changes in molecular crystals. **b** Mechanofluorochromic behavior before and after grinding. **c** Crystal-to-crystal domino phase transition. Reproduced from Ref. [11] with permission from The Royal Society of Chemistry

allows them to be processed into the desired shape. The difference between these materials and flexible molecular crystals is thought to be the relatively high refractive index caused by the electronic system of the constituent molecules (Fig. 5.3).

Unlike many one-dimensional materials, i.e., fibers, molecular crystals have various shapes formed by self-organization [28–30]. In the crystal, the molecules are arranged and oriented depending on various parameters of chemical structures, and the transition dipole moments (TDM, μ) are also arranged with anisotropy based on this. Light that diffuses perpendicular to the transition dipoles strongly interacts with them and is reabsorbed. On the other hand, if they are parallel, the photon dissipation by reabsorption is small. In other words, the orientation of the molecules correlates with the efficiency of this OWG [31]. The loss coefficient is calculated as the efficiency of the OWG by the method described later in Sect. 5.3.2. When this evaluation is performed on 2D crystals, different attenuation coefficients are often obtained depending on the waveguiding direction [31]. When placed in a 2D platelet crystal and arranged in 1D, the transition dipole moment is oriented in 1D. For example, in the alternating cocrystal of bipyridyl-appended phenylene acrylonitrile **1** and 1,2,4,5-tetrafluoro-3,6-diiodobenzene **2**, the 1D structure, poly(**1**-*alt*-**2**), is oriented along the long axis of the crystal, so there is a transition dipole moment that

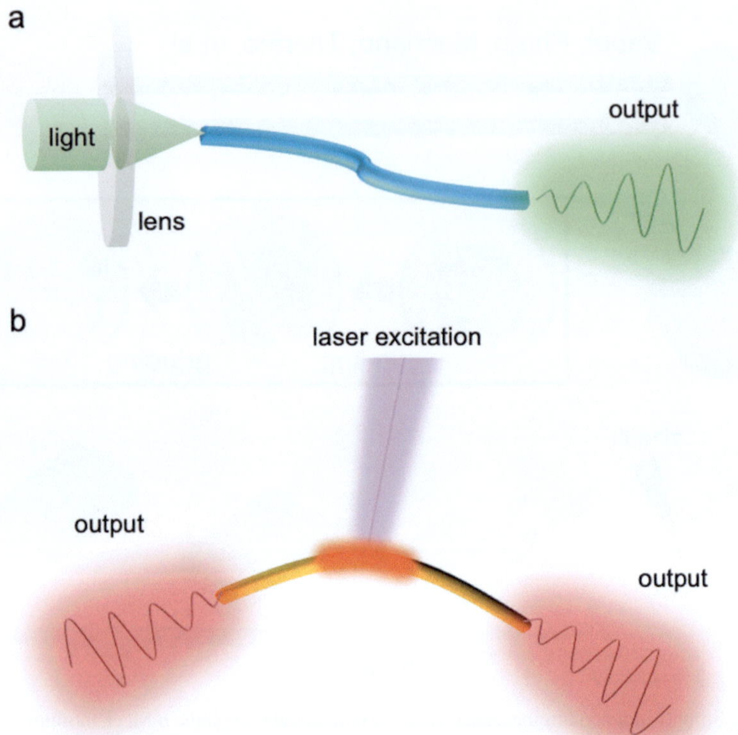

Fig. 5.3 Schematic representation of flexible optical waveguiding process. **a** Passive. **b** Active. Reprinted with permission from Ref. [11], The Royal Society of Chemistry

matches it (Fig. 5.4b) [32]. When the laser-excited luminescent photon is confined and diffused in the crystal, it interacts strongly or weakly depending on the orientation of TDM. As a result, the reabsorption efficiency differs depending on the orientation, resulting in a large difference in the loss coefficient (Fig. 5.4c). Thus, molecular crystals have the potential to function as anisotropic OWGs.

5.1.2.4 Optical Resonators

An optical cavity is a device that confines light inside itself and creates a standing wave of light because of photon interference (Fig. 5.5a) [33]. A Fabry-Pérot (F-P) mode cavity is a device in which mirrors are arranged so that light travels back and forth many times (Fig. 5.5b, c). When the optical distance of one round trip is an integer multiple of the wavelength of light incident on the cavity, the optical electric field inside the optical cavity is constructive. Due to this property, optical cavities induce phenomena such as nonlinear optical effects that occur only with

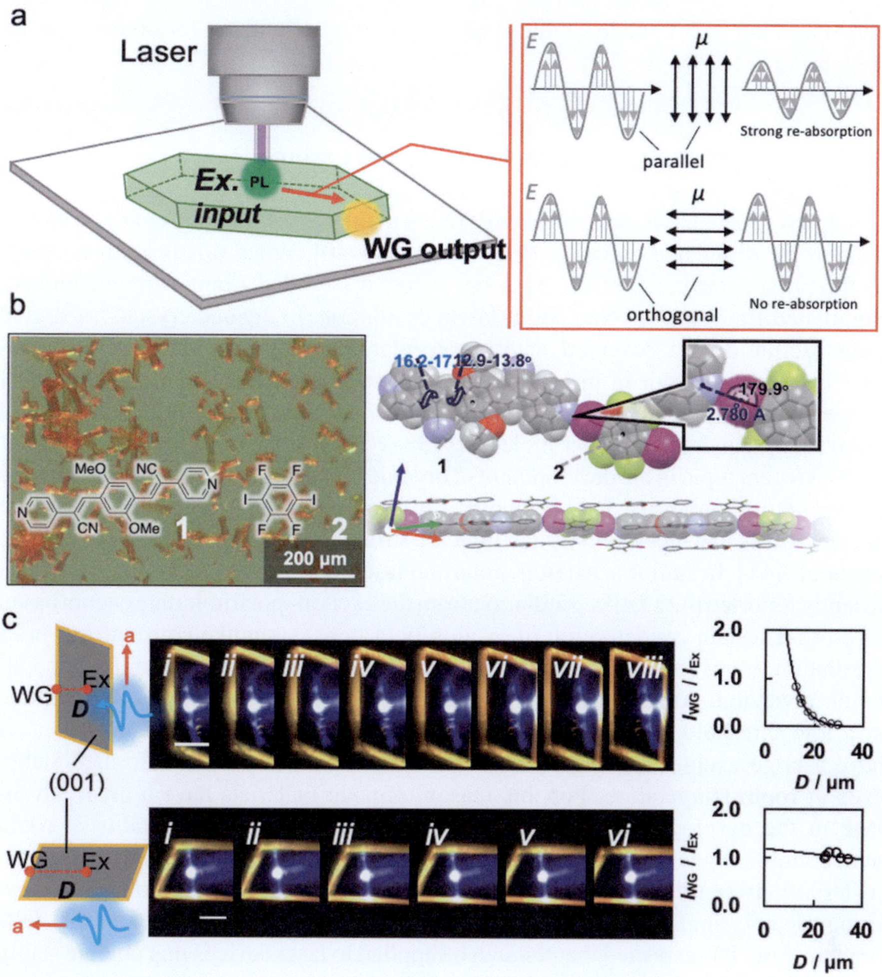

Fig. 5.4 **a** Illustration of measurement and relationships between electric field, E, and TDM, μ. **b** Chemical structure, microscope image, and crystal structure of the cocrystals. **c** Optical waveguiding characteristics. Reproduced from Ref. [32] with permission from Wiley–VCH

high-intensity light. In many cases, the confinement and leakage of light in a cavity are an anisotropic phenomenon, so the position-dependent enhancement of optical resonance is observed. One of the most efficient light confinements is achieved by exciting the edges of a spherical optical cavity in the whispering gallery mode (WGM). However, in this case, excitation at the top of the sphere is inefficient due to light leakage into the substrate. On the other hand, the efficiency of light confinement in a F-P mode cavity is expected to be improved by precise optimization of the excitation and detection positions. Organic crystals may have facing surfaces corresponding to F-P mode cavities and are potential candidates for optical cavities (Fig. 5.5b) [34, 35].

5.1.2.5 Amplified Spontaneous Emission and Lasing

Amplified spontaneous emission (ASE) refers to the process of light-emission amplification by stimulated emission (Fig. 5.6) [6, 36–40]. When this process is added to the photon feedback effect by a resonator, it is called Light Amplification of Stimulated Emissions of Radiation (laser) oscillation. In organic crystals, ASE and laser oscillation are observed using molecular materials that show high brightness and enhancement in the solid state. Representative molecular skeletons that become organic crystal lasers include phenylene/vinylene [6], fluorene [36, 37], and thiophene/phenylene oligomer molecules [38–40].

As a recent topic in the development of organic crystal lasers, we introduce organic exciton–polariton lasers. It is known that exciton-polaritons, which are composite quasiparticles in which excitons and light are strongly coupled, are formed in organic materials [41]. In addition, exciton–polariton lasers using organic crystals have been recently reported [42]. Laser oscillation from the exciton–polariton state occurs based on Bose–Einstein condensation of exciton-polaritons, so unlike conventional laser oscillation, it has the potential to achieve laser oscillation without forming a population inversion. Attention is being paid to developing laser devices that operate at a low threshold based on this principle. Furthermore, since organic materials have a large exciton binding energy, the exciton–polariton state is formed stably even at room temperature. For this reason, organic materials have a great advantage in the development of exciton–polariton lasers that operate stably at room temperature. Organic exciton–polariton lasers are made by sandwiching an organic material between dielectric multilayer films, but recently exciton–polariton lasers using organic single crystals alone have also been reported. With only simple self-organization, it is expected that this can be applied to laser devices that operate stably at room temperature with a low threshold [43, 44].

5.1.3 Mechanical Deformability in Molecular Crystals

It has long been known that some molecular crystals can change shape (deform) mechanically [45–47]. However, this deformation has never been understood systematically. Many crystals are known to be brittle, and they generally cleave when subjected to mechanical stimulation. This phenomenon is explained by the fact that many of the attractive forces between molecules are based on dipole–dipole interactions. It is more common for ionic crystals to be brittle, but ions arranged in an alternating order based on cation and anion charges are easily cleaved because stress converts them into repulsive forces.

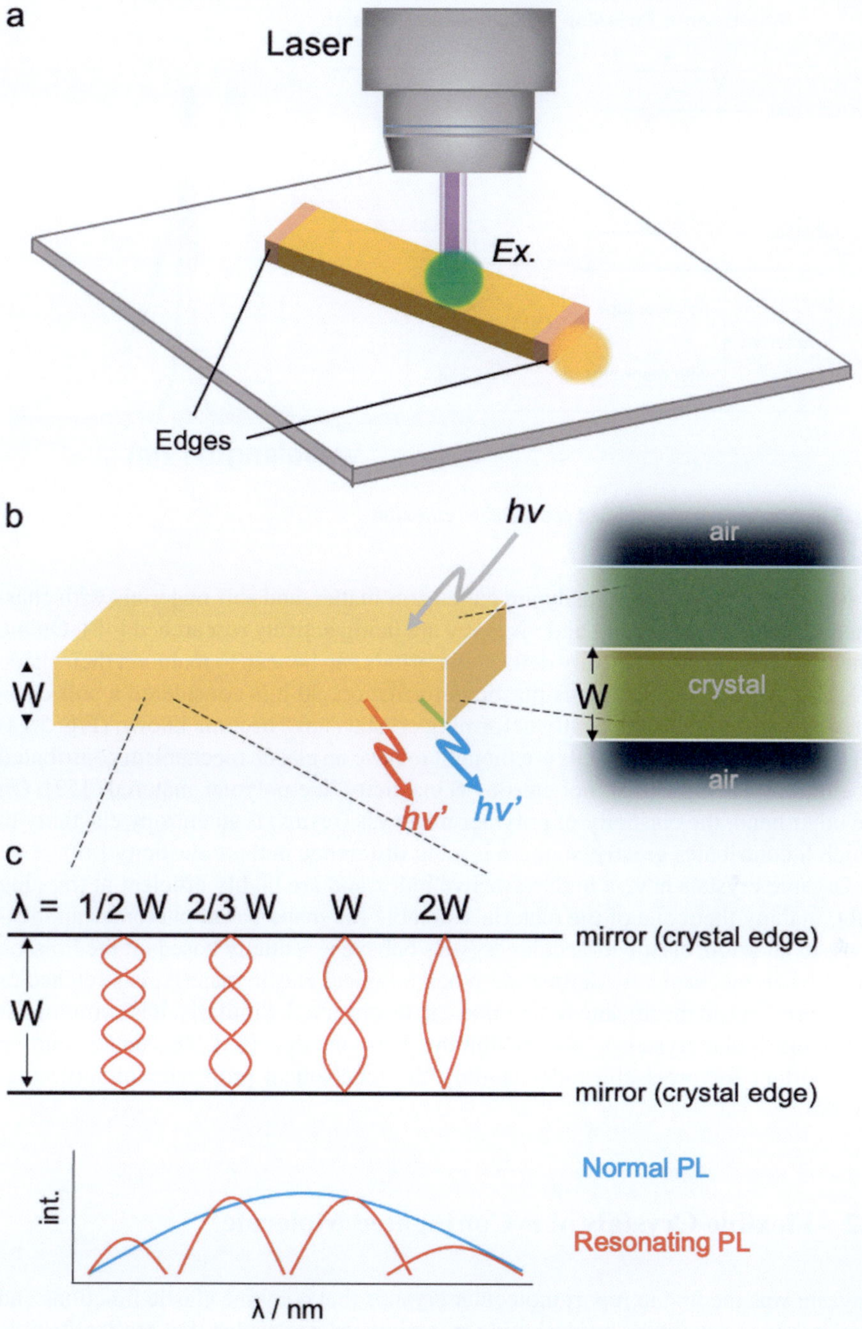

Fig. 5.5 **a** Illustration of crystal with laser excitation. **b** Representation of optical resonating process

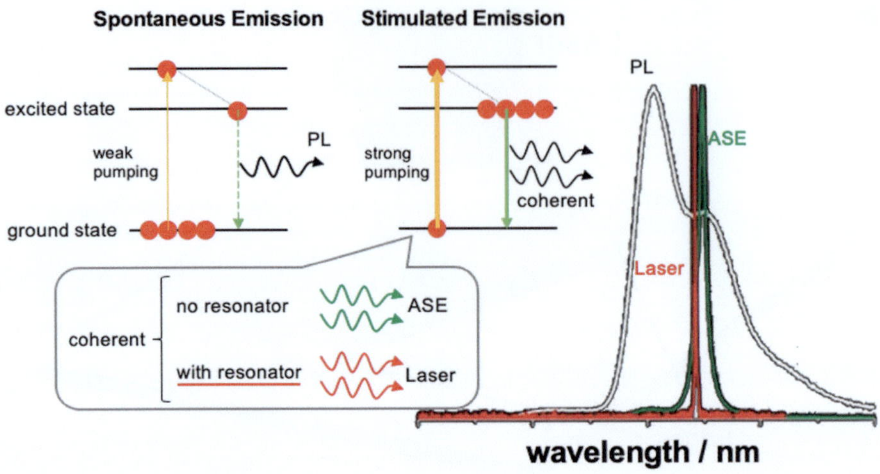

Fig. 5.6 Mechanism of amplified spontaneous emission

Materials that crumble easily are called soft matter, and soft materials with characteristics such as plasticity and viscosity are being actively researched [48]. On the other hand, some mechanically deformed crystals can be seen to maintain their shape [49, 50]. At first glance, this is plastic deformation, so it is considered a soft characteristic. In contrast, elastically deforming crystals have become known (Fig. 5.7a) [51]. Elastic molecular crystals are thought to have an elastic mechanism contributed by enthalpy H (Fig. 5.7b), not entropy S elasticity like polymer materials [52]. On the other hand, the elasticity of polymer materials (resins) is an entropy elasticity in which S contributes greatly, so there is a big difference in their elasticity [52].

Organic crystals have a high refractive index and are highly efficient at trapping light, making them one of the materials suitable for shape-based photonics applications. In addition, elastic molecular crystals behave essentially based on the Poisson effect when mechanically deformed. When a typical elastic material is stretched or bent, shrinkage in the thickness direction can be observed. Similarly, it is common for elastic molecular crystals to also exhibit this Poisson effect [53]. This understanding is important for precisely understanding the mechanical properties of molecular crystals.

5.2 Flexible Crystals of π-Conjugated Molecule

Hayashi was the first to report molecular crystals that combine elastic functions and luminescence properties by designing π-conjugated molecules that realize flexible molecular crystals (Fig. 5.7a) [51]. Polyaromatic hydrocarbons are often formed by interactions based on CH–π or π–π, but these can be controlled by introducing substituents, and the crystal structure can be intentionally formed. By introducing

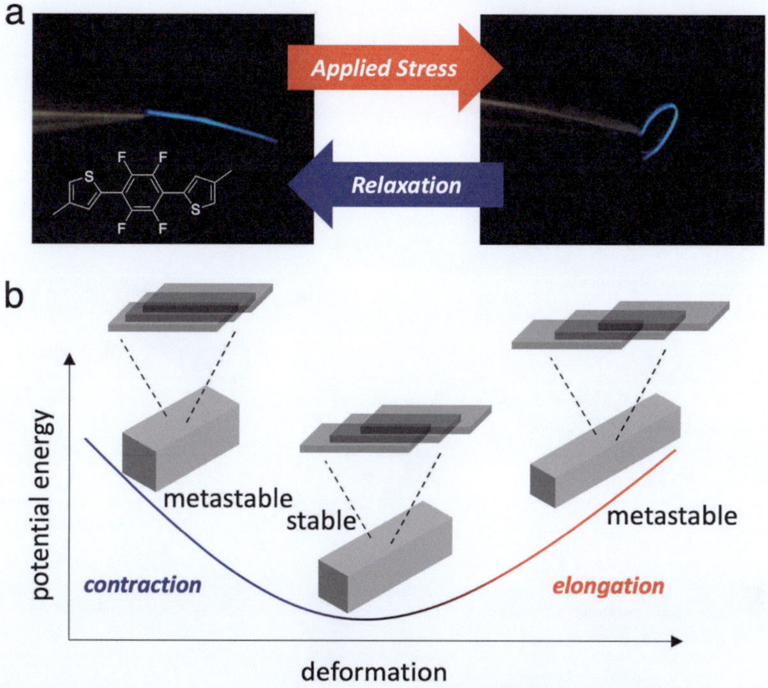

Fig. 5.7 **a** Photographic images of elastic bending. Fluorescent molecular crystal and its chemical structure. **b** Plausible potential energy curve of the elastic organic crystals under bending stress. Length changes at the elastic organic crystals' outside and inside energy curve under bending stress. Reproduced from Ref. [51] with permission from Wiley–VCH

halogen groups to the short axis of π-conjugated molecules, the strong dipole–dipole attractive network by CH–π can be suppressed, and the repulsive network by π–π can be enriched, allowing the design of elastically deformable crystals. Hayashi showed that elastic semiconductor crystals can be obtained by constructing π–π columnar structure crystals using a combination of perfluorophenylene and thienyl ring, 1,4-bis(thien-2-yl)-2,3,5,6-tetrafluorobenzene (Fig. 5.8). This approach may be developed into photonic functions due to the luminescence based on the semiconductor properties of the crystals and optical switching functions driven by intermolecular packing changes due to mechanical deformation.

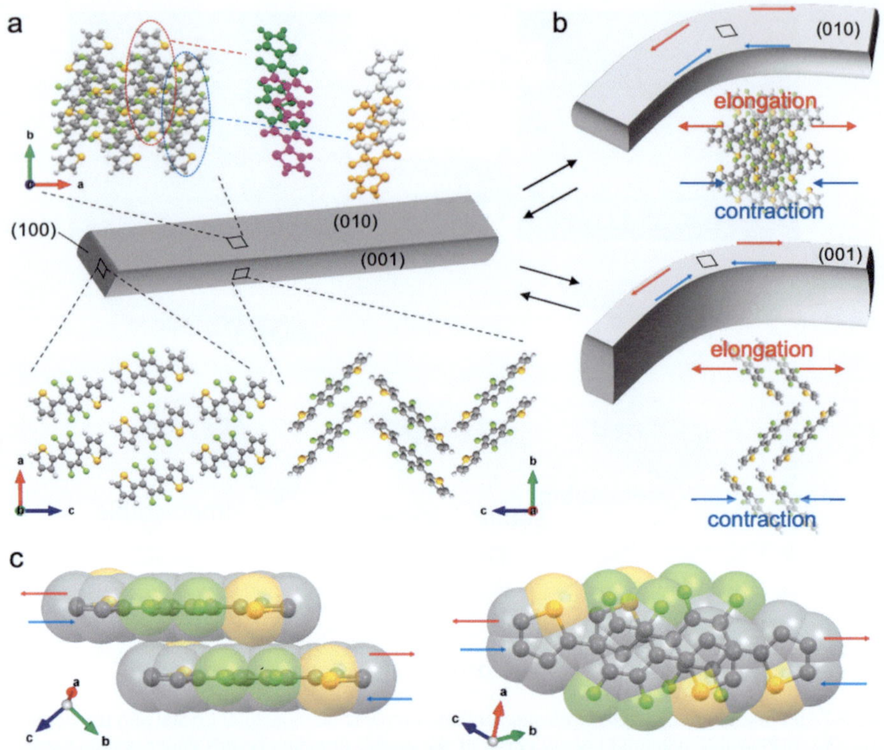

Fig. 5.8 **a** Schematic showing the columnar arrangement of the molecules based on a slip-stacked structure. **b** Schematic illustration of the elastic deformation. The red and blue arrows correspond to the elongation and contraction, respectively. **c** π-Stacked dimer and possible molecular sliding under elastic deformation. Reprinted with permission from Ref. [54], The Royal Society of Chemistry

5.3 Optical Characterization of π-Conjugated Molecular Crystals

5.3.1 Measurement Setup for Crystal Photonics

Optical analyses for materials and biological systems have been performed in various ways, such as two- or three-dimensional analysis or by changing the laser power, so far. This can also be applied to the photonics of organic crystals. Here, we will mention some of the measurement systems that have been used for crystal photonics.

5.3.1.1 Spatially Resolved μ-Photoluminescent Analysis

The optical waveguide properties of organic crystals can be evaluated using spatially resolved μPL spectral measurements (Fig. 5.9). While observing an isolated crystal using an objective lens, the crystal is focused and excited with a laser, and the light emission from the organic crystal itself is guided within the crystal to the end of the crystal. The light emission including the guided light is introduced into a spectrometer through the objective lens, and the light can be spatially resolved using a pinhole, for example. The position of the pinhole can be adjusted while observing the field of view of the objective lens with a camera, and the guided light from the end of the crystal can be spatially selectively acquired as a PL spectrum, allowing the waveguide properties of the organic crystal to be evaluated. A detailed analysis will be described later.

We will outline the measurement equipment used to evaluate the laser characteristics of organic crystals under optical excitation (Fig. 5.10). Measurements can be performed with essentially the same setup as the spatially resolved μPL spectrum measurement equipment used in the waveguide evaluation described above. An organic crystal is isolated on a substrate such as quartz, and the crystal is strongly excited by a pulsed laser or the like to measure the emission spectrum. Stimulated emission light from the crystal is guided within the crystal and emitted from the crystal end. Because of stimulated emission and optical resonance, the ASE light or laser light from the organic crystal has strong directionality, and gain is obtained. In contrast, the stimulated emission light is guided within the crystal. For this reason, it is desirable to capture the emission from the organic crystal end using a lens horizontally to the substrate.

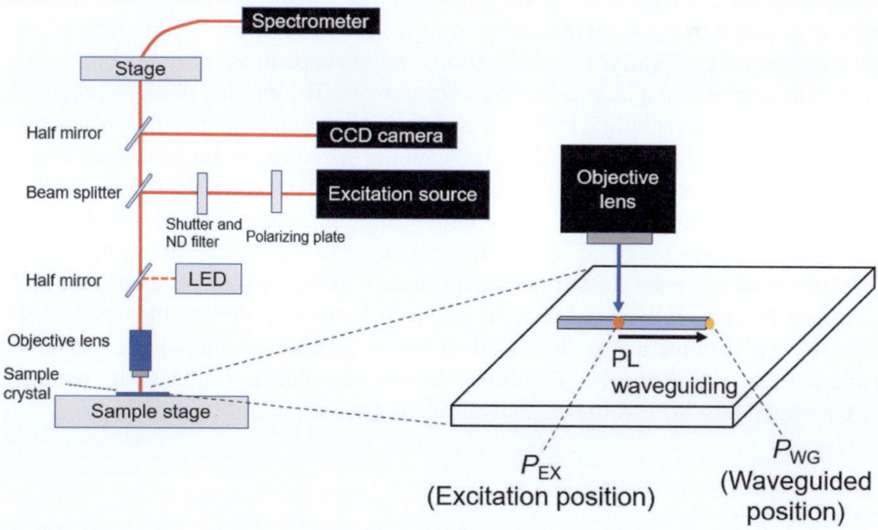

Fig. 5.9 Schematic depiction of the setup of spatially resolved μPL measurements

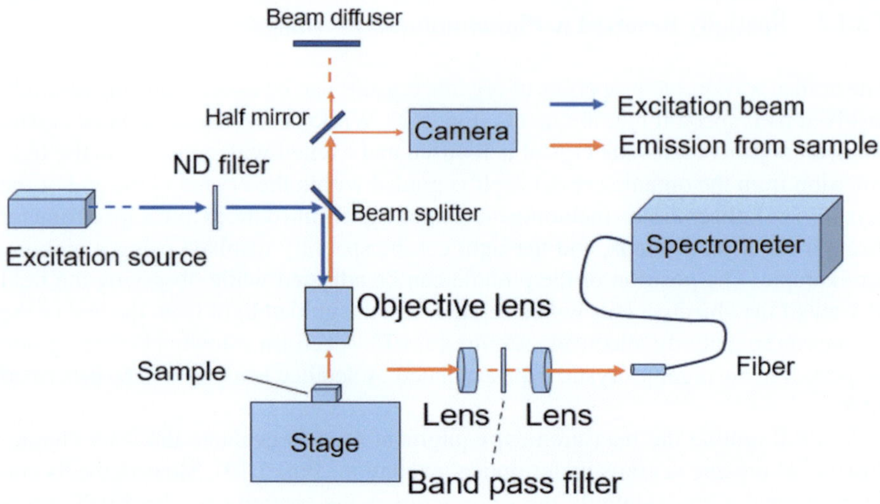

Fig. 5.10 Schematic depiction of the setup of lasing spectra measurements

5.3.1.2 Multi-photon Fluorescent Microscopy

In contrast to microscopy, which focuses a spotlight on the surface of a crystal, multi-photon fluorescent microscopy (MPFM) is known to be an important technique for observing the internal (bulk) emission of a crystal in living systems. The optical configuration is shown in Fig. 5.11 [55]. Here, a femtosecond fiber laser oscillating at 1030 nm was used as the light source, and a quarter-wave plate was used to generate circularly polarized light to avoid the polarization dependence of the crystal. The laser light is scanned using a galvanometer mirror system to obtain a two-dimensional microscope image. Similar two-dimensional microscope images are obtained using MPFM as conventional laser scanning microscopes. The emitted photons propagate through the crystal, including reflection and scattering. Since the speed of propagating light is sufficiently fast compared to the scanning speed of the galvanometer mirror, the emitted photons propagating through the crystal are integrated by a photodetector before moving to the next point in the laser scanning process. The two-dimensional microscope images obtained by moving the stage in the Z-axis direction are then superimposed to create a 3D microscope image using ImageJ. A dichroic mirror (cutoff at 780 nm) is used to focus the reflected beam onto the sample using a 20× objective with a numerical aperture of 0.45. Fluorescence from the sample is detected using a photomultiplier tube. The emission intensity plots in Fig. 5.11 are obtained by corresponding MPFM image analysis using ImageJ.

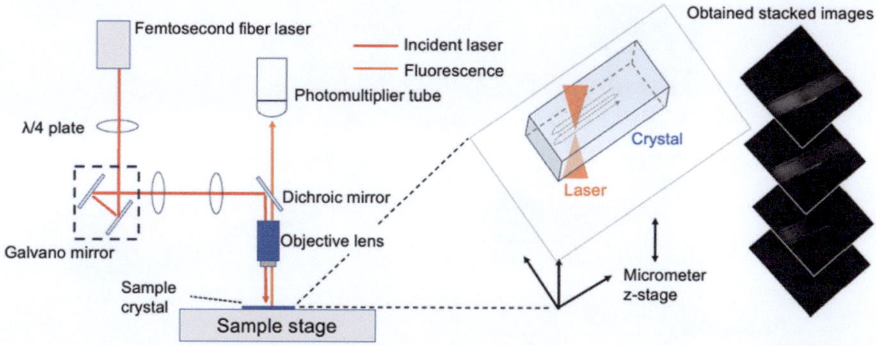

Fig. 5.11 Schematic depiction of setup of multi-photon fluorescent microscopy. Adapted from Ref. [55] with permission from American Chemistry Society

5.3.2 Characterization of Optical Waveguides

OWGs can be evaluated by spatially resolved μPL spectroscopy. In the setup shown in 5.3.1.1, the attenuation coefficient ε is obtained by plotting the distance ($l_{i\text{–}o}$) from the laser excitation position (input) to the crystal edge (output) and the intensity ratio (I_i/I_o) of the emission spectrum obtained from each position, and this value can be used to evaluate the efficiency (performance) [56]. For the example crystal shown in Fig. 5.12, when a laser is excited at any position, emission is observed from the edge of the crystal. The spectrum obtained from the edge of the crystal is generally observed as a characteristic emission band where the short wavelength band where the absorption spectrum and the fluorescence spectrum overlap appear to disappear due to reabsorption within the crystal (Fig. 5.12). The attenuation coefficient, ε, is calculated by plotting the distance ($l_{i\text{–}o}$) from the edge of the crystal to an arbitrary excitation position and the intensity ratio (I_i/I_o) and is expressed in dB/cm.

5.3.3 Characterization of Optical Resonators

It has been shown that micron-width molecular crystals can function as optical resonators [57–60]. The sharpness of the resonant peak at this time is evaluated as the Q value, expressed as $Q = v/\Delta v$. v and Δv represent the resonant peak's wavelength (cm^{-1}) and linewidth (cm^{-1}), respectively. The larger the value is the better the resonator performance (Fig. 5.13). The Q value is also greatly influenced by the reflectivity of the resonator. Therefore, a higher resonator Q value can be expected for WGM resonators based on total reflection than F-P resonant modes. Organic resonators fabricated by self-organization usually have a reflectivity below 50%, so the effect of WGM resonance in improving the Q value is very large. Improving the Q value by improving the reflectivity will lead to spectrum narrowing (band-narrowing)

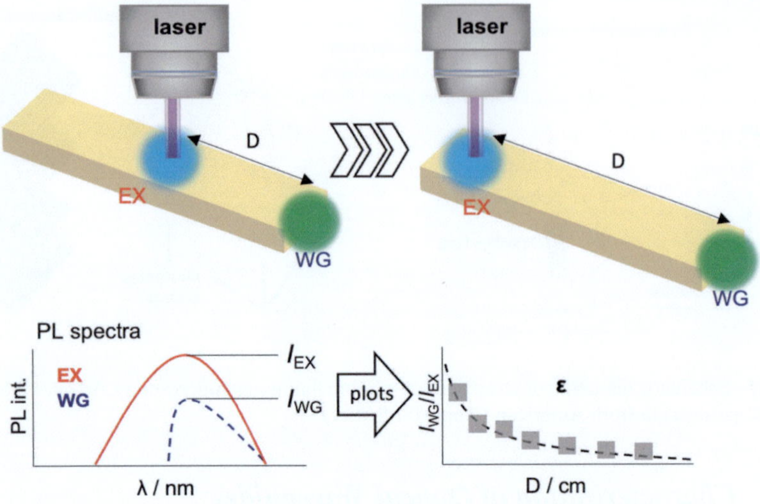

Fig. 5.12 Schematic depiction of PL spectra measured for optical waveguide

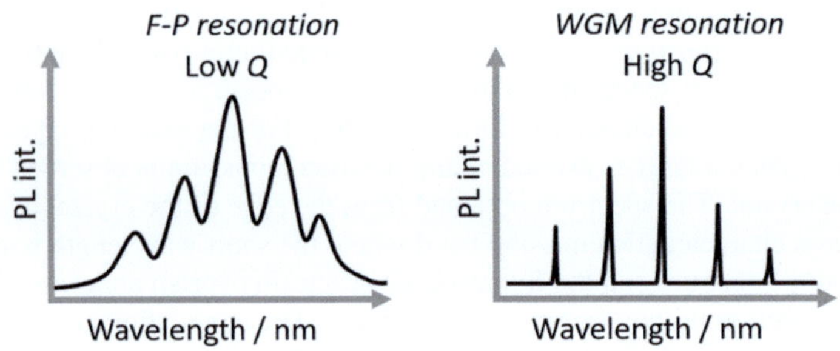

Fig. 5.13 Schematic depiction of PL spectra measured for **a** low Q resonator and **b** high Q resonator

and low threshold lasing. For example, laser oscillation has been achieved by fabricating a sandwich structure with an organic single crystal as the active layer using a DBR mirror to increase the reflectivity [58]. In a high Q resonator structure, a significant increase in the probability of transition due to the coupling between the optical transition and the resonator mode can occur, known as the Purcell effect [59].

5.3.4 Characterization of Lasing

Laser oscillation and ASE are easily confused, and the terms are sometimes misused in the literature. In the past, I. D. W. Samuel et al. published a commentary paper titled "How to recognize lasing" [61] in which they presented the conditions for laser oscillation. The conditions are: (1) the emission intensity must increase for the excitation energy in a threshold manner due to the stimulated emission process, and (2) there must be a feedback effect of photons on the active medium due to the resonator. These are the minimum criteria that must be met to define laser oscillation.

The most standard index for evaluating the quality of a laser medium is the oscillation threshold. To calculate the oscillation threshold, the emission spectrum is measured at each excitation energy while increasing the excitation energy (kA/cm^2 or $\mu J/cm^2$). Then, the excitation energy vs. emission intensity is plotted. In the low excitation energy region before the laser oscillation threshold, the linewidth of the emission intensity becomes sublinear, and the slope of the log–log plot is less than 1. However, in the high excitation energy region after the oscillation threshold, the slope of the log–log plot exceeds 1 and increases superlinearly. The boundary between the sublinear and superlinear regions is defined as the oscillation threshold. In addition, the excitation energy versus emission linewidth is also plotted at the same time, and the behavior of the emission linewidth narrowing rapidly is often investigated. This type of evaluation is also performed for organic single-crystal lasers, but various parameters determine the oscillation threshold. As already mentioned, there are three components of a laser: the excitation source, the active medium, and the resonator, and the oscillation threshold is also greatly influenced by the properties. For the excitation light, the wavelength, pulse duration, incident angle to the sample, polarization direction, and excitation beam shape affect the oscillation threshold, while for the active medium, the emission quantum yield, emission lifetime, presence or absence of a reabsorption process of emission in the ground and excited states, heat resistance, molecular orientation, sample size, morphology, etc. Furthermore, for the resonator, the oscillation threshold also changes significantly depending on the reflectivity, roughness, and resonance mode. Although the oscillation threshold is a standard indicator for evaluating the quality of an organic laser, it should be noted that the oscillation threshold does not strictly evaluate the superiority of the material.

In addition, there is a method called the variable stripe length (VSL) method for quantitatively evaluating the quality of organic laser materials as active media, which is known to be a method that can easily calculate the gain coefficient [62–64]. The VSL method is a method in which excitation light is shaped into a stripe shape and irradiated onto the active medium, and the emission spectrum from the active medium is measured while changing the length of the excitation light and adjusting the beam power to achieve the same excitation density ($\mu J/cm^2$) (Fig. 5.14).

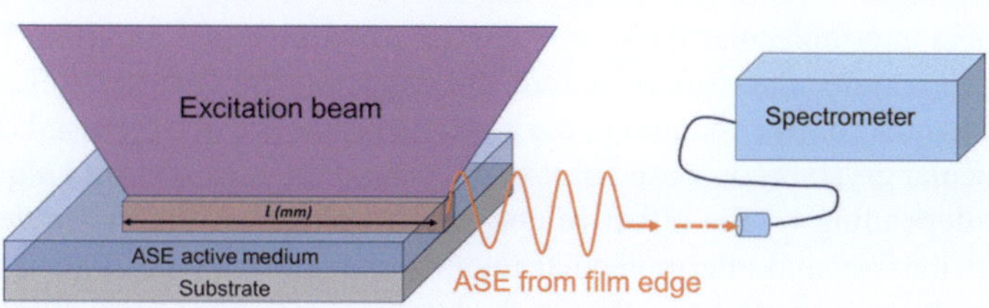

Fig. 5.14 Schematic depiction of PL measurement in VSL method

Here, the beam length of the excitation light is plotted against the emission intensity, and the gain coefficient can be calculated from the slope. The larger the slope, the larger the gain coefficient. In the VSL method, the relationship between the stripe the beam length, and output intensity is described by Eq. 5.1.

$$I(\lambda) = \frac{A(\lambda)I_p}{g(\lambda)}\left(e^{g(\lambda)l} - 1\right) \tag{5.1}$$

where $A(\lambda)$ is a constant, I_p is the incident intensity, $g(\lambda)$ is the gain coefficient, and l is the length of the beam stripe. The gain $g(\lambda)$ is calculated by fitting.

The evaluation of the gain coefficient by the VSL method is often performed on organic thin films, but a similar evaluation is possible for organic single-crystal thin films if the sample size is sufficiently larger than the excitation light [62–64]. Unlike amorphous or polycrystalline films, organic single crystals have uniform molecular orientation on a macroscopic scale, which makes stimulated emission more efficient and increases the gain coefficient accordingly. The group of S. Y. Park and J. Gierschner has reported a large gain coefficient of 77 cm^{-1} for a methyl-substituted distyrylbenzene single crystal [62].

5.4 Flexible Crystal Photonics

5.4.1 Mechano-Responsive Optical Properties

Mechanochromism is a phenomenon that has been known for a long time and refers to a phenomenon in which a color change is observed due to a change in solid structure caused by mechanical stress such as grinding [19]. Results based on various state changes such as crystal-to-crystal [20], crystal-to-amorphous [21], and amorphous-to-amorphous [22] have been reported. Mutai et al. reported that the crystals have superelasticity and that the mechanical deformation associated with it changes the luminescence color due to a crystal-to-crystal change [65]. Such a change without losing crystallinity is a characteristic of elastic molecular crystals, and it is possible to realize a robust luminescence color change [65].

The deformation of molecular crystals based on enthalpy elasticity is caused by changes in intermolecular packing or crystal structure transformation. Therefore, changes in intermolecular packing or molecular/crystal structure due to elastic deformation led to quantitative changes in the physical properties of the crystal (Fig. 5.15) [65]. Molecular crystals composed of π-conjugated molecules exhibit unique optical properties depending on the effective conjugation length of the molecules, charge density distribution, and intermolecular interactions in the crystal. The deformation of elastic molecular crystals can change the luminescence due to structural changes. Elastic crystal deformation into a curved shape causes the crystal to elongate on the outside and contract on the inside. In other words, the change in crystal structure is coupled with the macroscopic deformation of the crystal. This is due to slight arrangement changes in the intermolecular packing structure. It is important to obtain detailed information on the effects of these changes on the luminescence of organic crystals and to consider the related mechanisms. Therefore, spatially resolved physical property measurements are an important tool for understanding the changes caused by the bending of elastic molecular crystals [54].

In fact, for some elastic molecular crystals, it is possible to observe changes in optical properties due to stress, that is, changes in fluorescence due to deformation of the crystal [66]. The PL spectrum of 9,10-dibromoanthracene crystal had an emission

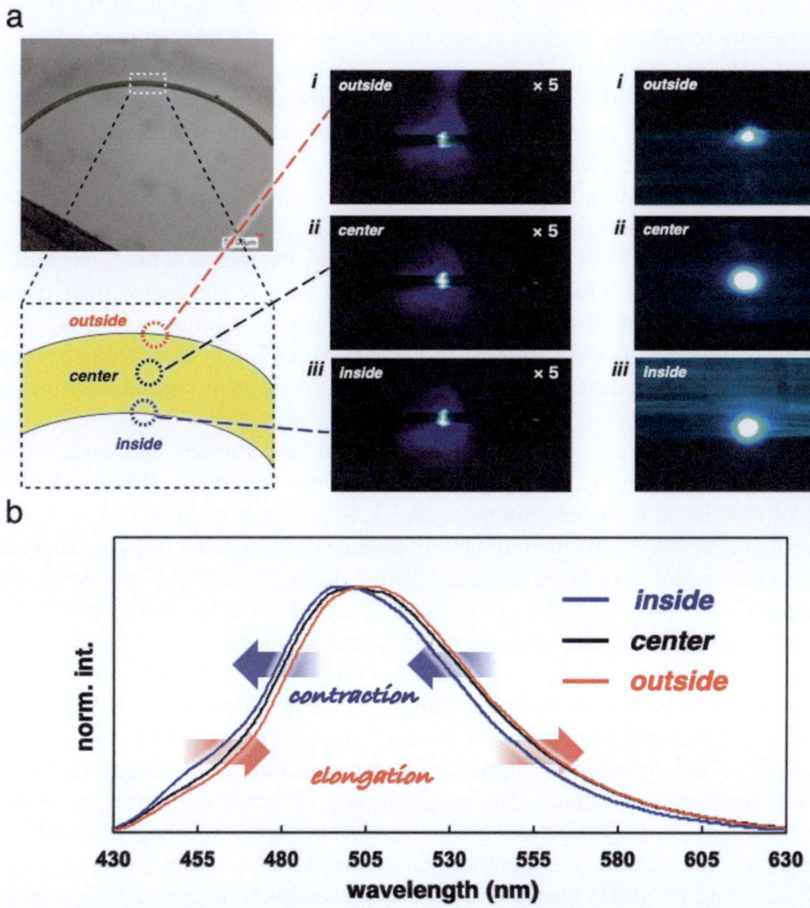

Fig. 5.15 Spatially resolved microscopic photoluminescence measurement of the crystal. **a** Microscopy image and schematic illustration of the μPL measurement of the straight crystal. **b** Emission spectra of the bent crystal at the inside (red), center (black), and outside (blue). Reproduced from Ref. [66] with permission from Wiley–VCH

maximum wavelength (λ^{PL}) of 500 nm (Fig. 5.15). When the crystal is bent along the *a*-axis in a *b*-face-up configuration, expansion and contraction occur simultaneously on the curvature's convex (outer) and concave (inner) sides, and no distortion occurs. Using a spatially resolved μPL measurement system, we attempted to measure the PL spectrum of micro-areas (inner, center, and outer positions) of the bent crystal. The PL spectrum of the straight crystal's inner, center, and outer positions was found to be $\lambda^{PL} = 500$ nm (Fig. 5.15b). However, at the outer and inner positions of the bent crystal under a bending strain of $\varepsilon_a = 1.3\%$, a red-shift ($\lambda^{PL} = 506$ nm) and blue-shift ($= 497$ nm) of the PL spectrum were observed, respectively, when compared with the center position ($\lambda^{PL} = 505$ nm). These observations suggested that indeed, the stretch-induced red-shift and the contraction-induced blue-shift occurred at the outer and inner positions of the bent crystal, respectively. The bending strain induces the slip of the π-planes, increasing the characteristics of the J-aggregates on the outer side and the H-aggregates on the inner side, resulting in the red and blue-shifts, respectively. As seen in the bending relaxation, the stretching of the crystals caused similar PL spectral changes to those observed on the outer side of the bent crystal (Fig. 5.15b). When stretched along the *a*-axis with $\varepsilon_a = 1.0\%$, a clear red-shift of the PL spectrum was observed, and λ^{PL} shifted to 505 nm [66]. Moreover, the PL spectrum returned to its original shape when the stretching stress was released. For $\lambda^{PL} = 511$ nm, a larger red-shift was observed when stretched along the *a*-axis with $\varepsilon_a = 2.0\%$. The spectral changes were fully reversible and could be repeated many times upon applying and releasing the stretching stress. These deformation-induced fluorescence changes probably result from changes in molecular orbitals associated with changes in molecular packing, and therefore, the crystals can function as stimuli-responsive materials.

Distinguishing luminescence contributions from the surface and bulk of a crystal is a long-standing challenge in crystalline materials. When MPFM imaging of the elastic molecular single crystal 1,4-bis(4-methylthien-2-yl)-2,3,5,6-tetrafluorobenzene was performed, the luminescence contributions from the surface and bulk of the crystal were experimentally distinguished [55]. Strong luminescence was observed only from the surface of the crystal, and the bulk did not emit strongly. In addition, the surface and bulk luminescence behavior responded well to the mechanical shape change of the crystal. MPFM images of a straight crystal on the (001) faces are shown in Fig. 5.16. In this technique, luminescence was detected point-by-point as the focus was scanned from the top to the bottom of the crystal. Thus, the waveguiding effect of the crystal was ignored. Compared to the luminescence intensity of the bulk of the crystal, stronger luminescence was observed at the surface of the crystal on the two-photon image along with the (001) face (Fig. 5.16a). Meanwhile, a bent crystal was observed from the (001) face using MPFM. The employed crystal can be bent along the (001) plane. From the cross-sectional images, luminescence was observed from the surface as well as in the case of not bending. Furthermore, luminescence was observed inside the bent crystal. This is evident from the comparison of the central images in Fig. 5.16b. Stronger luminescence was also observed on the surface of the curved crystal in response to the elongation stress.

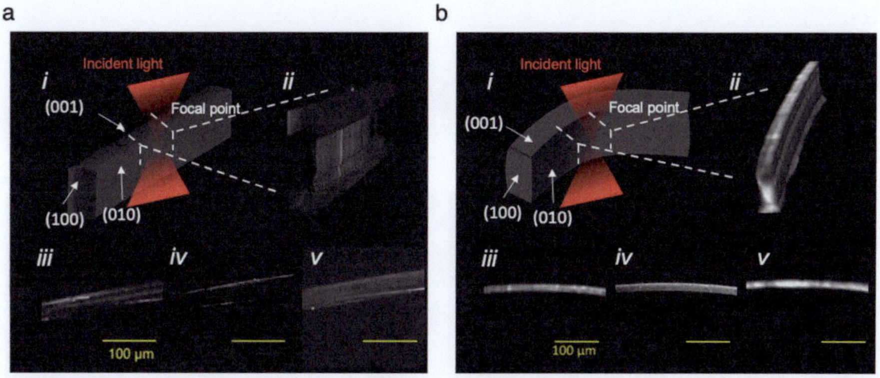

Fig. 5.16 **a** MPFM images of the original state observed along the (001) plane, and **b** MPFM images of the bent state observed from the (001) plane (*i*: illustration of overview crystal, *ii*: 3D microscope image clipped at the part of the crystal, *iii*–*v*: cross-sectional images, *iii*: top surface, *iv*: inside, *v*: bottom surface). Reproduced from Ref. [55] with permission from American Chemistry Society

5.4.2 *Flexible Optical Waveguides*

5.4.2.1 Flexible Optical Waveguides of Fluorescent Elastic Molecular Crystals

It has been found that single crystals of many luminescent molecules exhibit OWG characteristics [67]. Here, it is unsurprising that fluorescent elastic molecular crystals exhibit OWG characteristics in natural and curved crystal states [68]. For example, when a linear crystal of molecule (Fig. 5.17a) was irradiated with a focused laser (405 nm) at different positions, reddish-orange emission was always detected from the edge of the linear crystal regardless of the excitation position (Fig. 5.17b). Bent crystal, with a calculated strain value of 1.9%, also showed similar OWG performance (Fig. 5.17c). The emission band at the crystal edge peaked at 597 nm and showed a narrower full width at half maximum (FWHM: 34 nm) than at the illumination position (573 nm, FWHM: 56 nm). The spectral profile did not change substantially with the illumination position, but the emission intensity at the edge slightly decreased as the distance increased. On the other hand, in the case of long-distance OWG characterization, which is difficult to observe with a microscope, the calculation of the extinction coefficient from a crystal spanning several centimeters requires a great deal of effort due to the variation in values based on the crystal quality and the need to focus on the fluorescence spectrum at the illumination position (I_{in}) and the edge of the crystal (I_{out}). Therefore, future OWG characterization will likely require a system configuration that allows mapping analysis with clear spatial resolution and will yield more accurate values using short-distance OWGs.

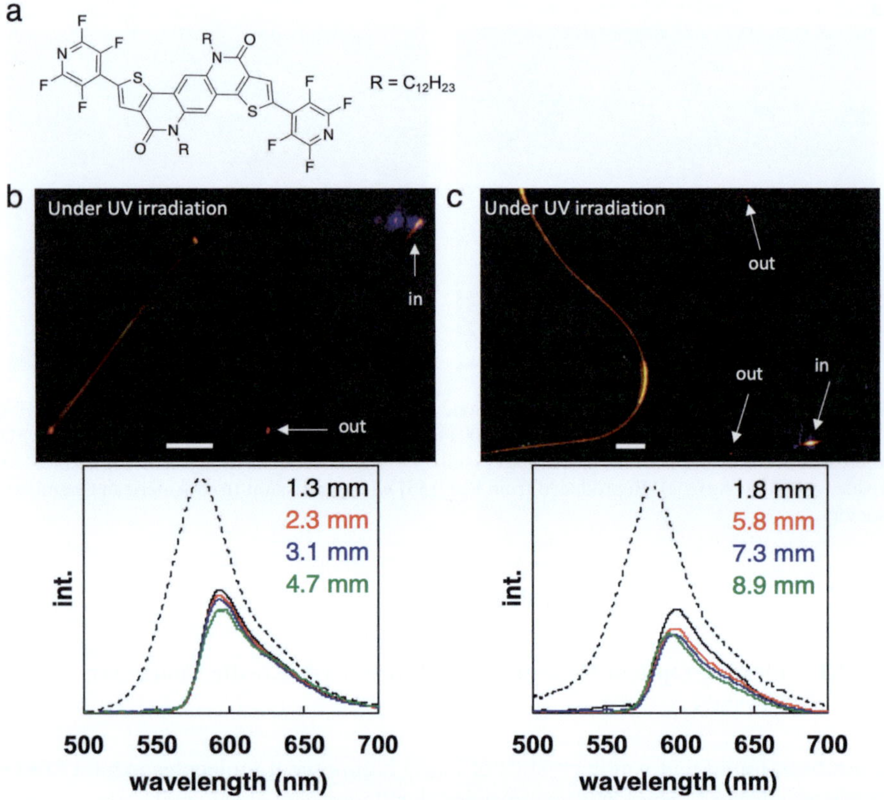

Fig. 5.17 **a** Chemical structure. **b** Microscope images of linear crystal under UV irradiation and upon excitation with a focused laser, respectively. μPL spectra at the end of the crystals. **c** Microscope image of a curved crystal under UV irradiation and upon excitation with a focused laser, respectively. μPL spectra at the end of the crystals. Reproduced from Ref. [68] with permission from Wiley–VCH

5.4.2.2 FRET-Assisted Flexible Optical Waveguides

Flexible molecular crystal waveguides based on elastic molecular crystals (EMCs) are essential for flexible and compact optical materials. The increase in loss coefficient α due to self-absorption is often a problem in fluorescent chemical material optical waveguides (OWGs) that guide photons in the active mode [69]. The fluorescent molecular crystal of 9,10-dibromoanthracene has an α value of 0.1258 dB/μm. In many cases, reabsorption in the crystal leads to attenuation of photon diffusion, which is related to the increase in α value. On the other hand, the development of elastic mixed molecular crystals (EMMCs) may realize a lower α value, i.e., a highly efficient OWG, as a Förster resonance energy transfer (FRET)-assisted OWG (Fig. 5.18).

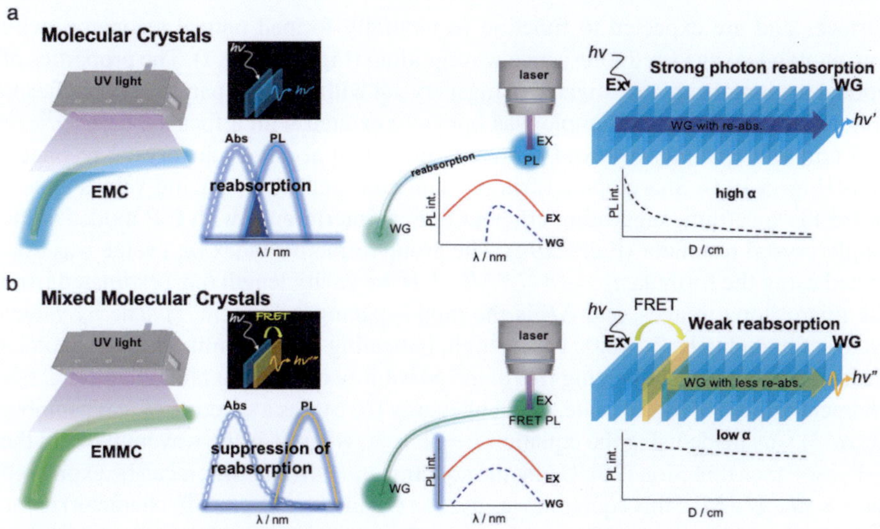

Fig. 5.18 Schematic illustration of fluorescence and optical waveguide in molecular crystals. **a** General OWG. **b** Elastic FRET-assisted OWG

To obtain a FRET crystal system based on elastic molecular crystals, fluorescent molecular crystals of 9,10-dibromoanthracene with 1–5% 9,10-diformylanthracene as the same regioisomer acceptor with electron-withdrawing groups as the dopant were successfully prepared. In addition to the conversion to a mixed system, there are differences in elastic modulus and hardness between EMC and EMMC. However, elastic behavior was also shown with a few percent doping of the acceptor. The α value (0.0077 dB/μm) of this EMMC, which consists of EMMC with 1% of 9,10-diformylanthracene, is significantly lower than that of EMC (0.1258 dB/μm) due to the reduced self-absorption in the FRET system. Selecting the appropriate acceptor and its low doping rate to make the mixed crystal structure leads to the development of an efficient and flexible OWG. This method is a practical approach in various functional and flexible crystal systems.

5.4.3 *Flexible Optical Resonators*

5.4.3.1 Flexible Fabry-Pérot Crystal Resonators

Elastic molecular crystals with fluorescence and smooth edges have potential as flexible crystalline optical resonators [70]. Crystals obtained from diketopyrrolopyrrole derivatives have good photoluminescence quantum efficiency ($\Phi^{PL} = 0.45$) and show mechanical deformation shape to a curved state (Fig. 5.19a, b). High-resolution microscope and scanning electron microscopy reveal that the crystals have smooth surfaces and are expected to function as naturally formed optical resonators with low photon leakage for fluorescence waveguiding (Fig. 5.19c, e, f). The properties of optical resonators are investigated using a crystal with a pair of parallel crystal facets (Fig. 5.19d). The crystal sample was optically excited with a focused laser (λ^{PL} = 405 nm) and spatially resolved μ-PL was measured at one of the crystal resonator ends (Fig. 5.19d). Sharp optical fringes with constant mode spacing values appear in the PL spectrum, suggesting efficient optical interference with F-P modes in the single-crystal resonator (Fig. 5.19g). The group refractive index (n_g) value was estimated using the formula $n_g = 1/(2L^*\Delta E)$. L is the cavity length (cm) estimated from the microscopy images, and ΔE is the mode spacing value (cm^{-1}). The n_g values were estimated to be 3.7–6.0. This is high, indicating strong confinement of photons inside the single crystal, giving rise to the possibility of efficient photon guiding. For characterization of the EP state, a plot of energy (E: eV) as a function of wavenumber (k: m^{-1}) was made using the equation $k = 2\pi n_g/\lambda$, where k is the wavenumber in the F-P resonance direction [70]. Using the equation $n_g = 1/(2L^*\Delta E)$, k can be expressed as: $k = \pi hc/(L\lambda\Delta E)$; this equation was used to obtain the E-k plot. To characterize the E-k plot according to the EP model, the exciton energy (E_{ex}: eV) was estimated as the midpoint between the 0–0 peaks of the fluorescence excitation and fluorescence spectra (i.e., 2.128 eV). The energy dispersion of the cavity photon (E_{ph}) is obtained from $E_{ph} = (\hbar c/n_p)k$, where $\hbar$, c, and n_p are the Dirac constant, the speed of light, and the phase refractive index, respectively. The energy dispersion of the upper polariton branch (EUPB) and the lower polariton branch (ELPB) can be calculated according to the following Eq. 5.2:

$$E_{\mathrm{UPB,LPB}} = \frac{1}{2}\left(E_{ex} + E_{ph}\right) \pm \frac{1}{2}\sqrt{\left(E_{ex} - E_{ph}\right)^2 + (\hbar\Omega)^2} \tag{5.2}$$

where $\hbar\Omega$ (eV) is the Rabi splitting energy, which indicates the coupling strength between the exciton and the cavity photon. The obtained E-k plot was well fitted as the ELPB, and the values of the Rabi splitting energy and n_p were estimated to be 1.38 and 2.03, respectively. The estimated value of $\hbar\Omega$ is larger than the exciton width, indicating emission from the EP state in the crystal. Thus, this crystal is a F-P crystal cavity formed naturally due to its flat crystal surface. The group refractive index (n_g) and Rabi splitting energy ($\hbar\Omega$), which indicate the coupling strength between the exciton and the cavity photons, are large (n_g = 3.7–6.0, $\hbar\Omega$ = 1.38 eV), suggesting strong confinement of the guided photons.

5.4.3.2 Tunable Fabry-Pérot Resonators Based on Poisson's Effect

Resonators fabricated from elastic molecular crystals may exhibit variable resonance peak changes due to the reversible deformation of the crystal width caused by the Poisson effect. Despite the excellent light guiding properties along the long axis of tiny rod-shaped single-crystal particles produced by slow precipitation of fluorescent π-conjugated molecules, cyano-substituted oligo(phenylenevinylene),

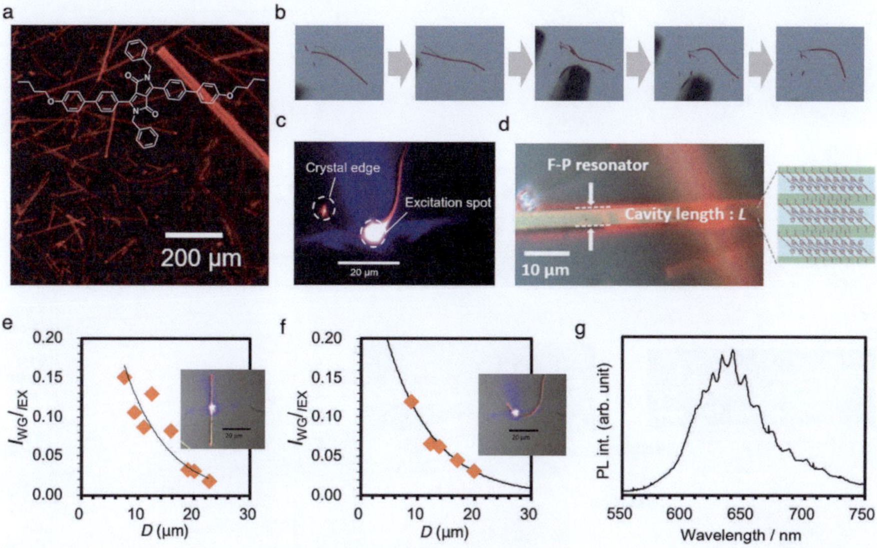

Fig. 5.19 **a** Fluorescence microscope image and chemical structure. **b** Elastic bending. **c** Microscope image upon laser excitation. **d** Single-crystal with parallel edges and alignment of molecules. **e**, **f** Optical waveguiding characteristics. **g** Resonating PL band. Reproduced from Ref. [70] with permission from American Chemistry Society

their PL spectra barely show the periodic peaks characteristic of optical resonators (Fig. 5.20a–c) [71]. This is attributed to the non-planar surfaces of the end faces and the optical losses during light propagation and reflection. However, when a much wider grain with $L = 120.4$ μm and $W = 22.3$ μm is excited with a focused laser, the emission is observed at the vertical end faces and the horizontal end faces, as shown in Fig. 5.20c, d. Interestingly, the PL spectrum collected at the side facets shows periodic PL peaks overlapping the broad PL band characteristic of F-P-type optical resonators (Fig. 5.20d). To further confirm the F-P mode resonance along the horizontal direction, μPL spectra are measured for a series of elastic molecular crystals with widths ranging from 8.6 to 30.6 μm. The mode spacing ($\Delta\lambda$) between adjacent resonant peaks decreases as W increases, which is consistent with Eq. 5.3.

$$\Delta\lambda = \frac{\lambda^2}{l[n - \lambda(dn/d\lambda)]} \tag{5.3}$$

where l is the optical path length ($l = 2W$ for the crystal) and n is the crystal's refractive index. A plot of $\Delta\lambda$ as a function of $1/l$ shows a linear relationship, indicating that the optical confinement and resonance occur along the crystal's lateral direction. Since the optical confinement occurs independent of the excitation point, both fluorescence spectra detected at the side facets feature periodic peaks that can be assigned to the FP optical resonances along the lateral direction.

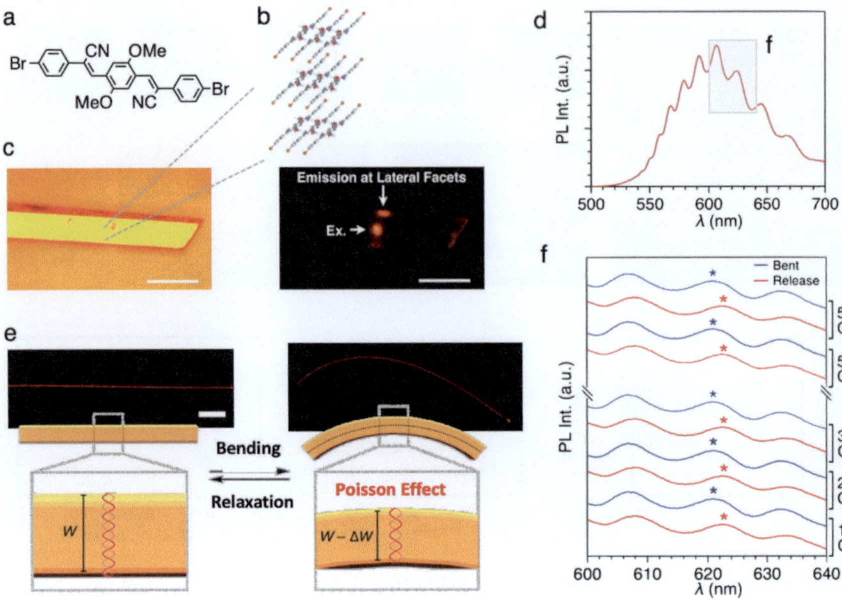

Fig. 5.20 **a** Chemical structure. **b** Crystal structure. **c** Microscopic images under white light irradiation and upon excitation with a focused laser, respectively. **d** μPL spectrum of the crystal collected at its lateral facet upon excitation with a focused laser. **e** Illustration and images of bent and relaxed crystals. **f** μPL spectra of the crystal upon multiple mechanical bending and relaxation. Reproduced from Ref. [71] with permission from Wiley–VCH

To study the change in optical properties due to bending, one end of an elastic molecular crystal with L 1.5 mm is glued to a glass substrate and bent in the in-plane direction relative to the substrate by applying a constant force in a stepwise manner. The other end of the crystal is attached to a sharp tungsten needle mounted on a micromanipulator. Meanwhile, an excitation laser is focused on the same position of the crystal, and the emission from the side facet is collected. The bending is quantitatively evaluated by calculating the bent crystal arc's r and elastic strain ε_n. The smaller the r, the larger the bending. Remarkably, the resonance peak at 606.4 nm observed in the raw specimen shifts monotonically toward the shorter wavelength region upon bending. The mechanical bending and the associated spectral shift can be repeated more than 50 times (Fig. 5.20e, f). The μ-PL spectrum of the crystal after 52 cycles of bending and releasing is nearly identical to that of the pristine sample. Both the peak positions in the released and bent states show significantly narrower deviations. The crystals' mechanical and spectral features indicate their potential utility as flexible light emitters, bending sensors, and mechanically robust optical components for optical fibers.

The shift of the peak to the shorter wavelength region upon bending is attributed to the decrease in lateral width and the change in refractive index due to the Poisson effect. Bending a crystal generates stress along the long axis, which causes expansion and contraction in the outer and inner arcs of the crystal, respectively. Assuming an ideal neutral plane, the change in the overall width at a given curvature has traditionally been essentially zero. Although this assumption works well for a variety of macroscopic objects, it is speculated that slight contraction/expansion may occur in the lateral dimensions, causing a spectral shift. This is negligible in mechanical measurements but detectable in optical measurements. At the same time, the contraction and expansion of the crystal are usually accompanied by a change in the refractive index, which also affects the spectral position of the resonant peak. The peak shift due to the change in refractive index and optical path length is given by the following approximate Eq. 5.4:

$$\frac{\delta\lambda}{\lambda} = \frac{\delta l}{l} + \delta n/n \tag{5.4}$$

where $\delta\lambda$ is the shift in the resonant wavelength, δl is the change in the optical path length, and δn is the change in the refractive index. The peak shift $\delta\lambda$ is a function of δl and δn, which means that an experimentally observed $\delta\lambda$ cannot be used to find a unique solution that simultaneously solves for δl and δn, unless the relationship between n and l at a particular stress is provided elsewhere.

5.5 Conclusions

Elastic molecular crystals are flexible materials with mechanical deformation ability and robustness. Creating elastic crystals of π-conjugated molecules gives them electrochemical or photochemical properties based on their semiconducting performance. Therefore, their application to photonics will be important fundamental research in developing flexible and wearable optical devices. Expected photonic functions include switching, waveguiding, resonators, and lasers. These results depend on various systems such as microspectroscopy, lasers, and detectors. However, quantitative understanding and comparison between studies will not be possible without standardization of the measurement method. In this chapter, we have summarized how to measure flexible molecular crystals by showing the measurement method. Furthermore, we have described the possibility of developing these photonic applications.

Acknowledgements S. H. acknowledges the JST FOREST Program (no. JPMJFR211W) and a KAKENHI (Grant-in-Aid for Scientific Research B: no. 24K01574) of the Japan Society for the Promotion of Science (JSPS). T. M. acknowledges a KAKENHI (Grant-in-Aid for Young Scientific Research: no. 22K14671) of JSPS.

References

1. D. Barman, K. Narang, P. Parui, N. Zehra, M.N. Khatun, L.R. Adil, P.K. Iyer, Review on recent trends and prospects in π-conjugated luminescent aggregates for biomedical applications. Aggregate **3**, e172 (2022)
2. Y. Hino, T. Matsuo, S. Hayashi, Structural phase transition in anthracene crystals. ChemPlusChem **87**, e202200157 (2022)
3. Q. Zhang, W. Tao, J. Huang, R. Xia, J. Cananillas-Gonzalez, Toward electrically pumped organic lasers: a review and outlook on material developments and resonator architectures. Adv. Photonics Res. **2**, 2000155 (2021)
4. S. Hayashi, T. Koizumi, N. Kamiya, 2,5-Dimethoxybenzene-1,4-dicarboxaldehyde: an emissive organic crystal and highly efficient fluorescent waveguide. ChemPlusChem **84**, 247–251 (2019)
5. T. Matsuo, H. Tanikubo, S. Hayashi, Facile wet-process to free-standing whispering gallery mode resonators mixed with spherical silica gel and π-conjugated molecules. Adv. Opt. Mater. **13**, 2401119 (2024)
6. T. Matsuo, S. Hayashi, Lasing in low-dimensional crystals of a fumaronitrile-based luminogen. J. Phys. Chem. Lett. **15**, 3968–3974 (2024)
7. S. Ghosh, M.K. Mishra, Elastic molecular crystals: from serendipity to design to applications. Cryst. Growth Des. **21**, 2566–2580 (2021)
8. S. Hayashi, Elastic organic crystals of π-conjugated molecules: anisotropic densely packed supramolecular 3D polymers exhibit mechanical flexibility and shape tunability. Polym. J. **51**, 813–823 (2019)
9. S. Hayashi, A. Asano, N. Kamiya, Y. Yokomori, T. Maeda, T. Koizumi, Fluorescent organic single crystals with elastic bending flexibility: 1,4-bis(thien-2-yl)-2,3,5,6-tetrafluorobenzene derivatives. Sci. Rep. **7**, 9453 (2017)
10. Y. Makino, M. Yoshida, S. Hayashi, T. Sasaki, S. Takamizawa, A. Kobayashi, M. Kato, Elastic and bright assembly-induced luminescent crystal of platinum(II) complexes with near-unity emission quantum yield. Dalton Trans. **52**, 8864–8872 (2023)
11. T. Seki, N. Hoshino, Y. Suzuki, S. Hayashi, Functional flexible molecular crystals: intrinsic and mechanoresponsive properties. CrystEngComm **23**, 5686–5696 (2021)
12. V.C. Sundar, J. Zaumseil, V. Podzorov, E. Menard, R.L. Willett, T. Someya, M.E. Gershenson, J.A. Rogers, Elastomeric transistor stamps: reversible probing of charge transport in organic crystals. Science **303**, 1644–1646 (2004)
13. H. Gercek, Poisson's ratio values for rocks. Int. J. Rock Mecha. Min. Sci. **44**, 1–13 (2007)
14. T. Ono, T. Hisaeda, Vapochromism of organic crystals based on macrocyclic compounds and inclusion complexes. Symmetry **12**, 1903 (2020)
15. L. Huang, L. Liu, X. Li, H. Hu, M. Chen, Q. Yang, Z. Ma, X. Jia, Crystal-state photochromism and dual-mode mechanochromism of an organic molecule with fluorescence, room-temperature phosphorescence, and delayed fluorescence. Angew. Chem. Int. Ed. **58**, 16445–16450 (2019)
16. Z. Jiang, H. Zhao, W. Wu, K. Chen, H. Yu, T. Wang, X. Huang, N. Wang, L. Zhou, H. Hao, Multi-stimuli responsive organic polymorphic crystals: anisotropic elasticity and plasticity, mechanochromism and photomechanical motions. J. Mater. Chem. C **11**, 4375–4383 (2023)
17. Y. Morita, S. Suzuki, K. Fukui, S. Nakazawa, H. Kitagawa, H. Kishida, H. Okamoto, A. Naito, A. Sekine, Y. Ohashi, M. Shiro, K. Sasaki, D. Shiomi, K. Sato, T. Takui, K. Nakasuji, Thermochromism in an organic crystal based on the coexistence of σ- and π-dimers. Nat. Mater. **7**, 48–51 (2008)
18. Y.-A. Lee, R. Eisenberg, Luminescence tribochromism and bright emission in gold(I) thiouracilate complexes. J. Am. Chem. Soc. **125**, 7778–7779 (2003)
19. Y. Sagara, T. Kato, Mechanically induced luminescence changes in molecular assemblies. Nat. Chem. **1**, 605–610 (2009)
20. K. Nagura, S. Saito, H. Yusa, H. Yamawaki, H. Fujihisa, H. Sato, Y. Shimoikeda, S. Yamaguchi, Distinct responses to mechanical grinding and hydrostatic pressure in luminescent chromism of tetrathiazolylthiophene. J. Am. Chem. Soc. **135**, 10322–10325 (2013)

21. Z. Chi, X. Zhang, B. Xu, X. Zhou, C. Ma, Y. Zhang, S. Liua, J. Xu, Recent advances in organic mechanofluorochromic materials. Chem. Soc. Rev. **41**, 3878–3896 (2012)
22. H. Tanikubo, T. Matsuo, S. Hayashi, Alkyl-chain bridged acrylonitrile-appended anthracene chromophore toward mechanically-induced bathochromic shift. Bull. Chem. Soc. Jpn **96**, 178–182 (2023)
23. N. Zhang, W. Sun, Y. Zhang, H. Jang, R. Xiong, S. Dong, H. Zhang, Organic radical ferroelectric crystals with martensitic phase transition. Nat. Commun. **14**, 5854 (2023)
24. Y. Hino, S. Hayashi, Thermo-triggered domino-like single-crystal-to-single-crystal phase transition from face-to-edge to face-to-face packing of anthracene. Chem. Eur. J. **27**, 17595–17600 (2021)
25. T. Seki, K. Sakurada, H. Ito, Controlling mechano- and seeding-triggered single-crystal-to-single-crystal phase transition: molecular domino with a disconnection of aurophilic bonds. Angew. Chem. Int. Ed. **52**, 12828–12832 (2013)
26. K. Takazawa, J. Inoue, K. Mitsuishi, Y. Yoshida, H. Kishida, P. Tinnemans, H. Engelkamp, P.C.M. Christianen, Phase-transition-induced jumping, bending, and wriggling of single crystal nanofibers of coronene. Sci. Rep. **11**, 3175 (2021)
27. S. Chen, M.P. Zhuo, X.D. Wang, G. Wei, L. Liao, Optical waveguides based on one-dimensional organic crystals. PhotoniX **2**, 2 (2021)
28. Y. Hino, T. Matsuo, S. Hayashi, Unique 2D face topologies in naphthyl-appended naphtho [k] fluoranthene-based 3D crystals for optical waveguide. Adv. Opt. Mater. **12**, 2201810 (2023)
29. S. Hayashi, T. Koizumi, Direction-specific fluorescence of an engineered organic crystal and its mechanically induced shaping-caused new face appearance. CrystEngComm **21**, 5990–5994 (2019)
30. S. Hayashi, T. Koizumi, Mechanically induced shaping of organic single crystals: facile fabrication of fluorescent and elastic crystal fibers. Chem. Eur. J. **24**, 8507–8512 (2018)
31. Y. Liu, H. Hu, L. Xu, B. Qiu, J. Liang, F. Ding, K. Wang, M. Chu, W. Zhang, M. Ma, B. Chen, X. Yang, Y.S. Zhao, Orientation-controlled 2D anisotropic and isotropic photon transport in co-crystal polymorph microplates. Angew. Chem. Int. Ed. **59**, 4456–4463 (2020)
32. M. Nakabayashi, T. Matsuo, S. Hayashi, Non-covalent supramolecular 1D alternating copolymer in crystal toward 2D anisotropic photon transport. Chem. Eur. J. **29**, e202302351 (2023)
33. A. Qiagedeer, H. Yamagishi, S. Hayashi, Y. Yamamoto, Polymer optical microcavity sensor for volatile organic compounds with distinct selectivity toward aromatic hydrocarbons. ACS Omega **6**, 21066–21070 (2021)
34. J. Wu, H. Gao, R. Lai, M. Zhuo, J. Feng, X. Wang, Y. Wu, L. Liao, L. Jiang, Near-infrared organic single-crystal nanolaser arrays activated by excited-state intramolecular proton transfer. Matter **2**, 1233–1243 (2020)
35. H. Fang, J. Yang, J. Feng, T. Yamao, S. Hotta, H. Sun, Functional organic crystals for laser applications. Laser Photo. Rev. **8**, 687–715 (2014)
36. L. Chen, J. Cai, Y. Zhen, C. Ou, X. Ding, J. Lin, Ultraviolet organic laser from rhombus microcrystal: benefits of single-molecule emission from twisted structure. J. Phys. Chem. Lett. **15**, 1028–1033 (2024)
37. D. Liu, J. De, H. Gao, S. Ma, Q. Ou, S. Li, Z. Qin, H. Dong, Q. Liao, B. Xu, Q. Peng, Z. Shuai, W. Tian, H. Fu, X. Zhang, Y. Zhen, W. Hu, Organic laser molecule with high mobility, high photoluminescence quantum yield, and deep-blue lasing characteristics. J. Am. Chem. Soc. **142**, 6332–6339 (2020)
38. T. Matsuo, C. Rössiger, J. Herr, R. Göttlich, D. Schlettwein, H. Mizuno, F. Sasaki, H. Yanagi, Synthesis and characterization of methoxy- or cyano-substituted thiophene/phenylene co-oligomers for lasing application. RSC Adv. **10**, 24057–24062 (2020)
39. T. Matsuo, F. Sasaki, H. Yanagi, Optically pumped lasing in a single crystal cavity of thiophene/phenylene co-oligomers grown via improved crystal growth methods in solution. Appl. Phys. Express **15**, 051002 (2022)
40. T. Matsuo, Y. Ueda, H. Mizuno, F. Sasaki, K. Yamashita, H. Yanagi, Optically pumped lasing based on vibrationally dressed exciton polaritons in a single-crystal molecular cavity at room temperature. ACS Photonics **9**, 2015–2023 (2022)

41. D. Lidzey, D.D.C. Bradley, M.S. Skolnick, T. Virgili, S. Walker, D.M. Whittaker, Strong exciton–photon coupling in an organic semiconductor microcavity. Nature **395**, 53–55 (2023)
42. S. Kéna-Cohen, S. Forrest, Room-temperature polariton lasing in an organic single-crystal microcavity. Nat. Photon. **4**, 371 (2010)
43. J. Gautier, M. Li, T.W. Ebbesen, C. Genet, Planar chirality and optical spin–orbit coupling for chiral Fabry-Perot cavities. ACS Photonics **9**, 2015 (2022)
44. J. Tang, J. Zhang, Y. Lv, H. Wang, F.F. Xu, C. Zhang, L. Sun, J. Yao, Y.S. Zhao, Room temperature exciton–polariton Bose-Einstein condensation in organic single-crystal microribbon cavities. Nat. Commun. **12**, 3265 (2021)
45. S. Saha, M.K. Mishra, C.M. Reddy, G.R. Desiraju, From molecules to interactions to crystal engineering: mechanical properties of organic solids. Acc. Chem. Res. **51**, 2957–2967 (2018)
46. C.M. Reddy, R.C. Gundakaram, S. Basavoju, M.T. Kirchner, K.A. Padmanabhan, G.R. Desiraju, Structural basis for bending of organic crystals. Chem. Commun. **31**, 3945–3947 (2002)
47. C.M. Reddy, K.A. Padmanabhan, G.R. Desiraju, Structure-property correlations in bending and brittle organic crystals. Cryst. Growth Des. **6**, 2720–2731 (2006)
48. J. Li, W. Wong, X. Tao, Recent advances in soft functional materials: preparation, functions and applications. Nanoscale **12**, 1281–1306 (2020)
49. T.T. Nguyen, S.G. Arkhipov, D.A. Rychkov, Simple crystallographic model for anomalous plasticity of L-leucinium hydrogen maleate crystals. Mater. Today Proc. **25**, 412–415 (2020)
50. T. Seki, S. Kobayashi, R. Ishikawa, K. Yano, T. Matsuo, S. Hayashi, Preparation of intrinsically fragile bent crystals. Chem. Sci. **15**, 12258–12263 (2024)
51. S. Hayashi, T. Koizumi, Elastic organic crystal of a fluorescent π-conjugated molecule. Angew. Chem. Int. Ed. **55**, 2701–2704 (2016)
52. S. Hayashi, Elastic organic crystals of π-conjugated molecules: new concept for materials chemistry. Symmetry **12**, 2022 (2020)
53. S. Hayashi, Elastic molecular crystals: their deformation-induced reversible unit cell changes with specific Poisson effect. Bull. Chem. Soc. Jpn **95**, 721–727 (2022)
54. S. Hayashi, Curving deformation-induced photoluminescence changes and anisotropy analysis in elastic organic crystals. CrystEngComm **23**, 5763–5767 (2021)
55. Y. Suzuki, S. Koga, K. Kitaura, J. Kawamata, K. Yano, N. Hoshino, T. Akutagawa, S. Hayashi, Noninvasive three-dimensional assessment of single molecular crystals using multiphoton microscopic observation and their deformation-induced emission characteristics. Langmuir **39**, 11646–11652 (2023)
56. T. Matsuo, S. Hayashi, F. Sasaki, H. Yanagi, Efficient optical waveguiding and lasing in needle crystals of thiophene/phenylene co-oligomers grown in solution phase. Jpn. J. Appl. Phys. **62**, 030903 (2023)
57. S. Zhao, H. Yamagishi, Y. Norikane, S. Hayashi, Y. Yamamoto, Optical control of aggregation-induced emission shift by photoisomerizable precipitant in a liquid droplet microresonator. Adv. Opt. Mater. **11**, 2202134 (2023)
58. S. Kera-Cohen, S.R. Forrest, Room-temperature polariton lasing in an organic single-crystal microcavity. Nat. Photon. **4**, 371 (2010)
59. M. Ghulinyan, D. Navarro-Urrios, A. Pitanti, A. Lui, G. Pucker, L. Pavesi, Whispering-gallery modes and Purcell effect in a Si-nanocrystal-based single microdisk resonator, in *Proceedings of the 5th IEEE International Conference on Group IV Photonics* (Sorrento, Italy--19 September 2008), p. 17
60. A.S.D. Sandanayaka, K. Yoshida, M. Inoue, C. Qin, K. Goushi, J.-C. Ribierre, T. Matsushima, C. Adachi, Quasi-continuous-wave organic thin-film distributed feedback laser. Adv. Opt. Mater. **4**, 834–839 (2016)
61. I.D.W. Samuel, E.B. Namdas, G.A. Turnbull, How to recognize lasing. Nat. Photon. **3**, 546–549 (2009)
62. S. Varghese, S.K. Park, S. Casado, R.C. Fischer, R. Resel, B. Milián-Medina, R. Wannemacher, S.Y. Park, J. Gierschner, Stimulated emission properties of sterically modified distyrylbenzene-based H-aggregate single crystals. J. Phys. Chem. Lett. **4**, 1597 (2013)

63. S.-W. Choi, S. Yamamoto, T. Iwata, H. Kikuchi, Optically isotropic liquid crystal composite incorporating in-plane electric field geometry. Appl. Phys. Lett. **90**, 141110 (2007)
64. M. Zeb, M. Tahir, F. Muhammad, S.M. Said, M.F.M. Sabri, M.R. Sarker, S.H.M. Sli, F. Wahab, Amplified spontaneous emission and optical gain in organic single crystal quinquethiophene. Crystals **9**, 609 (2019)
65. S. Hayashi, T. Koizumi, N. Kamiya, Elastic bending flexibility of fluorescent organic single crystal: new aspects of commonly used building block "4,7-dibromo-2,1,3-benzothiadiazole." Cryst. Growth Des. **17**, 6158–6162 (2017)
66. S. Hayashi, F. Ishiwari, T. Fukushima, M. Tashiro, Y. Imamura, M. Katouda, Anisotropic Poisson's effect and deformation-induced fluorescence change of elastic 9,10-dibromoanthrathene single crystals. Angew. Chem. Int. Ed. **59**, 16195–16201 (2020)
67. C. Zhang, Y.S. Zhao, J. Yao, Optical waveguides at micro/nanoscale based on functional small organic molecules. Phys. Chem. Chem. Phys. **13**, 9060–9073 (2011)
68. S. Hayashi, S. Yamamoto, D. Takeuchi, Y. Ie, K. Takagi, Creating elastic organic crystals of π-conjugated molecules with bending mechanofluorochromism and flexible optical waveguide. Angew. Chem. Int. Ed. **57**, 17002–17008 (2018)
69. T. Matsuo, K. Ikeda, S. Hayashi, Flexible Förster resonance energy transfer-assisted optical waveguide based on elastic mixed molecular crystals. Aggregate **4**, e378 (2023)
70. T. Matsuo, J. Kuwabara, T. Kanbara, S. Hayashi, Flexible and red-emissive organic single crystal micro-resonator for efficient active waveguides. J. Phys. Chem. Lett. **14**, 6577–6582 (2023)
71. S. Zhao, H. Yamagishi, O. Oki, Y. Ihara, N. Ichiji, A. Kubo, S. Hayashi, Y. Yamamoto, Mechanically flexible and optically tunable organic crystal resonator. Adv. Opt. Mater. **10**, 2101808 (2022)

Part II
Luminescent, Electrical, and Magnetic Properties

Chapter 6
Tunable Mechanochromic Luminescence of Organic Crystals

Suguru Ito

Abstract Luminescent organic crystals often exhibit mechanochromic luminescence (MCL), which refers to the reversible color change of photoluminescence induced by mechanical stimuli such as grinding, crushing, and pressing. Initial developments of mechanoresponsive luminescent crystals relied on serendipity, making the rational design of MCL a challenging task. This chapter reviews recent advancements in tunable MCL of organic crystals. One fundamental strategy involves tuning molecular conformation in the crystalline state by introducing sterically different substituents, which controls the luminescence color and alters MCL properties. Another significant approach is the preparation of two-component organic crystals. Formation of cocrystals and solvates is a fundamental method for tuning the luminescence properties of organic crystals. Additionally, the formation of segregated crystals has proven to be an effective method for achieving MCL with a substantial shift in emission wavelength. Furthermore, the luminescent properties of organic crystals can be finely tuned by preparing mixed crystals containing trace amounts of dopant compounds, and this approach can extend the range of mechanoresponsive luminescence shifts.

Keywords Mechanochromism · Fluorescence · Solid-state emission · Luminescent crystal · Energy transfer

S. Ito (✉)
Department of Chemistry and Life Science, Graduate School of Engineering Science, Yokohama National University, 79-5 Tokiwadai, Hodogaya-Ku, Yokohama 240-8501, Japan
e-mail: suguru-ito@ynu.ac.jp

S. Kobatake and H. Uekusa (eds.), *Advances in Organic Crystal Chemistry*,
https://doi.org/10.1007/978-981-96-4479-7_6

6.1 Introduction

6.1.1 Luminescent Organic Crystals

Photoluminescence of organic molecules refers to the emission of light energy from excited states [1]. When an organic molecule in a singlet ground state (S_0) absorbs photons, one electron is excited to an anti-bonding orbital. The resulting excited state molecule undergoes vibrational relaxation to reach the lowest excited singlet state (S_1), from which it can emit fluorescence or return to S_0 through non-radiative decay processes. In some cases, the molecule may undergo intersystem crossing to the lowest excited triplet state (T_1), resulting in phosphorescence. The efficiency of photoluminescence is assessed by the fluorescence or phosphorescence quantum yield (Φ_F or Φ_P), defined as the ratio of emitted photons to absorbed photons.

Crystals of luminescent aromatic hydrocarbons often exhibit lower Φ_F compared with their dilute solutions due to impurities, structural defects, and reabsorption, although their high-purity single crystals maintain relatively high Φ_F [2]. Many luminescent organic molecules, especially polar molecules, typically experience severe concentration quenching in their crystalline state, resulting in significantly diminished luminescence [3]. Recently, there have been increasing reports of luminescent organic crystals achieved through design strategies that either suppress intermolecular interactions [4] or exploit aggregation-induced emission [5]. Moreover, organic crystals that switch their emission color in response to external stimuli are actively being investigated owing to their promising applications, such as in sensor materials [6].

6.1.2 Mechanochromic Luminescence (MCL) of Organic Crystals

Manual grinding or crushing of organic crystals by using spatulas or pestles can induce phase transitions to polymorphic crystalline or amorphous states. Such transitions alter molecular arrangements and conformations, leading to changes in luminescent properties for organic crystals. The first reported mechanoresponsive organic crystal that switches its luminescence intensity is 4-*tert*-butyl-1-(4′-dimethylaminobenzylideneamino)pyridinium perchlorate, as described by Gawinecki and coworkers in 1993 [7]. Although this pioneering work did not immediately spur continuous research, from 2007 to 2009, several researchers reported organic and organometallic materials exhibiting mechanochromic luminescence (MCL), where mechanical stimuli change the emission color [8–12]. These reports have initiated a rapid advancement in the research on MCL.

In typical MCL of organic crystals, mechanical stimuli induce a bathochromic shift in the emission wavelength (Fig. 6.1) [13–15]. The mechanically changed emission color recovers to the original color by heating or exposing to organic solvents,

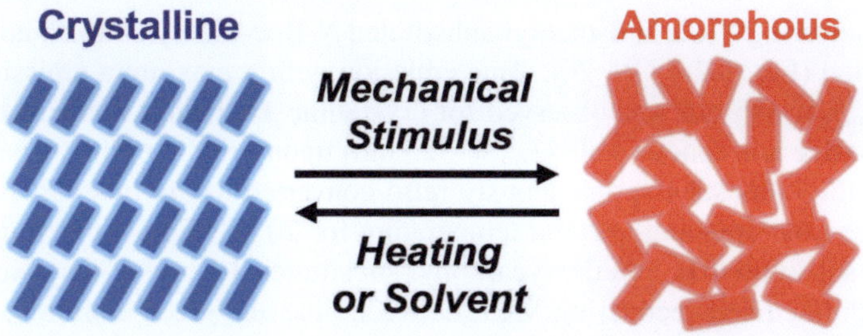

Fig. 6.1 Schematic illustration for typical bicolor MCL of organic crystals. Reprinted with permission from Ref. [14]. Copyright 2021, The Chemical Society of Japan

resulting in a bicolor emission switching. Early reports of organic MCL crystals were often serendipitous discoveries. However, recent research has begun to establish methods for controlling MCL properties. This chapter first discusses the control of MCL through the steric effects of substituents. Following this, the effectiveness of creating two-component crystals, such as cocrystals, solvates, segregated crystals, and doped crystals, in tuning MCL properties is highlighted.

6.2 Steric Effects of Substituents

With advancements in the study of organic MCL crystals, it has been found that the steric effects of substituents, in addition to their electronic effects, play a significant role in tuning MCL properties. This section presents examples of how the steric effects of substituents attached to luminescent organic molecules can control self-recovery behavior in MCL. Moreover, it covers cases where the direction of the emission wavelength shift upon mechanical stimulation has been effectively controlled by the steric differences of the substituents.

6.2.1 Tuning of Self-recovering MCL

Unlike typical MCL crystals that require a secondary stimulus to revert to their original state after mechanical stimulation, organic crystals exhibiting self-recovering MCL spontaneously return to their initial emission color after the mechanical stimulus is removed. Early reports on self-recovering MCL include findings on difluoroboron avobenzone [16] and 4-[bis(4-methylphenyl)amino]benzaldehyde [17]. In addition, among anthracene derivatives with propyl, pentyl, and dodecyl side chains, the dodecyl-substituted derivative has been observed to exhibit self-recovering MCL. This behavior is attributed to the low cold-crystallization temperature of the dodecyl-substituted compound [18].

Self-recovering MCL of 3-methyl-substituted *N*-Boc-indolylbenzothiadiazole **1a** was reported (Fig. 6.2a) [19, 20]. Blue emission with a maximum emission wavelength (λ_{em}) of 485 nm was observed for crystalline **1a**, which was converted to a green-emissive amorphous state (λ_{em} = 525 nm) upon grinding with a spatula. The ground amorphous state spontaneously retro-converted to the blue-emissive crystalline state after standing at room temperature for 20 s. 3-Ethyl-substituted **1b**, 3-isopropyl-substituted **1c**, *N*-ethoxycarbonyl-substituted **1d**, and *N*-tosyl-substituted **1e** also exhibited self-recovering MCL between blue and green emission, although the recovery time varied from 1 min to 1.5 h (Fig. 6.2b–e). For **1f** (R^2 = PhCO), **1g** (R^2 = ^{t}BuCO), and **1h** (R^2 = Me), the emission wavelength shifted in the bathochromic direction with a progressive decrease in the electron-withdrawing property of the substituent on the nitrogen atom (Fig. 6.2f–h). A longer recovery time was required in the self-recovering MCL of **1f** (2 h) and **1g** (40 min), whereas the mechanically changed color of **1h** immediately recovered to the original color in 10 s. The differences in self-recovery behavior can be explained by the glass transition temperature (T_g) determined through differential scanning calorimetry. For instance, the T_g of *N*-benzoyl derivative **1g** was 45 °C, whereas the T_g of *N*-methyl derivative **1h** was 16 °C. When the T_g is higher than room temperature, the material remains in a glassy state at room temperature, resulting in reduced molecular mobility and thus a slower recrystallization process.

6.2.2 Controlling the Direction of MCL

Most organic MCL crystals exhibit a bathochromic shift (red-shift) in emission wavelength in response to mechanical stimulation, with instances of hypsochromic shift (blue-shift) being relatively rare [21]. Especially, achieving both red-shifted and blue-shifted MCL within derivatives of the same molecular framework is uncommon. A strategy for designing the direction of mechanoresponsive luminescence shift has been reported, which is based on controlling the dipole moment of donor–acceptor molecules [22]. Specifically, both bathochromic and hypsochromic shifts were observed for 2,7-diaryl-[1,2,4]triazolo[1,5-*a*]pyrimidines by interchanging the electron-donating and electron-withdrawing groups on the 2- and 7-positions. Another strategy utilizes cation–anion interactions [23]. The direction of mechanoresponsive luminescence shift was controlled by changing the anion of tributyl(perylene-3-ylmethyl)phosphonium fluorophore.

In addition to the more commonly observed red and blue shifts in emission color induced by mechanical stimuli, there are also a few reported cases where the emission color changes in a stepwise manner in response to varying intensities of mechanical stimuli. For example, a two-step MCL was achieved by a complex composed of BH_3 and an amine linker, which connects a tetraphenylethylene unit and rhodamine B [24]. In this case, the initial emission color change is attributed to the phase transition to the amorphous phase, while further strong grinding induces the ring-opening reaction of

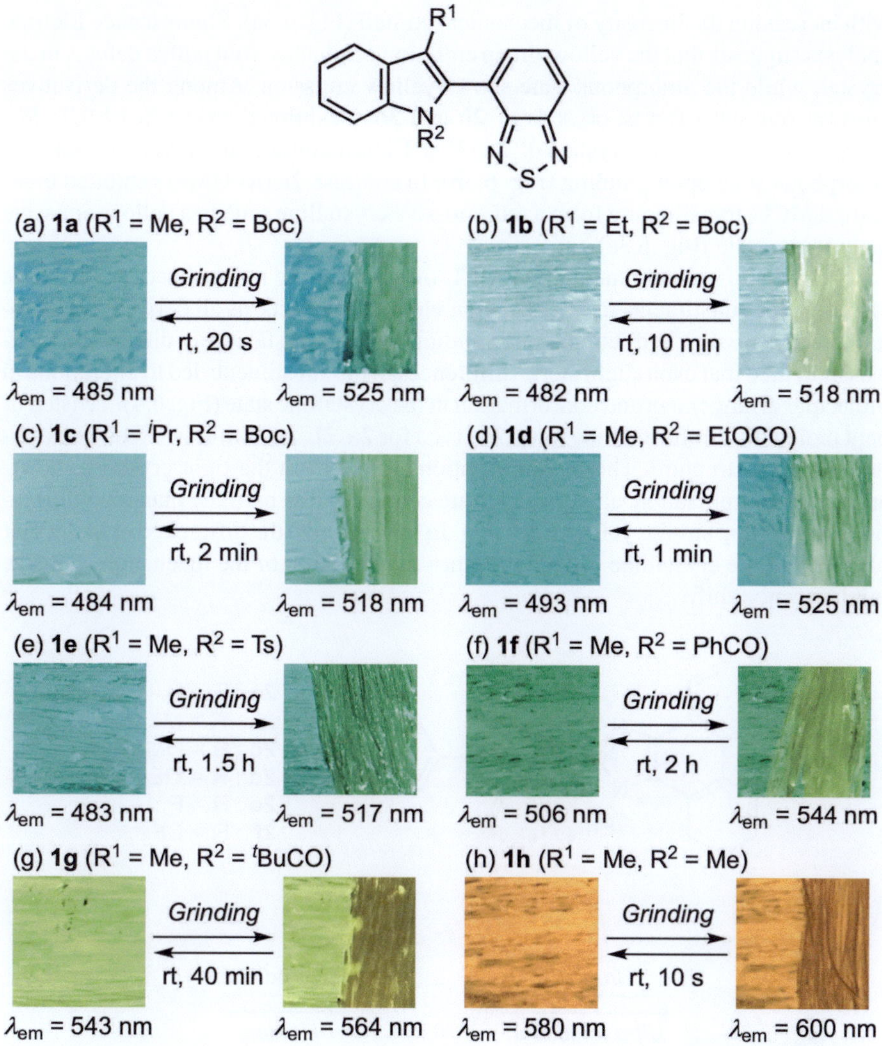

Fig. 6.2 **a–h** Photographs for the self-recovering MCL of indolylbenzothiadiazoles **1a–h** under UV light. Reproduced with permission from Ref. [20]. Copyright 2017, The Royal Society of Chemistry

the rhodamine moiety. For another example, a biindole derivative exhibited a two-step MCL based on changes in molecular conformation and arrangement without altering the molecular structure [25].

Phenanthroimidazole derivatives **2a–f** exhibited not only typical red-shifted MCL but also blue-shifted MCL and two-step MCL, depending on the steric bulk of the substituted phenyl groups [26]. The unsubstituted derivative **2a** (R = H) exhibited a transition from green-emissive crystals to yellow-green and yellow-emissive states with increasing the intensity of mechanical stimuli (Fig. 6.3a). Fluorescence lifetime analysis suggests that the yellow-green emission originates from lattice defects in the crystal, while the amorphous state shows yellow emission. Among the derivatives with various substituents, crystals of **2b** and **2d–f** exhibited red-shifted MCL. For example, blue-emissive crystals of **2e** (R = F) transformed into a yellow-emissive amorphous state upon grinding (Fig. 6.3b). In contrast, **2c** (R=OMe) exhibited blue-shifted MCL, transitioning from a red-emissive crystalline state to a yellow-emissive amorphous state (Fig. 6.3c).

The red-shifted and blue-shifted MCL of **2a–f** can be rationalized by the steric effects of the substituents rather than their electronic effects, as all derivatives exhibited green emission in dilute toluene solution. Single-crystal X-ray diffraction analysis indicated that even slight steric differences in the substituents led to alterations in molecular arrangement and conformation in the crystalline state (Fig. 6.4). Twisted or bent molecular conformations were observed for **2a**, **2b**, and **2d–f**, whereas **2c** exhibited a planar structure. These conformations result from the tight crystal packing, and upon amorphization, all derivatives are converted into partially planar conformations, exhibiting similar yellow emission. In other words, the difference in molecular planarity in the crystalline state determines the direction of the mechanoresponsive luminescence shift.

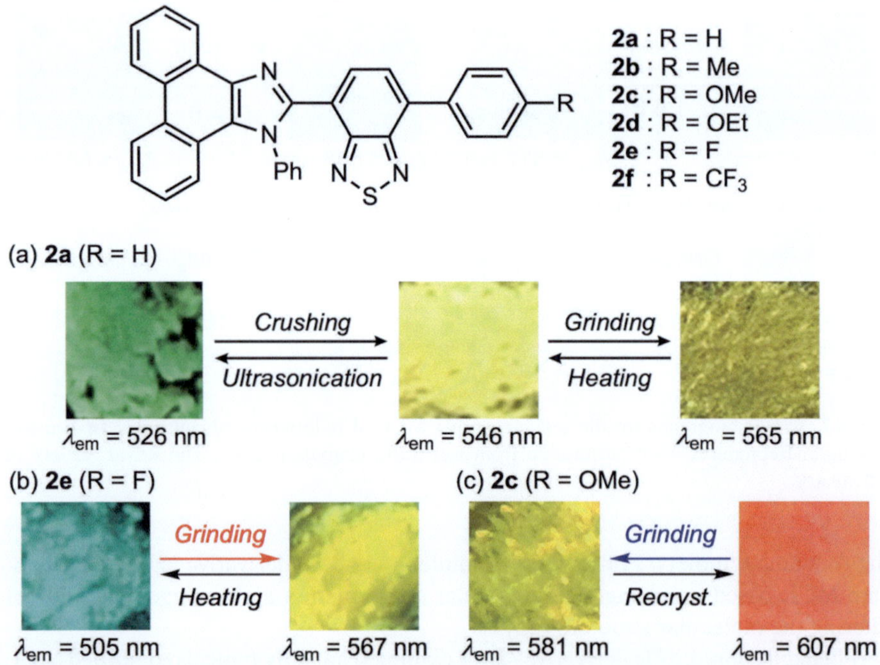

Fig. 6.3 **a** Two-step MCL of **2a**. **b** Red-shifted MCL of **2e**. **c** Blue-shifted MCL of **2c**. Reproduced with permission from Ref. [26]. Copyright 2019, The Royal Society of Chemistry

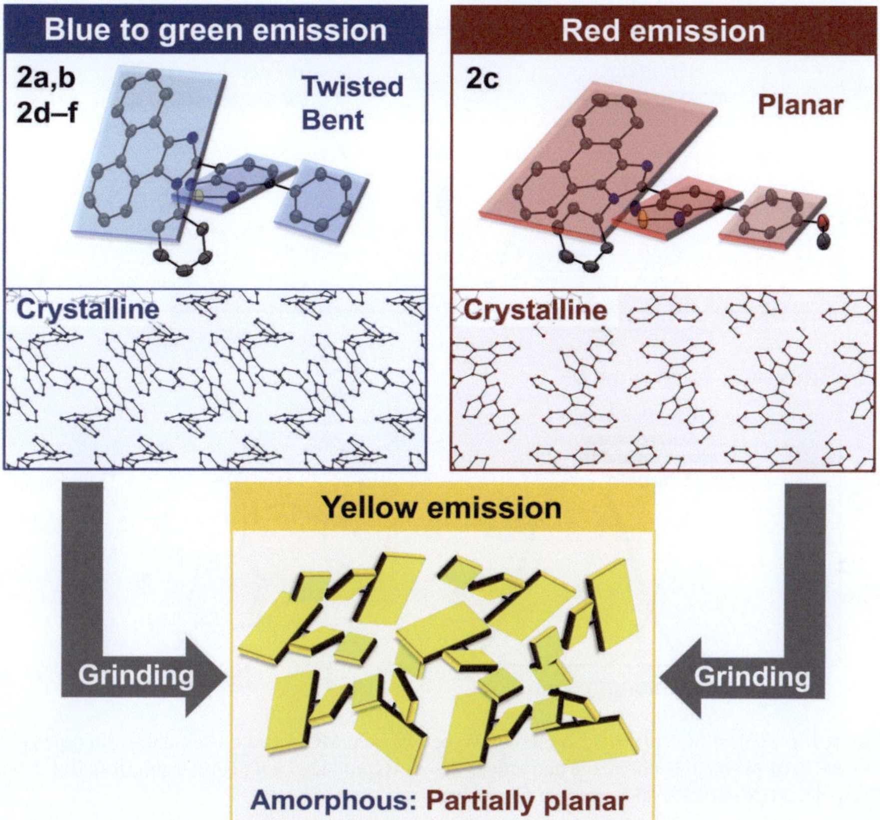

Fig. 6.4 Probable mechanism of the red-shifted and blue-shifted emission color change of **2a–f**. Reproduced with permission from Ref. [26]. Copyright 2019, The Royal Society of Chemistry

Two-step MCL can be categorized into two types. The first is the one-way type, exemplified by the unsubstituted derivative **2a**, where initial crushing with low intensity followed by more intense grinding results in a unidirectional shift in the emission wavelength. The second is the back-and-forth type, where crushing induces a shift in the emission wavelength in one direction, and subsequent grinding causes a shift in the opposite direction. Among the few examples of two-step MCL, the back-and-forth type is particularly rare [27].

Phenanthroimidazole derivatives **2g–m**, each featuring different substituents on the phenyl group attached to the nitrogen atom, displayed diverse MCL behaviors [28]. Derivatives **2g–j** (R = *p*-Me, *p*-OMe, *p*-F, and *p*-CF_3) and **2l** (R = *m*-Br) exhibited typical bicolor MCL (Fig. 6.5a). Additionally, **2k** (R = *p*-Br) demonstrated one-way type two-step MCL (Fig. 6.5b), while **2m** (R = *o*-Br) achieved the rare back-and-forth type two-step MCL (Fig. 6.5c). In the one-way type MCL of **2k**, the emission color change of the initial crystal upon crushing was attributed to the phase transition to another crystalline state. For **2m**, the formation of defects on the fracture

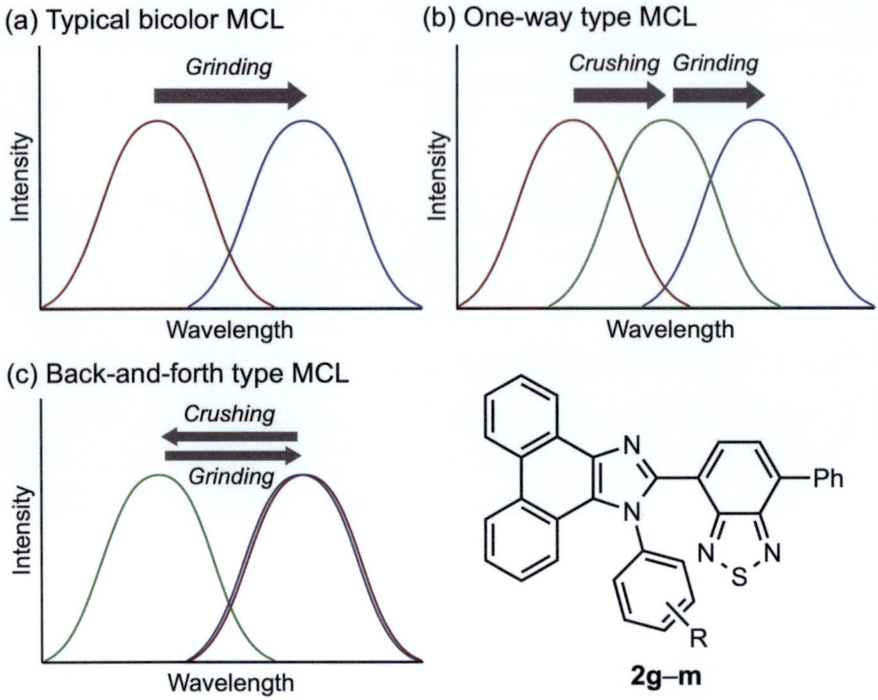

Fig. 6.5 **a** Typical bicolor MCL, **b** one-way type two-step MCL, and **c** back-and-forth type two-step MCL of phenanthroimidazole derivatives **2g–m**. Reproduced with permission from Ref. [28]. Copyright 2020, The Royal Society of Chemistry

surface accounts for the initial hypsochromic shift of the emission upon crushing. The second color change of both **2k** and **2m** was attributed to amorphization by strong grinding.

6.3 Cocrystals and Solvates

Crystallizing a compound with another solid or liquid compound to form cocrystals or solvates is an effective method for significantly altering the properties of crystals [29–31]. While the majority of MCL crystals are single-component, several instances have been reported for the MCL of cocrystals and solvates [32–36]. This section will discuss the tunable MCL by forming cocrystals or solvates.

6.3.1 Tunable MCL of Cocrystals

The MCL properties of pyridyl-substituted phenanthroimidazole derivative **3** were tunable by forming cocrystals with substituted benzoic acids **4a–d** [37]. The crystals of **3** exhibited green luminescence ($\lambda_{em} = 521$ nm) (Fig. 6.6a), whereas cocrystals with **4a–d** exhibited bathochromically shifted emission wavelengths with increasing acidity of **4a–d**. This phenomenon can be attributed to the enhanced electron-withdrawing ability of the pyridyl group when hydrogen bonded to the more acidic carboxyl groups, facilitating intramolecular charge-transfer absorption and subsequent emission at lower energies. When these cocrystals were ground with a spatula, they demonstrated MCL with further bathochromic shifting of the emission wavelength. For instance, the λ_{em} of cocrystal with **4a** shifted from 551 to 594 nm (Fig. 6.6b), while the λ_{em} of cocrystal with **4d** shifted from 593 to 627 nm (Fig. 6.6c). Remarkably, the emission color of the mechanically generated state gradually shifted in the hypsochromic direction upon heating. At lower temperatures, the cocrystals recrystallized, and at higher temperatures, the benzoic acid derivatives volatilized, resulting in the green emission of crystalline **3**.

6.3.2 Multicolor MCL of Solvates

Dianisyl-substituted phenanthroimidazole derivative **5** formed four types of solvates with different emission colors [38]. In response to grinding and exposure to different solvents, the emission colors were switchable between blue-green to yellow-green ($\lambda_{em} = 505$–542 nm) for the solvates and yellow-orange ($\lambda_{em} = 576$ nm) for the amorphous states (Fig. 6.7a). In the crystalline state, adjacent molecules of **5** in a head-to-tail orientation form π-stacks through their anisyl groups on the nitrogen atom. The spaces between these dimers incorporate solvent molecules, resulting in the formation of solvates.

Crystals containing pyridine molecules exhibited nearly identical molecular conformations and arrangements as those containing benzene molecules. However, in addition to exhibiting MCL, the pyridine-containing crystals also showed a remarkable bathochromic shift in emission wavelength in response to HCl vapor (Fig. 6.7b). Typically, changes in emission color in response to acids are achieved by the protonation of basic groups directly attached to the emissive molecules [39–42]. In contrast, this system uniquely realizes acid-responsive luminescence through the protonation of non-emissive pyridine molecules surrounding the luminescent molecules.

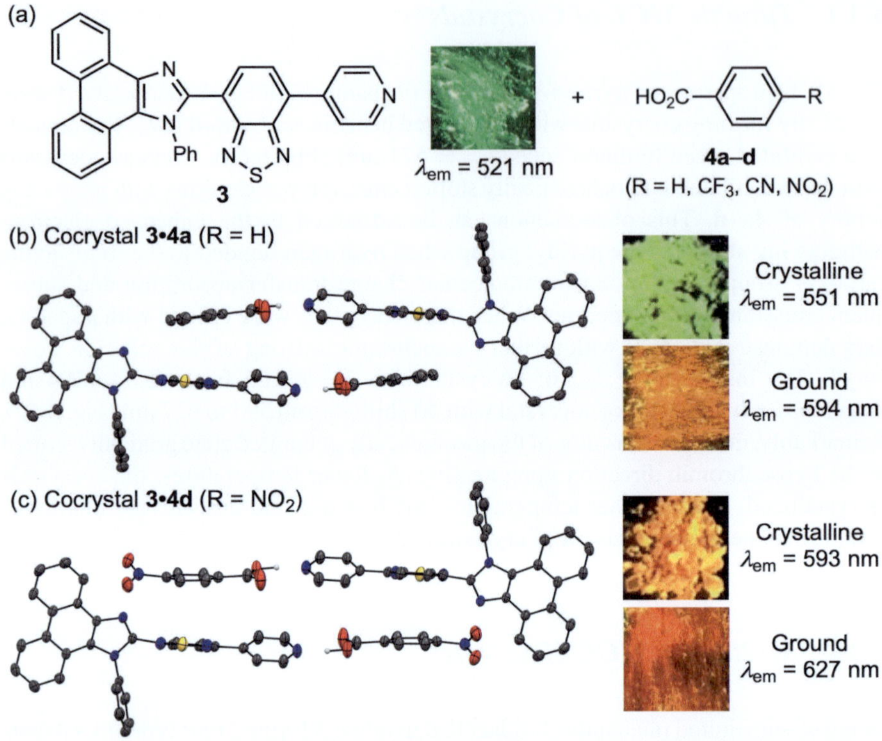

Fig. 6.6 **a** Structures of **3** and **4a**–**d**. Photograph of crystalline **3** under UV light. **b** and **c** Crystal structures and MCL behaviors of cocrystals **3•4a** and **3•4d**. Reproduced with permission from Ref. [37]. Copyright 2019, The Royal Society of Chemistry

6.4 Two-Component Segregated Crystals

As shown in Sect. 6.3, equimolar amounts of two organic compounds often form cocrystals and solvates. However, in some cases, two compounds independently form their own crystals. The preparation of such segregated crystals has become a useful strategy to control the MCL properties of organic crystals [43–47]. This section covers tricolor MCL, self-recovering MCL, wide-range and two-step MCL, and mechanoresponsive on–off luminescence switching by two-component segregated crystals.

6.4.1 Tricolor MCL

Dipyrenylbithiophene **6** exhibited a typical bicolor MCL between blue and blue-green with a maximum emission wavelength shift of 13 nm, which could be

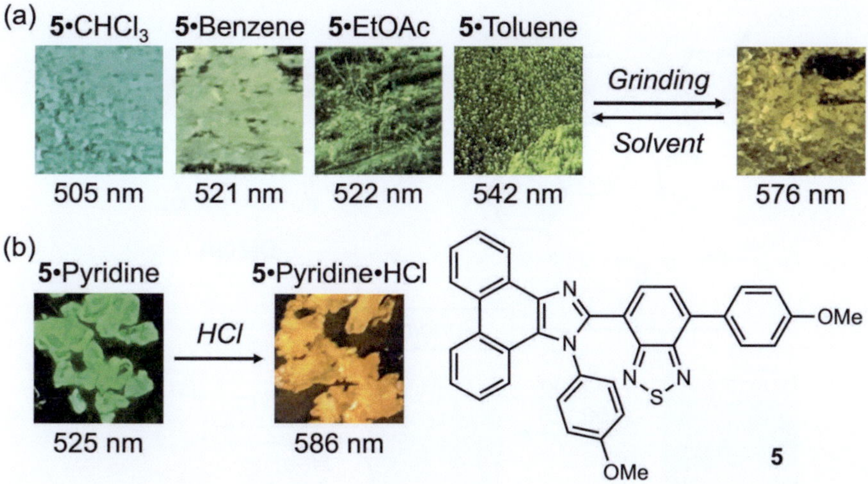

Fig. 6.7 **a** Photographs for the MCL of solvates composed of **5** and solvent molecules under UV light. **b** Photographs for the acid-responsive emission color change of **5**•pyridine under UV light. Reproduced with permission from Ref. [38]. Copyright 2023, The Royal Society of Chemistry

improved to show tricolor MCL by forming segregated crystals with *N*,*N*′-dimethylquinacridone (**DMQA**) (Fig. 6.8) [43]. Because the absorption band of solid-state **DMQA** is in the same region as the luminescence band of ground amorphous **6**, Förster resonance energy transfer (FRET) from **6** to **DMQA** was expected by mixing these two compounds. A green-emissive molten mixture **BD-M** (bithiophene/**DMQA**-molten; λ_{em} = 531 nm) was prepared from a 1:1 molar mixture of **6** and **DMQA** by heating at 150 °C and then cooling to room temperature. Upon grinding with a spatula, the emission color of the mixture changed to orange (λ_{em} = 605 nm). The resulting **BD-G** (bithiophene/**DMQA**-ground) state retro-converted into **BD-M** upon heating to 150 °C. Meanwhile, upon heating to 110 °C, **BD-G** converted into a blue-emissive **BD-C** (bithiophene/**DMQA**-crystallized) state.

The proposed mechanism for the tricolor MCL of **6/DMQA** is summarized in Fig. 6.9. The blue-emissive **BD-C** is composed of independent crystals of **6** and **DMQA** (Fig. 6.9a). As crystalline **DMQA** is almost non-emissive, the blue emission of crystalline **6** is observed from the segregated crystals. In the **BD-M** state, **DMQA** molecules are partially dissolved in perfectly amorphous **6** (Fig. 6.9b). The energy transfer from **6** to dissolved **DMQA** results in the green emission of **BD-M**. Since **BD-G** crystallizes into **BD-C**, **6** in the **BD-G** state should exist as protocrystalline amorphous state (Fig. 6.9c). Although **DMQA** molecules are dissolved in protocrystalline amorphous **6**, they interact more efficiently with each other to exhibit orange luminescence due to the solvatofluorochromic property.

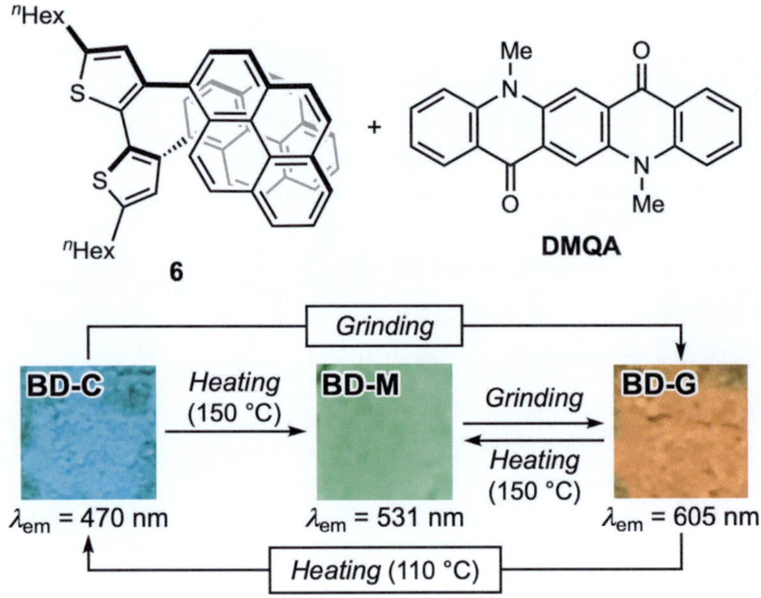

Fig. 6.8 Photographs for the tricolor MCL of **6/DMQA** under UV light. Reproduced with permission from Ref. [43]. Copyright 2019, The Royal Society of Chemistry

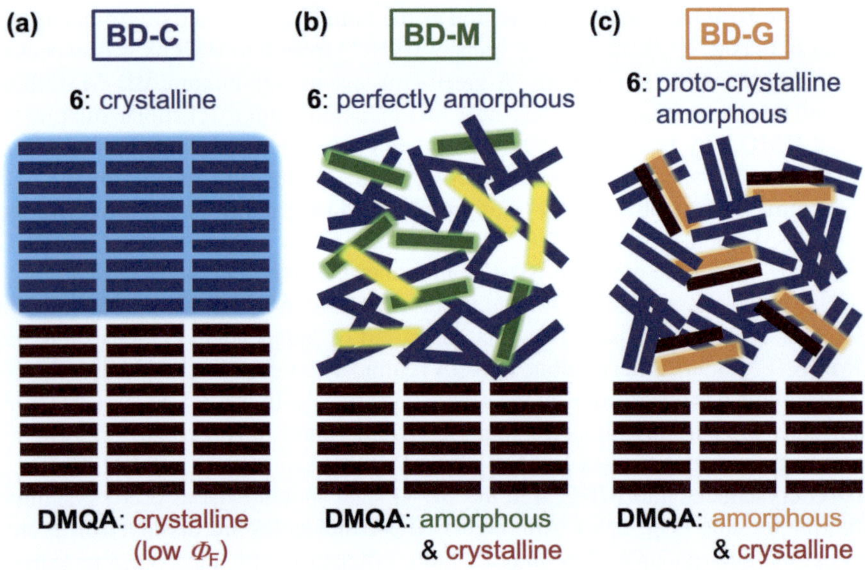

Fig. 6.9 Proposed mechanism for the tricolor MCL of **6/DMQA**. **a BD-C** state. **b BD-M** state. **c BD-G** state. Reproduced with permission from Ref. [43]. Copyright 2019, The Royal Society of Chemistry

6.4.2 Controlling the Self-recovering Properties and the MCL Range

The MCL properties of 2-alkyl-4-(pyren-1-yl)thiophenes **7** were tunable by the length of the alkyl group and/or by mixing with **DMQA** [44]. Crystalline samples of **7a** and **7b** with shorter alkyl groups showed typical MCL between the blue monomer emission and light blue excimer emission of the pyrene groups (Fig. 6.10a). Partial self-recovering MCL was observed for **7c–f** with longer alkyl groups (Fig. 6.10b). The T_g values of **7c–f** were below room temperature, which accounts for their self-recovering behaviors from the ground amorphous states. Meanwhile, the MCL range of **7a** and **7b** could be expanded by mixing with **DMQA**. The emission color of the resulting two-component mixtures was switchable between violet and orange (Fig. 6.10c). Furthermore, a high-contrast, self-recovering MCL was achieved by combining these two approaches (Fig. 6.10d). Similar to the mechanism discussed in Sect. 6.4.1, segregated crystals of **7/DMQA** exhibited violet emission, and the energy transfer from **7** to **DMQA** in the ground amorphous mixture accounts for the orange luminescence.

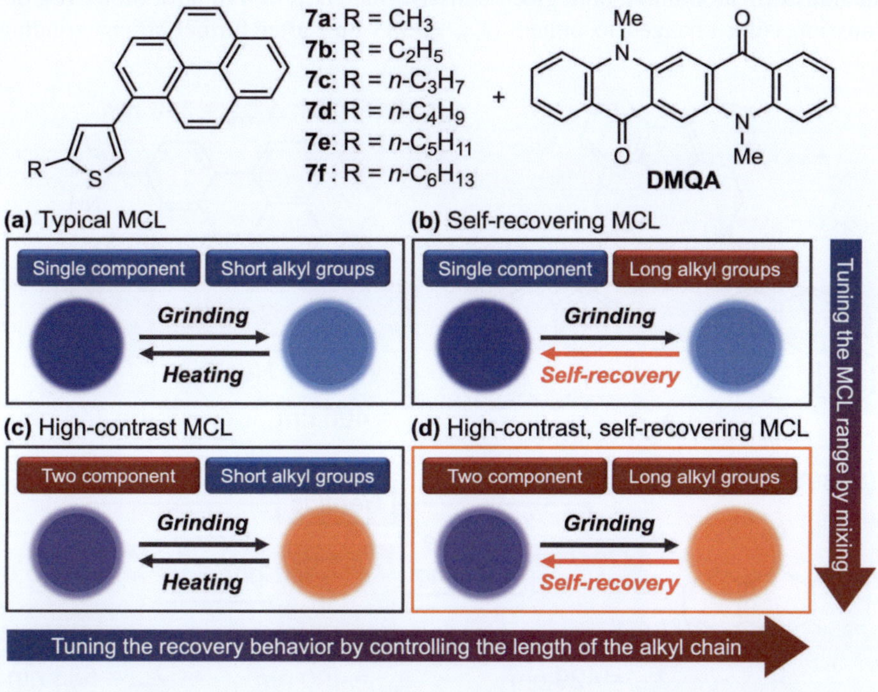

Fig. 6.10 Tunable MCL of **7** by controlling the length of the alkyl chain and/or by mixing with **DMQA**. **a** Typical MCL. **b** Self-recovering MCL (partial self-recovery was observed). **c** High-contrast MCL. **d** High-contrast self-recovering MCL. Reproduced with permission from Ref. [44]. Copyright 2019, The Royal Society of Chemistry

6.4.3 Wide-Range MCL and Two-Step MCL

Several two-component segregated crystals were prepared by mixing a 1:1 molar amount of dibenzofuran-based bispyrene **8** and commercially available organic fluorophores [45]. Among these, a two-component mixture with 3,4,9,10-perylenetetracarboxylic diimide (**PTCDI**) exhibited a remarkable shift of the maximum emission wavelength by 340 nm (Fig. 6.11a). Specifically, the violet-emissive segregated crystals of **8/PTCDI** (λ_{em} = 406 nm) were switched to an amorphous mixture exhibiting far-red luminescence (λ_{em} = 746 nm) upon grinding. Achieving wide-range MCL over 300 nm is quite challenging. Previous examples of wide-range MCL are based on the switching between the stacked and isolated structures of an aminobenzopyranoxanthene [48] and the switching between blue fluorescence and infrared phosphorescence of a gold complex [49]. The present two-component system is based on the different mechanisms.

When segregated crystals were prepared from **8** and 9,10-anthraquinone (**AQ**), two-step MCL in response to mechanical stimuli of different intensities was achieved (Fig. 6.11b). Blue emission was observed from the two-component crystals **8/AQ** (λ_{em} = 429 nm). Gentle grinding (5 N cm^{-2}) of **8/AQ** with a spatula triggered the transformation into a blue-green-emissive state (λ_{em} = 490 nm). Moreover, the emission color changed to orange (λ_{em} = 583 nm) after further strong grinding

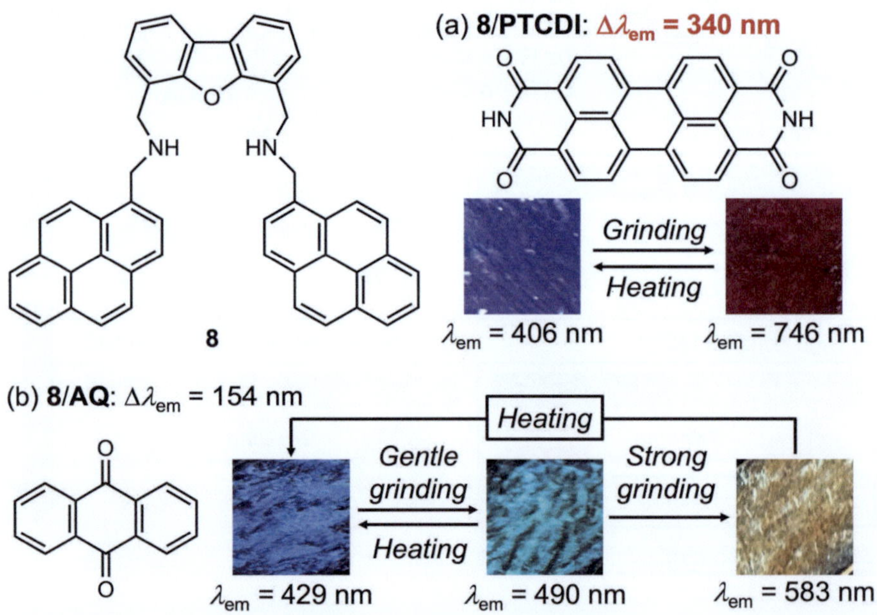

Fig. 6.11 **a** Photographs for the wide-range MCL of **8/PTCDI** under UV light. **b** Photographs for the two-step MCL of **8/AQ** under UV light. Reproduced with permission from Ref. [45]. Copyright 2021, Wiley–VCH Verlag GmbH & Co. KGaA, Weinheim

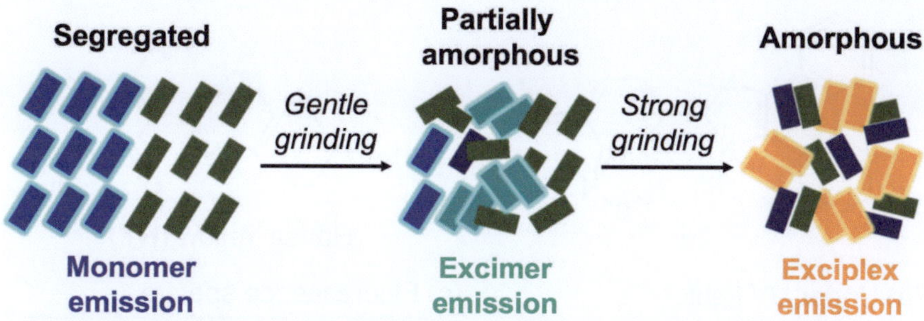

Fig. 6.12 Schematic illustration for the two-step MCL of **8/AQ**. Reproduced with permission from Ref. [45]. Copyright 2021, Wiley–VCH Verlag GmbH & Co. KGaA, Weinheim

(11 N cm^{-2}). Upon heating to 135 °C, the orange-emissive state was retro-converted to the original blue-emissive state.

The tricolor luminescence in the two-step MCL of **8/AQ** can be attributed to the monomer emission of **8**, excimer emission of **8**, and exciplex emission of **8/AQ** (Fig. 6.12). Upon gentle grinding, the amorphization of **8** should occur, resulting in blue-green excimer emission between pyrenyl groups. Strong grinding induces the amorphization of **AQ**, leading to the formation of charge-transfer complexes between **8** and **AQ**. This interaction results in orange exciplex emission.

6.4.4 Mechanoresponsive Fluorescence On–Off Switching

Changes in the luminescent properties of organic crystals upon mechanical stimulation are not limited to shifts in emission wavelength; emission intensity can also be affected [7, 50–53]. Just as the rational design of organic MCL crystals that exhibit color changes is challenging, creating crystals that show significant changes in emission intensity in response to mechanical stimuli is also difficult.

By preparing segregated crystals from an MCL-active organic crystal and a non-luminescent organic pigment, mechanoresponsive on–off switching of luminescence intensity was achieved [47]. When MCL crystal **2a**, discussed in Sect. 6.2, was mixed with Hansa Yellow (HY), the intensity of the blue-green emission ($\lambda_{em} = 523$ nm) in the segregated crystals significantly decreased upon grinding (Fig. 6.13a). However, the apparent color under room light remained largely unchanged, appearing yellow before and after grinding (Fig. 6.13b). Heating the mechanically stimulated state restored the emission intensity, with a luminescence intensity ratio of 45:1 between the heated and ground states (Fig. 6.13c). Using different MCL compounds allowed the emission color to be altered, while the use of other pigments enabled changes in the apparent color.

The feasibility of utilizing the two-component emitter **2a/HY** for encrypting and decrypting luminescent displays was demonstrated (Fig. 6.14). A 5 × 3 glass

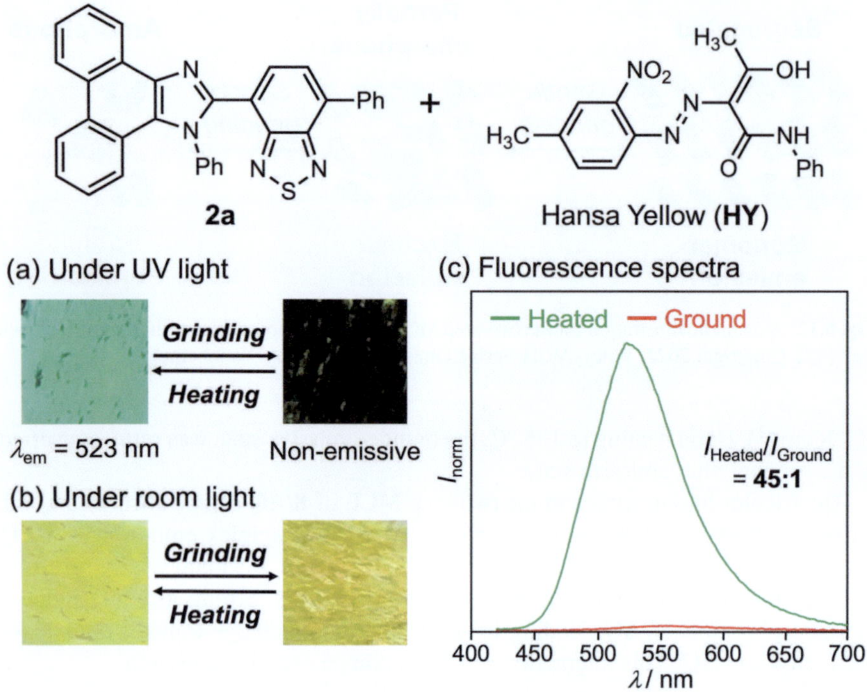

Fig. 6.13 Mechanoresponsive fluorescence on–off switching of **2a/HY**. **a** Photographs under UV light. **b** Photographs under room light. **c** Fluorescence spectra. Reproduced with permission from Ref. [47]. Copyright 2023, Wiley–VCH Verlag GmbH & Co. KGaA, Weinheim

microplate was set up with non-luminescent yellow pigment **HY**, green-emissive heated **2a/HY**, and non-luminescent ground **2a/HY**. Under room light, the apparent colors of **HY**, heated **2a/HY**, and ground **2a/HY** were similar yellow, displaying the numeral "8". Under UV light, only the heated **2a/HY** showed green emission, revealing the numeral "4". When the plate was heated, the non-emissive ground **2a/HY** transformed into green-emissive **2a/HY**, changing the numeral "4" to "9" under UV light. Meanwhile, the apparent colors under room light remained unchanged, continuing to display the numeral "8". Thus, the numeral "4", which was visible only under UV light and served as dummy information, could be decrypted to reveal the actual numeral "9" through heating.

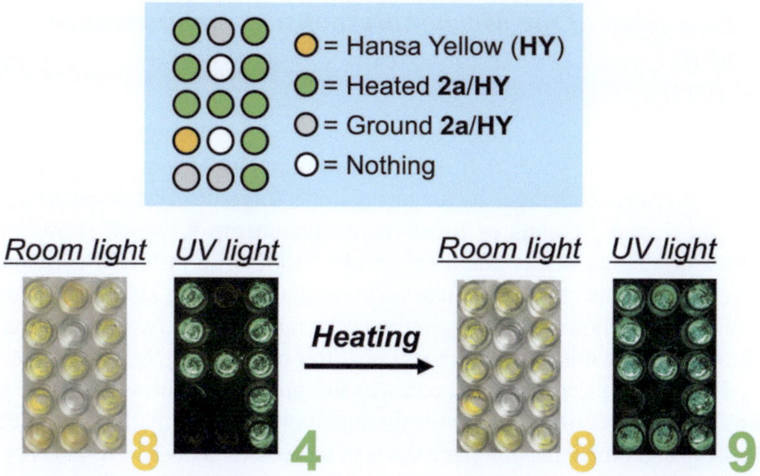

Fig. 6.14 Top: Schematic illustration of the arrangement of **HY**, heated **2a/HY**, and ground **2a/HY** in a glass microplate. Bottom: Photographs of the glass microplate under room light and UV light before and after heating. Reproduced with permission from Ref. [47]. Copyright 2023, Wiley–VCH Verlag GmbH & Co. KGaA, Weinheim

6.5 Doped Organic Crystals

6.5.1 Luminescence Tuning by Molecular Doping

In Sects. 6.3 and 6.4, the MCL properties of organic crystals are controlled by forming cocrystals, solvates, and segregated crystals, which require equimolar amounts of another compound. Meanwhile, the luminescent properties of organic molecular crystals can also be controlled by doping with trace amounts of another compound. The control of luminescent properties in organic crystals through molecular doping has been practiced since the 1960s [54]. However, the structural constraints of doped organic crystals, which require the substitution of heteromolecules, have limited such examples. In the 2000s, with the growing interest in applications for organic electronics and luminescent sensors, there has been a renewed focus on controlling luminescence through molecular doping [55, 56].

In the crystals of a fluorescent organic host molecule doped with another guest molecule that exhibits fluorescence at a longer wavelength, energy transfer occurs from the excited host molecules to the ground-state guest molecules. This process excites the guest molecules, resulting in their emission being observed. For instance, doping 1,4-bis(4-methylstyryl)benzene crystals with pentacene as the guest molecule allows the emission color to be tuned from blue to white, pink, and red, depending on the concentration of the guest molecule [57].

6.5.2 *Extension of the Mechanoresponsive Luminescence Shift*

Controlling the luminescent properties of organic crystals through molecular doping is a useful technique. However, there had been no examples of controlling mechanoresponsive luminescence changes by preparing doped organic crystals. In 2022, it was reported that doping crystals of thienylbenzothiadiazole **9**, which show minimal emission color change before and after grinding [58], with dithienyl compound **10** resulted in a significant change in emission color upon grinding [59]. Prior to doping, the emission wavelength shift upon grinding was only 10 nm between blue-green and blue (Fig. 6.15a), but this expanded to 69 nm after doping (Fig. 6.15b). In the doped crystals, FRET from the host molecules to the guest molecules results in yellow emission. When the crystal structure is disrupted upon grinding, the energy transfer is suppressed, and blue emission from the host molecules is observed. This approach represents a promising new method for rationally controlling the mechanoresponsive luminescence shift of organic crystals.

6.6 Conclusions

This chapter has provided an overview of recent advancements in controlling the response behavior of luminescent organic crystals to mechanical stimuli. For crystals composed of single-component luminescent molecules, the steric effects of substituents significantly influence their luminescent properties. This, in turn, affects the recovery behavior of emission color after mechanical stimulation and the direction of mechanoresponsive emission wavelength. Furthermore, mixing organic MCL molecules with other molecules to form cocrystals, solvates, or segregated crystals has proven to be a useful method for obtaining organic crystalline materials with diverse stimuli-responsive behaviors. In particular, the preparation of segregated crystals is an excellent strategy, as it enables significant changes in emission wavelength and substantial decreases in emission intensity upon mechanical stimulation by utilizing intermolecular energy transfer. Additionally, doping with small amounts of heteromolecules can lead to a red-shift of the emission color through the FRET mechanism, thereby extending the wavelength shift induced by mechanical stimuli.

These methods for controlling the luminescent color and mechanoresponsive behavior of organic crystals are becoming increasingly established. However, for practical applications of organic crystals as materials, it is desirable to develop homogeneous crystals and crystals amenable to molding and processing. With further advancements in this field, it is anticipated that luminescent organic crystals will be implemented in societal applications such as pressure sensors.

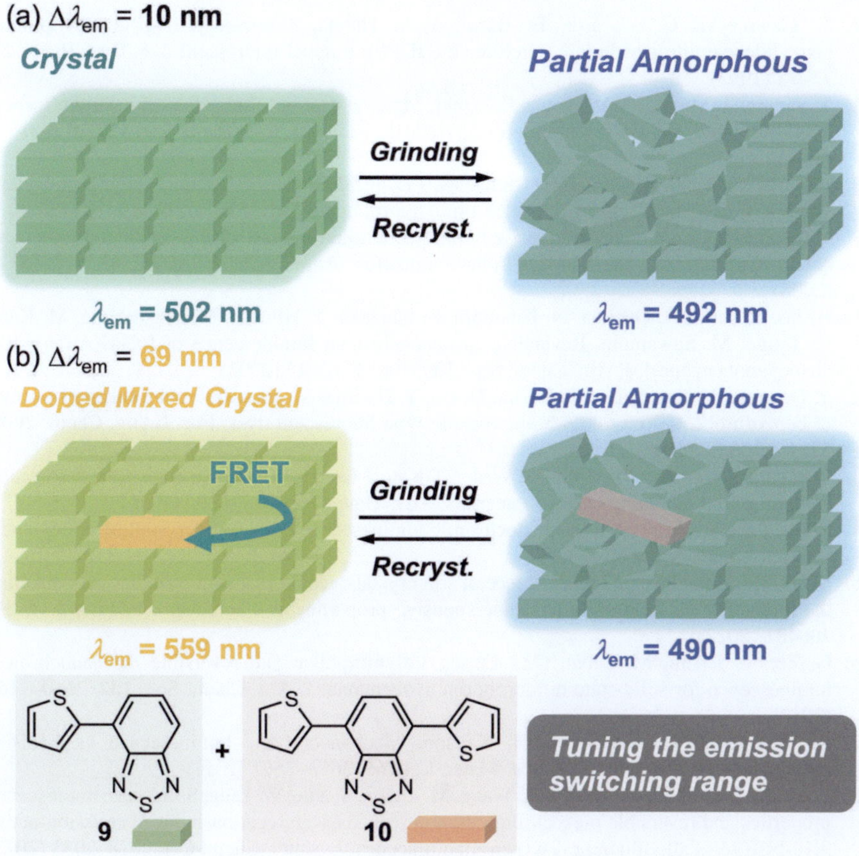

Fig. 6.15 **a** Mechanoresponsive emission color switching of **9**. **b** Mechanoresponsive emission color switching of doped mixed crystals of **9** and **10**. Reproduced with permission from Ref. [59]. Copyright 2022, The Royal Society of Chemistry

References

1. B. Wardle, *Principles and Applications of Photochemistry* (Wiley-VCH, Weinheim/New York, 2009)
2. R. Katoh, K. Suzuki, A. Furube, M. Kotani, K. Tokumaru, Fluorescence quantum yield of aromatic hydrocarbon crystals. J. Phys. Chem. C **113**, 2961–2965 (2009)
3. A.P. Green, A.R. Buckley, Solid state concentration quenching of organic fluorophores in PMMA. Phys. Chem. Chem. Phys. **17**, 1435–1440 (2015)
4. M. Shimizu, T. Hiyama, Organic fluorophores exhibiting highly efficient photoluminescence in the solid state. Chem. Asian J. **5**, 1516–1531 (2010)
5. J. Mei, N.L.C. Leung, R.T.K. Kwok, J.W.Y. Lam, B.Z. Tang, Aggregation-induced emission: together we shine, united we soar! Chem. Rev. **115**, 11718–11940 (2015)
6. M. Kato, H. Ito, M. Hasegawa, K. Ishii, Soft crystals: flexible response systems with high structural order. Chem. Eur. J. **25**, 5105–5112 (2019)

7. R. Gawinecki, G. Viscardi, E. Barni, M.A. Hanna, 4-*Tert*-butyl-1-(4'-dimethylamino-benzylideneamino)pyridinium perchlorate (BDPP): a novel fluorescent dye. Dyes Pigm. **23**, 73–78 (1993)
8. Y. Sagara, T. Mutai, I. Yoshikawa, K. Araki, Material design for piezochromic luminescence: hydrogen-bond-directed assemblies of a pyrene derivative. J. Am. Chem. Soc. **129**, 1520–1521 (2007)
9. J. Kunzelman, M. Kinami, B.R. Crenshaw, J.D. Protasiewicz, C. Weder, Oligo(*p*-phenylene vinylene)s as a "new" class of piezochromic fluorophores. Adv. Mater. **20**, 119–122 (2008)
10. Y. Sagara, T. Kato, Stimuli-responsive luminescent liquid crystals: change of photoluminescent colors triggered by a shear-induced phase transition. Angew. Chem. Int. Ed. **47**, 5175–5178 (2008)
11. H. Ito, T. Saito, N. Oshima, N. Kitamura, S. Ishizaka, Y. Hinatsu, M. Wakeshima, M. Kato, K. Tsuge, M. Sawamura, Reversible mechanochromic luminescence of $[(C_6F_5Au)_2(\mu\text{-}1,4\text{-}$diisocyanobenzene)]. J. Am. Chem. Soc. **130**, 10044–10045 (2008)
12. Y. Ooyama, Y. Kagawa, H. Fukuoka, G. Ito, Y. Harima, Mechanofluorochromism of a series of benzofuro[2,3-*c*]oxazolo[4,5-*a*]carbazole-type fluorescent dyes. Eur. J. Org. Chem. **2009**, 5321–5326 (2009)
13. Y. Sagara, S. Yamane, M. Mitani, C. Weder, T. Kato, Mechanoresponsive luminescent molecular assemblies: an emerging class of materials. Adv. Mater. **28**, 1073–1095 (2016)
14. S. Ito, Recent advances in mechanochromic luminescence of organic crystalline compounds. Chem. Lett. **50**, 649–660 (2021)
15. S. Ito, Mechanochromic luminescence of soft crystals: recent systematic studies in controlling the molecular packing and mechanoresponsive properties. J. Photochem. Photobiol., C **51**, 100481 (2022)
16. G. Zhang, J. Liu, M. Sabat, C.L. Fraser, Polymorphism and reversible mechanochromic luminescence for solid-state difluoroboron avobenzone. J. Am. Chem. Soc. **132**, 2160–2162 (2010)
17. K. Mizuguchi, H. Kageyama, H. Nakano, Mechanochromic luminescence of 4-[bis(4-methylphenyl)amino]benzaldehyde. Mater. Lett. **65**, 2658–2661 (2011)
18. L. Bu, M. Sun, D. Zhang, W. Liu, Y. Wang, M. Zheng, S. Xue, W. Yang, Solid-state fluorescence properties and reversible piezochromic luminescence of aggregation-induced emission-active 9,10-bis[(9,9-dialkylfluorene-2-yl)vinyl]anthracenes. J. Mater. Chem. C **1**, 2028–2035 (2013)
19. S. Ito, T. Yamada, T. Taguchi, Y. Yamaguchi, M. Asami, *N*-Boc-indolylbenzothiadiazole derivatives: efficient full-color solid-state fluorescence and self-recovering mechanochromic luminescence. Chem. Asian J. **11**, 1963–1970 (2016)
20. S. Ito, T. Taguchi, T. Yamada, T. Ubukata, Y. Yamaguchi, M. Asami, Indolylbenzothiadiazoles with varying substituents on the indole ring: a systematic study on the self-recovering mechanochromic luminescence. RSC Adv. **7**, 16953–16962 (2017)
21. F. Khan, A. Ekbote, G. Singh, R. Misra, Mechanochromic luminogens with hypsochromically shifted emission switching property: recent advances and perspectives. J. Mater. Chem. C **10**, 5024–5064 (2022)
22. J. Wu, Y. Cheng, J. Lan, D. Wu, S. Qian, L. Yan, Z. He, X. Li, K. Wang, B. Zou, J. You, Molecular engineering of mechanochromic materials by programmed C-H arylation: making a counterpoint in the chromism trend. J. Am. Chem. Soc. **138**, 12803–12812 (2016)
23. G. Li, Y. Xu, Q. Kong, W. Zhuang, Y. Wang, Cation–anion interaction directed dual-mode switchable mechanochromic luminescence. J. Mater. Chem. C **5**, 8527–8534 (2017)
24. Z. Ma, Z. Wang, X. Meng, Z. Ma, Z. Xu, Y. Ma, X. Jia, A mechanochromic single crystal: turning two color changes into a tricolored switch. Angew. Chem. Int. Ed. **55**, 519–522 (2016)
25. S. Ito, T. Yamada, M. Asami, Two-step mechanochromic luminescence of N,N'-bis-Boc-3,3'-di(pyren-1-yl)-2,2'-biindole. ChemPlusChem **81**, 1272–1275 (2016)
26. S. Nagai, M. Yamashita, T. Tachikawa, T. Ubukata, M. Asami, S. Ito, Efficient and versatile mechanochromic luminescence of phenanthroimidazolylbenzothiadiazoles: tricolor switching and directional control over the chromism. J. Mater. Chem. C **7**, 4988–4998 (2019)

27. A. Adak, T. Panda, A. Raveendran, K.S. Bejoymohandas, K.S. Asha, A.P. Prakasham, B. Mukhopadhyay, M.K. Panda, Distinct mechanoresponsive luminescence, thermochromism, vapochromism, and chlorine gas sensing by a solid-state organic emitter. ACS Omega **3**, 5291–5300 (2018)
28. S. Takahashi, S. Nagai, M. Asami, S. Ito, Two types of two-step mechanochromic luminescence of phenanthroimidazolylbenzothiadiazoles. Mater. Adv. **1**, 708–719 (2020)
29. C.B. Aakeröy, N.R. Champness, C. Janiak, Recent advances in crystal engineering. CrystEngComm **12**, 22–43 (2010)
30. S. Ito, Luminescent polymorphic crystals: mechanoresponsive and multicolor-emissive properties. CrystEngComm **24**, 1112–1126 (2022)
31. Y. Ding, Y. Zhao, Y. Liu, Organic cocrystals: from high-performance molecular materials to multi-functional applications. Aggregate, e626 (2024)
32. Y. Liu, Q. Zeng, B. Zou, Y. Liu, B. Xu, W. Tian, Piezochromic luminescence of donor–acceptor cocrystals: distinct responses to anisotropic grinding and isotropic compression. Angew. Chem. Int. Ed. **57**, 15670–15674 (2018)
33. S. Ito, S. Nagai, T. Ubukata, M. Asami, Switching the direction of mechano-responsive emission color change of triphenylimidazolylbenzothiadiazole. Chem. Lett. **48**, 1492–1495 (2019)
34. G. Huang, Y. Jiang, S. Yang, B.S. Li, B.Z. Tang, Multistimuli response and polymorphism of a novel tetraphenylethylene derivative. Adv. Funct. Mater. **29**, 1900516 (2019)
35. Z. Wang, F. Yu, W. Chen, J. Wang, J. Liu, C. Yao, J. Zhao, H. Dong, W. Hu, Q. Zhang, Rational control of charge transfer excitons toward high-contrast reversible mechanoresponsive luminescent switching. Angew. Chem. Int. Ed. **59**, 17580–17586 (2020)
36. Y. Sagara, K. Takahashi, A. Seki, T. Muramatsu, T. Nakamura, N. Tamaoki, Two-step mechanoresponsive luminescence and mechanical stimuli-induced release of small molecules exhibited by a luminescent cyclophane. J. Mater. Chem. C **9**, 1671–1677 (2021)
37. S. Ito, C. Nishimoto, S. Nagai, Sequential halochromic/mechanochromic luminescence of pyridyl-substituted solid-state emissive dyes: thermally controlled stepwise recovery of the original emission color. CrystEngComm **21**, 5699–5706 (2019)
38. T. Yagi, T. Tachikawa, S. Ito, Solvates of a dianisyl-substituted donor–acceptor-type benzothiadiazole: mechanochromic, vapochromic, and acid-responsive multicolor luminescence. CrystEngComm **25**, 2379–2389 (2023)
39. J. Jia, L. Wu, Reversible fluorescent switching properties of pyrene-substituted acylhydrazone derivatives toward mechanical force and acid vapor with aggregation-induced emission. J. Photochem. Photobiol., A **399**, 112640 (2020)
40. Y. Liu, Z. Ma, J. Liu, M. Chen, Z. Ma, X. Jia, Robust white-light emitting and multi-responsive luminescence of a dual-mode phosphorescence molecule. Adv. Opt. Mater. **9**, 2001685 (2021)
41. R. Kubota, Y. Yuan, R. Yoshida, T. Tachikawa, S. Ito, Tunable mechanochromic luminescence *via* surface protonation of pyridyl-substituted imidazole crystals. Mater. Adv. **3**, 5826–5835 (2022)
42. X. Han, X. Liu, N. Liu, W. Gao, Y. Lei, Y. Zhou, M. Liu, H. Wu, X. Huang, 1,7/8-Substituted isoquinoline derivatives: position isomerism caused by HIO_3-induced dehydrogenation and solid-state fluorescence stimulus-responsive properties. J. Mater. Chem. C **10**, 9875–9881 (2022)
43. S. Ito, G. Katada, T. Taguchi, I. Kawamura, T. Ubukata, M. Asami, Tricolor mechanochromic luminescence of an organic two-component dye: visualization of a crystalline state and two amorphous states. CrystEngComm **21**, 53–59 (2019)
44. M. Ikeya, G. Katada, S. Ito, Tunable mechanochromic luminescence of 2-alkyl-4-(pyren-1-yl)thiophenes: controlling the self-recovering properties and the range of chromism. Chem. Commun. **55**, 12296–12299 (2019)
45. S. Ito, R. Sekine, M. Munakata, M. Yamashita, T. Tachikawa, Mechanochromic luminescence (MCL) of purely organic two-component dyes: wide-range MCL over 300 nm and two-step MCL by charge-transfer complexation. Chem. Eur. J. **27**, 13982–13990 (2021)
46. M. Munakata, S. Ito, Wide-range tricolor mechanochromic luminescence of organic two-component dyes: different responses toward thermal and solvent treatments. Chem. Lett. **52**, 276–279 (2023)

47. R. Kubota, S. Takahashi, S. Nagai, S. Ito, Mechanoresponsive fluorescence on–off switching of two-component emitters composed of mechanochromically luminescent dyes and non-emissive pigments. Chem. Asian J. **18**, e202300124 (2023)
48. M. Tanioka, S. Kamino, A. Muranaka, Y. Ooyama, H. Ota, Y. Shirasaki, J. Horigome, M. Ueda, M. Uchiyama, D. Sawada, S. Enomoto, Reversible near-infrared/blue mechanofluorochromism of aminobenzopyranoxanthene. J. Am. Chem. Soc. **137**, 6436–6439 (2015)
49. T. Seki, N. Tokodai, S. Omagari, T. Nakanishi, Y. Hasegawa, T. Iwasa, T. Taketsugu, H. Ito, Luminescent mechanochromic 9-anthryl gold(I) isocyanide complex with an emission maximum at 900 nm after mechanical stimulation. J. Am. Chem. Soc. **139**, 6514–6517 (2017)
50. X. Luo, J. Li, C. Li, L. Heng, Y.Q. Dong, Z. Liu, Z. Bo, B.Z. Tang, Reversible switching of the emission of diphenyldibenzofulvenes by thermal and mechanical stimuli. Adv. Mater. **23**, 3261–3265 (2011)
51. Y. Lim, I. Choi, H. Lee, I.W. Kim, J.Y. Chang, Constitutional isomers of a C_3-symmetric molecule showing different piezochromic behaviours: on–off switching and colour tuning. J. Mater. Chem. C **2**, 5963–5968 (2014)
52. X. Su, Y. Wang, X. Fang, Y.-M. Zhang, T. Zhang, M. Li, Y. Liu, T. Lin, S.X.-A. Zhang, A high contrast tri-state fluorescent switch: properties and applications. Chem. Asian J. **11**, 3205–3212 (2016)
53. P.S. Hariharan, P. Gayathri, A. Kundu, S. Karthikeyan, D. Moon, S.P. Anthony, Synthesis of tunable, red fluorescent aggregation-enhanced emissive organic fluorophores: stimuli-responsive high contrast off–on fluorescence switching. CrystEngComm **20**, 643–651 (2018)
54. R.C. Powell, R.G. Kepler, Evidence for long-range exciton-impurity interaction in tetracene-doped anthracene crystals. Phys. Rev. Lett. **22**, 636–639 (1969)
55. H. Nakanotani, M. Saito, H. Nakamura, C. Adachi, Emission color tuning in ambipolar organic single-crystal field-effect transistors by dye-doping. Adv. Funct. Mater. **20**, 1610–1615 (2010)
56. J. Li, S. Takaishi, N. Fujinuma, K. Endo, M. Yamashita, H. Matsuzaki, H. Okamoto, K. Sawabe, T. Takenobu, Y. Iwasa, Enhancement of luminescence intensity in TMPY/perylene co-single crystals. J. Mater. Chem. **21**, 17662–17666 (2011)
57. R. Ding, X.-P. Wang, J. Feng, X.-B. Li, F.-X. Dong, W.-Q. Tian, J.-R. Du, H.-H. Fang, H.-Y. Wang, T. Yamao, S. Hotta, H.-B. Sun, Clarification of the molecular doping mechanism in organic single-crystalline semiconductors and their application in color-tunable light-emitting devices. Adv. Mater. **30**, 1801078 (2018)
58. R. Yoshida, T. Tachikawa, S. Ito, Mechano- and thermo-responsive luminescence of crystalline thienylbenzothiadiazole derivatives: stepwise hypsochromic switching of near-infrared emission. Cryst. Growth Des. **22**, 547–558 (2022)
59. R. Yoshida, T. Tachikawa, S. Ito, Extension of the mechanoresponsive luminescence shift *via* formation of a doped organic crystal. Chem. Commun. **58**, 6781–6784 (2022)

Chapter 7
Development of Organic Semiconductors with Exceptional Layered Crystallinity

Satoru Inoue, Seiji Tsuzuki, and Tatsuo Hasegawa

Abstract A class of unsymmetric rod-like organic molecules, which simply combine extended π-electron skeletons (π-cores) with straight-chain alkyl groups, exhibits a remarkable tendency to self-organize into crystalline molecular layers—referred to as layered crystallinity. These molecules facilitate the formation of highly uniform, two-dimensional organic semiconductor (OSC) thin films through simple solution processing, which enables the production of high-performance printed organic field-effect transistors (OFETs). In this chapter, we first illustrate the unique appearance of these crystals to characterize the layered crystallinity. We then explore the origins of this crystallinity using crystal structure analysis and intermolecular interaction calculations, with particular emphasis on how these characteristics are influenced by the length of the alkyl chains. Next, we discuss the development of various layered-crystalline OSCs, achieved through strategic molecular design modifications of both the π-cores and substituents, while maintaining their high degree of layered crystallinity. Finally, we highlight the enhanced device performance with showing high carrier mobility and sharp on/off switching at low voltages, achieved by forming exceptionally clean semiconductor–insulator interfaces that fully leverage the layered crystallinity.

Keywords Organic semiconductor · Field-effect transistor · Layered herringbone · Crystal engineering

---

S. Inoue (✉) · S. Tsuzuki · T. Hasegawa
Department of Applied Physics, The University of Tokyo, 7-3-1 Hongo, Tokyo 113 8656, Japan
e-mail: satoru.inoue@yz.yamagata-u.ac.jp

S. Inoue
Innovation Center for Organic Electronics, Yamagata University, 4-3-16 Jonan, Yonezawa, Yamagata 992 8510, Japan

S. Kobatake and H. Uekusa (eds.), *Advances in Organic Crystal Chemistry*,
https://doi.org/10.1007/978-981-96-4479-7_7

7.1 Introduction

Extended π-conjugated systems are the source of highly active π-electrons and serve as fundamental components of organic semiconductors (OSCs). These systems provide a wide range of electronic functions, stemming from low-lying electronic excitation and intermolecular carrier transport under the application of weak external fields, such as light irradiation or electric fields [1–3]. The electronic functions of OSCs often depend on their intermolecular packing arrangements. The most significant case is achieving efficient carrier transport in OSCs, which is realized through nearly perfect intermolecular ordering with translational symmetry (i.e., crystals) associated with band-like conduction [4–6]. The challenge in developing organic field-effect transistors (OFETs) lies in producing both homogeneous semiconductor layers and high-quality semiconductor interfaces, where molecules are arranged (or crystallized) at interfaces with translational symmetry. Therefore, the fundamental issue in developing materials and device processing is to achieve a strong tendency to form self-organized crystalline molecular layers (layered crystallinity), which is ideal for obtaining uniform channel layers in OFETs.

In this chapter, we outline the unique characteristics of a class of OSCs that exhibit exceptional layered crystallinity. These materials are derived from unsymmetric, rod-like organic molecules composed of extended π-electron skeletons (or π-cores) linked with alkyl chains. Due to their pronounced layered crystallinity, these materials enable the production of homogeneous channel semiconductor layers via solution processing, leading to high-performance OFETs. We begin by illustrating the distinct characteristics exhibited by these compounds due to their high layered crystallinity. Then, we discuss the origin of their high layered crystallinity and excellent thin-film formability, based on the crystal structure analyses of compounds with varying alkyl-chain lengths and the results of intermolecular interaction calculations within the crystals. We also discuss ongoing studies aimed at enhancing OSC materials through molecular design modifications while preserving their high layered crystallinity. Lastly, we show the device characteristics that leverage these unique properties.

7.2 Origin of Layered Crystallinity in π-Core–Alkyl-Chain Linked Molecule

A conjugated π-electron system in OSCs is typically derived from a rigid, ring-fused molecules in which five- and six-membered rings are connected (or condensed), known as π-electron skeletons or π-cores [7–9]. One notable example is the organic semiconductor C8-BTBT, which consists of benzothienobenzothiophene (BTBT) with octyl groups at both ends [10–14]. Professor Takimiya and his colleagues discovered that high-performance OFETs could be achieved using this compound through simple thin-film formation via spin coating [10, 11]. This finding has spurred the development of related molecular materials aimed at enhancing OSC performance.

However, the fundamental reasons behind the improved solubility and film-forming characteristics resulting from the introduction of alkyl groups remained unclear. We focused on understanding how the solubility, thermal properties, and crystal structures of π-core–alkyl-chain linked molecular systems, particularly those featuring alkyl groups connected to one side of the BTBT skeleton, vary with the length of the substituted alkyl chains. Our findings reveal that alkyl substitution is crucial not only for improving solubility but also for enhancing layered crystallinity and film formability [15–19].

Figure 7.1 shows optical micrographs for Ph-BTBT-C_n and *mono*-C_n-BTBT crystals precipitated from solutions with different alkyl-chain lengths n [15, 16]. Molecules with short alkyl chains primarily form needle-shaped crystals, whereas molecules with longer chains ($n = 5$ or more) precipitate as thin, plate-like crystals with rhombic or hexagonal shapes. Notably, molecules with even longer alkyl chains ($n \geq 10$) yield ultrathin crystals with molecular-level thickness from the solution. This apparent change in crystal morphology is attributed to variations in molecular packing, highlighting the role of alkyl chains in enhancing layered crystallinity. Figure 7.2 illustrates the crystal structures of Ph-BTBT-C_n at different chain lengths [15]. Regardless of the alkyl-chain length, the long axes of the rod-like molecules are approximately parallel to each other. For the molecules substituted with short alkyl groups ($n = 3$ and 4), there are pairs of laterally adjacent molecules aligned antiparallel to each other. In contrast, molecules substituted with alkyl groups at $n = 5$ or higher form a layered structure in which adjacent molecules are aligned in the same direction along the long axis and are connected laterally to each other, which eventually increases the contact between the π-cores. These π-cores form layered herringbone (LHB) packing within the molecular layer, where adjacent skeletal planes form a moderate dihedral angle (50–60°), as presented in Fig. 7.3. This LHB arrangement of π-cores is similar to that found in pentacene [20, 21], an early OSC used in OFETs, where overlap between π-electron orbitals facilitates efficient two-dimensional carrier transport within the molecular layer. Notably, the all-trans chain planes of the alkyl substituents also form T-shaped contacts with adjacent molecules, similar to the π-core planes, eventually leading to LHB packing, as depicted in Fig. 7.3. To investigate the origin of the layered molecular packing, we calculated the intermolecular interaction energy between adjacent molecules within the layers using the dispersion-corrected density functional theory (DFT) method [16]. Figure 7.3 summarizes the results, showing intermolecular interaction energy at the six nearest neighbor contacts in the herringbone layers. The intermolecular interaction energy remains almost constant for the BTBT core but systematically increases with the length of the alkyl chain (n). This enhancement is clearly attributed to the contributions from the attractive forces between the alkyl chains, indicating that the stability of layered crystals is increasingly enhanced by the alignment of longer alkyl chains. The results show that intermolecular attractive force is approximately proportional to the total number of carbon and sulfur atoms. It implies that the contribution of intermolecular π–π interactions is not special as to the attractive force, but that dispersion forces dominate intermolecular interactions for all the molecular components.

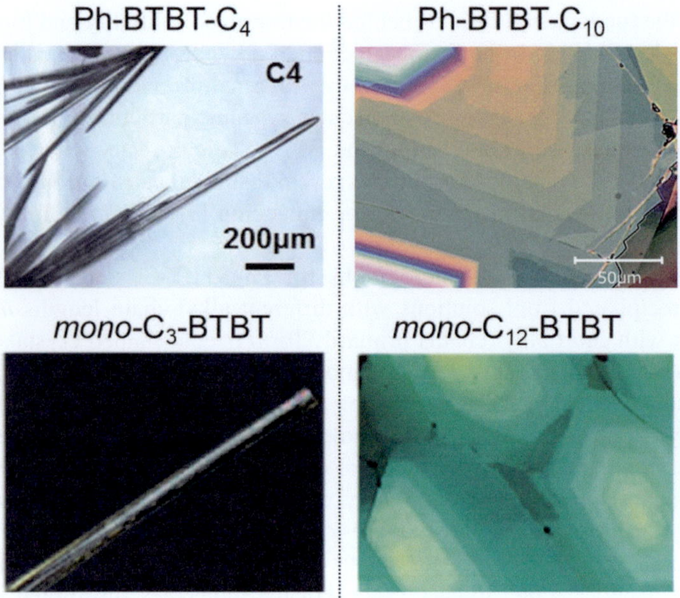

Fig. 7.1 Optical micrographs of OSC crystals obtained by recrystallization from solution. Needle-like crystals precipitate in OSCs with short-chain al-kyl groups (left), while ultrathin flake-like crystals precipitate in OSCs with long-chain alkyl groups (right). The step thickness of the crystal seen in the micrograph corresponds to one crystal lattice layer. Reprinted (adapted or reprinted in part) with permission from Refs. [15, 16]. Copyright 2015, 2017 American Chemical Society

In terms of the crystal stability, the intermolecular arrangement within a crystal should be determined to maximize the cohesive energy of the entire crystal. It can be regarded as an empirical rule that the size ratios between the *T*-shaped and slipped parallel contacts in the stable LHB packing are consistently about 3:2, which is associated with the number ratio between the *T*-shaped and slipped parallel contacts (i.e., 2:1). All-trans alkyl chains form dense intermolecular arrangements, where longer chains generate stronger self-cohesive forces. This interchain interactions eventually strengthen the layered molecular packing between the π-cores, ultimately leading to high layered crystallinity. Additionally, phenyl substituents form tight herringbone-type packing within the molecular layer, further contributing to the enhancement in layered crystallinity.

The respective molecular layers with aligned molecular long axes are stacked in head-to-head (and tail-to-tail) arrangements (Fig. 7.2). The feature resembles bilayer lipid membranes that make up cell membranes and is referred to as bilayer-type herringbone (b-LHB) packing. A similar layered ordering by alkyl substitutions is observed in a number of π-core–alkyl-linked molecules as described later. Alkyl-substituted molecules empirically exhibit high solubility in organic solvents, likely due to the conformational flexibility of alkyl chains in solution. Indeed, π-core–alkyl-chain linked molecular systems show increased solubility with short alkyl

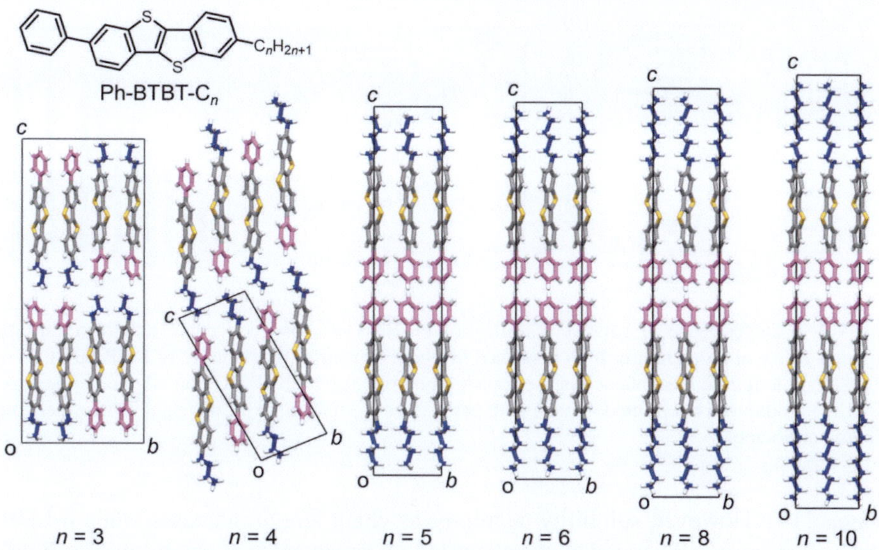

Fig. 7.2 Changes in crystal structure of Ph-BTBT-C_n, depending on the number of alkyl carbons n. The b-LHB packing is formed at $n = 5$ or more. Reprinted (adapted or reprinted in part) with permission from Ref. [15]. Copyright 2015 American Chemical Society

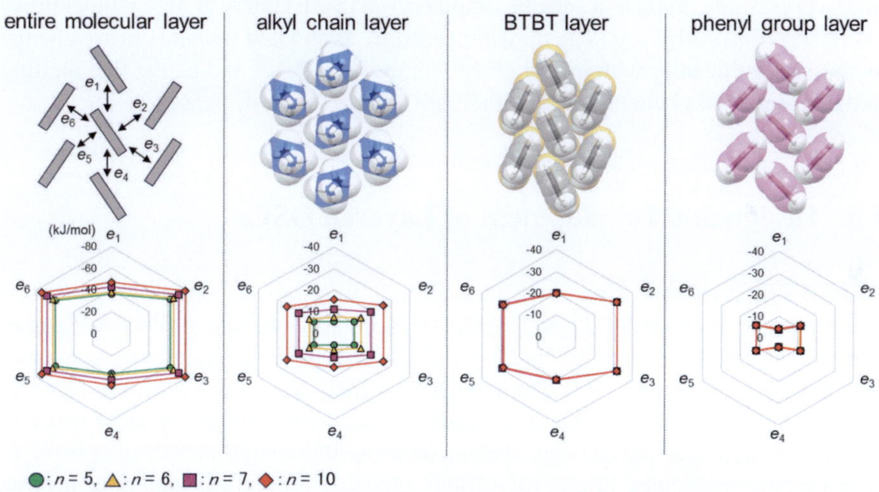

Fig. 7.3 Molecular packing structure of each layer in the b-LHB packing of Ph-BTBT-C_n (space filling model: top) and intermolecular interaction energy within the molecular layer (bottom). Calculations are conducted using dispersion-corrected DFT (B97D/6-311G**). Reprinted (adapted or reprinted in part) with permission from Ref. [16]. Copyright 2017 American Chemical Society

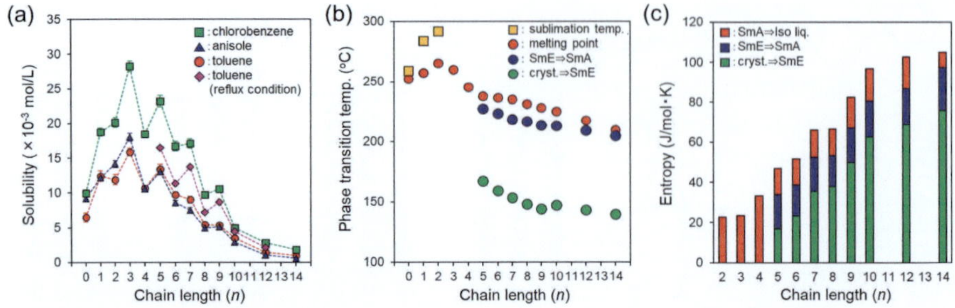

Fig. 7.4 **a** Dependence of solvent solubility of Ph-BTBT-C_n on alkyl chain length in various solvents at room temperature. **b** Dependence of phase transition temperature of Ph-BTBT-C_n on alkyl chain length. **c** Dependence of the transition entropy of Ph-BTBT-C_n on alkyl chain length. Reprinted (adapted or reprinted in part) with permission from Ref. [15]. Copyright 2015 American Chemical Society

chains [16]. However, solubility decreases as chain length increases when b-LHB packing forms, as illustrated by the chain-length dependence of solubility (Fig. 7.4a) [15]. This suggests that OSC solubility correlates more closely with the stability of layered molecular packing than with molecular conformational flexibility. Furthermore, compounds forming b-LHB packing at room temperature exhibit a transition from a crystalline phase to a smectic liquid crystal (SmE) phase at high temperatures (~150 °C) (Fig. 7.4b) [15]. The transition entropy associated with this phase change increases significantly with longer chain lengths (Fig. 7.4c), indicating that melting of the all-trans alkyl chains triggers the transition to the SmE phase.

7.3 Design and Development of Layered OSCs

Ph-BTBT-C_n is unsymmetric along the molecular long axis, in contrast to C8-BTBT, where both ends of the BTBT are substituted with alkyl groups of the same length. These unsymmetric molecules tend to exhibit higher layered crystallinity compared to symmetric molecules. This characteristic is likely associated with the fact that unsymmetric molecules can be more easily arranged with a high degree of freedom within the molecular layer. This section presents the design and development of unsymmetric molecules to control various physical properties, including layered crystallinity. One of the advantages of these OSCs is that the materials can be designed and optimized with a high degree of freedom by controlling the type, size, and substitution position of the π-cores and substituents. The molecules developed so far are shown in Fig. 7.5 [22–29].

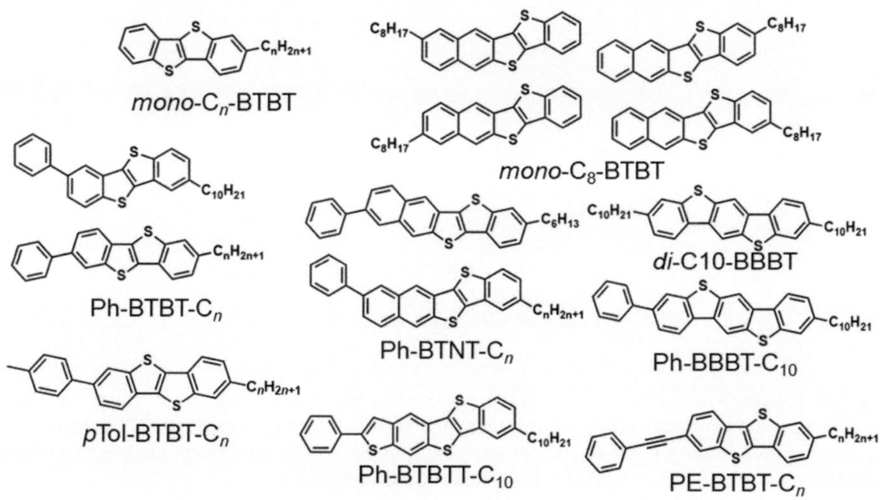

Fig. 7.5 Chemical structures of unsymmetric π-core–alkyl-chain linked molecules developed so far.

7.3.1 Ph-BTNT-C_n

For the application of solution-processable OSCs, a significant challenge is to achieve high thermal stability or durability in OSC characteristics. This is because the solvent solubility required for printing processes is closely correlated with the degradation of thermal stability, as it is associated with a decrease in the cohesive energy of the molecular materials. The solvent solubility of OSCs can be controlled or should be determined, by several factors, such as the size and shape of the π-electron skeletons, as well as the side-group substitutions. Notable examples include the comparison between BTBT and dinaphtho[2,3-*b*:2′,3′-*f*]thieno[3,2-b]-thiophene (DNTT) [30] frameworks, both of which are well-known small-molecule OSCs that form LHB packing and thus afford high-performance OFETs. The BTBT, with a relatively small π-electron skeleton, exhibits high solvent solubility but relatively low thermal stability, while the DNTT, with a larger π-electron skeleton, shows high thermal stability but extremely low solvent solubility [31, 32]. The increase in molecular weight as well as the enhanced interactions between the cores provides sufficient thermal durability, but the trade-off for this is inadequate solubility. It is necessary to address this dilemma toward practical application in printed electronics technology.

To address this issue, we focus on the use of benzothieno[3,2-b]naphtha[2,3-b]thiophene (BTNT) [22, 23], a hybrid π-electron framework between BTBT and DNTT. It is expected that BTNT should demonstrate a unique, delicate balance between high solvent solubility and thermal durability in device characteristics, as it lies between DNTT and BTBT. Figure 7.6 shows the chemical structures of the

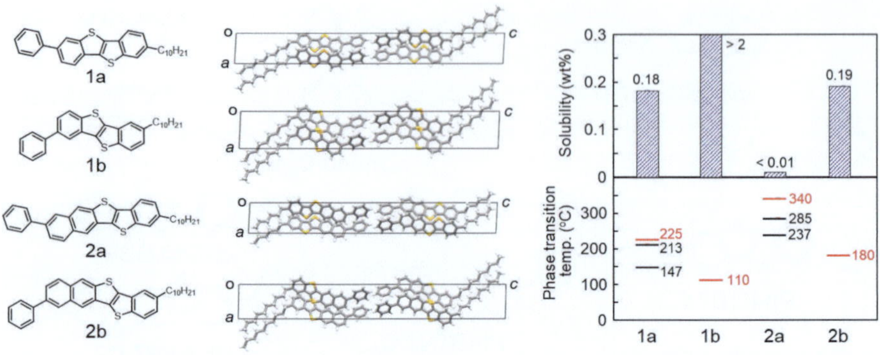

Fig. 7.6 Chemical structures (left), crystal packings (center), solubilities in *o*-xylene at room temperature, and phase transition temperatures of 1 (BTBT derivatives) and 2 (BTNT derivatives). Red line shows melting point of the compounds. Reprinted (adapted or reprinted in part) with permission from Ref. [22]. Copyright 2018 American Chemical Society

family of compounds used to investigate the effect of unsymmetric core substitutions with phenyl group and normal alkyl chains: 2-decyl-7-phenyl-BTBT (Ph-BTBT-C_{10}) (1a), 3-decyl-8-phenyl-BTBT (1b), 2-hexyl-8-phenyl-BTNT (2a), and 3-decyl-9-phenyl-BTNT (2b). The solvent solubility of substituted BTNT at the usual 2,8-substituting positions with alkyl chains and phenyl groups (2a) remains extremely low. In contrast, 3,9-substituted BTNT (2b) presents a unique balance between sufficiently high solvent solubility and thermal durability in device characteristics. These results indicate that tuning the substitution position slightly bends the rod-like molecular shapes, thereby decreasing the cohesive energy of the crystals while retaining the b-LHB packing for the unsymmetric rod-like molecules. This modification ultimately leads to increased solvent solubility, a decrease in phase transition temperature, and suppression of liquid-crystalline phases at high temperatures with excellent layered crystallinity. Ph-BTNT-C_n exhibits improved heat resistance and solubility compared to Ph-BTBT-C_n, making it an excellent candidate for OSCs suitable for multiple-purpose coating processes, including spin coating.

7.3.2 *Mono-C_n-BTNT*

To further investigate the effects of the regioisomeric structure–property relationship, we developed four different positional isomers of *mono*-octyl-substituted benzothieno[3,2-b]naphtho[2,3-b]thiophene (*mono*-C_8-BTNT), each with a linear alkyl-chain (octyl) group substituted at different positions (2, 3, 8, and 9) near the end of the BTNT core, as depicted in Fig. 7.7 [25]. All isomers form b-LHB packing with nearly isomorphous intralayer arrangements between the BTNT cores, indicating the rigid nature of the LHB arrangement in BTNTs compared to those in BTBTs. Nonetheless, the effect of the substitution position causes drastic variations

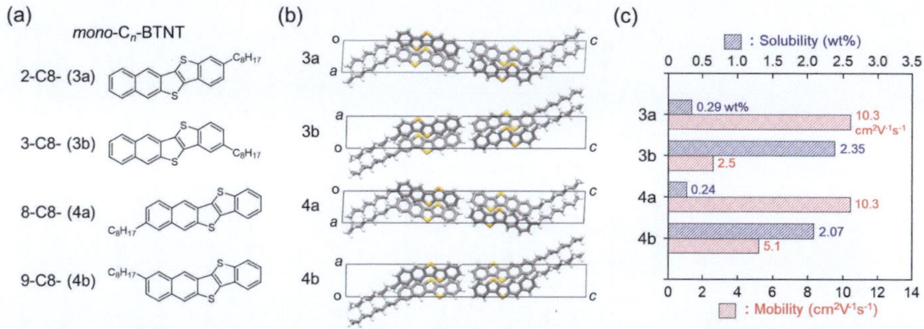

Fig. 7.7 **a** Chemical structures of the four positional isomers of *mono*-C_8-BTNTs. **b** Molecular packing motifs of the isomers viewed perpendicular to the molecular long axis (shown by the projection of *a*-*c* plane). **c** Solvent solubility of the isomers in *o*-xylene (blue bars), and estimated field-effect mobility of single-crystal OFETs of the isomers. Reproduced from Ref. [25] with permission from the Royal Society of Chemistry

in solvent solubility and liquid crystal phase transitions at high temperatures due to differences in crystal stability. The crystals of 3a and 4a are more stable than those of 3b and 4b, owing to the higher ordering of alkyl-chain layers in the crystals. The 3a and 4a allow the production of highly homogeneous single-crystal thin films using blade coating. Detailed structural analysis revealed that while the intermolecular arrangements of the BTNT skeletons are consistent across all isomers, the independence of the π-electron skeleton layer and the alkyl-chain layer varies due to the substitution position (Fig. 7.7b, c). Layered crystallinity is significantly enhanced when the single bond of the alkyl chain near the substitution position is nearly perpendicular to the molecular layer. The single-crystal OFETs of 3a and 4a exhibit mobilities of approximately 10 $cm^2 V^{-1} s^{-1}$, which are much higher than those of 3b and 4b. The correlation between layered crystallinity and semiconductor properties provides a clear understanding of the effects of alkyl substitution.

7.3.3 PE-BTBT-C_n

We developed a series of PE-BTBT-C_n ($n = 6, 8, 10, 12$), replacing the phenyl group in Ph-BTBT-C_n with a phenylethynyl (PE) group, which is relatively long and has rotational degrees of freedom, as shown in Fig. 7.8 [27]. This unique substituent effect influences both layered crystallinity and solubility. The layered crystallinity of PE-BTBT-C_n is higher than that of Ph-BTBT-C_n, resulting in the growth of ultrathin flake-like crystals through recrystallization in a solution droplet. The solubility of PE-BTBT-C_n at each n is approximately three times higher than that of Ph-BTBT-C_n. Full crystal structure analysis revealed that the compounds with n = 8, 10, 12 form b-LHB packing, which is fundamentally the same as in Ph-BTBT-C_n. In contrast, a distinctive layered solid crystalline phase emerges in PE-BTBT-C_n at n =

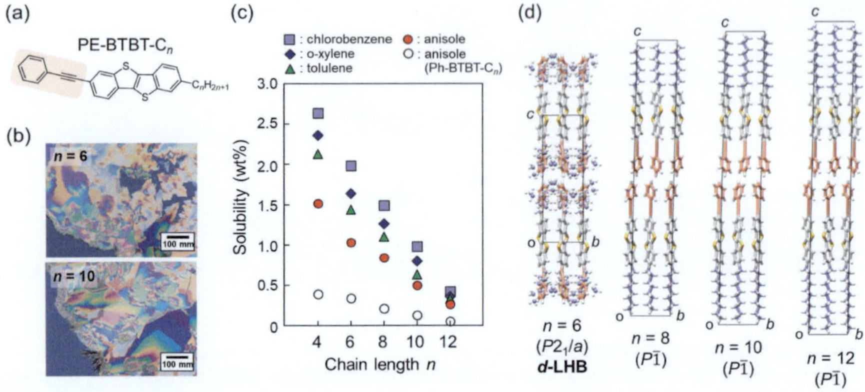

Fig. 7.8 **a** Chemical structures of PE-BTBT-C_n. **b** Optical micrographs of solution-grown crystals by drop casting at $n = 6$ and 10. **c** Comparison of solvent solubility at room temperature between PE-BTBT-C_n and Ph-BTBT-C_n. **d** Molecular packing motifs of PE-BTBT-C_n at $n = 6, 8, 10, 12$. Reprinted (adapted or reprinted in part) with permission from Ref. [27]. Copyright 2022 American Chemical Society

6, as determined by electron diffraction crystal structure analysis using cryo-electron microscopy [33–35]. In this crystal, the BTBT moiety maintains a rigid LHB packing, whereas molecules with opposite orientations are randomly distributed within the layer. Although the orientation of the molecular long axes is random, the herringbone arrangement of BTBT cores is maintained within the molecular layer. This packing motif is referred to as disordered LHB (d-LHB). The d-LHB phase originates from the nearly equal length of PE and –C_6 chains, stabilizing a unique phase through the random orientation of the molecular long axis. Despite this randomness, the intermolecular core—core interactions stabilize the d-LHB packing, allowing a relatively high field-effect mobility of approximately 3 $cm^2 V^{-1} s^{-1}$. Conversely, PE-BTBT-C_n at $n = 8, 10, 12$ forms b-LHB packing due to the unsymmetrical lengths of the substituents, exhibiting higher mobility of approximately 7 $cm^2 V^{-1} s^{-1}$.

The b-LHB and d-LHB phases in PE-BTBT-C_n compete with each other. The d-LHB phase at $n = 6$ disappears upon introducing a small amount of longer chain molecules ($n = 8, 10, 12$) in solution, leading to the emergence of the b-LHB phase. A temperature–composition phase diagram obtained through thermal analysis is presented in Fig. 7.9 [36]. The emergence of the b-LHB phase leads to improved film uniformity and carrier transport characteristics. Although the SmE phase is absent in the compound with $n = 6$, the SmE phase appears in the mixed system at high temperatures. Analysis of the transition entropy demonstrates that the entropy change from the b-LHB phase to the SmE phase decreases with a decreasing molar fraction of the long-chain molecules, approaching zero. From this, it is concluded that the unique d-LHB phase is closely associated with the emergence of liquid crystallinity, proving to be a frozen state of the SmE phase. These features are significant because molecular arrangement structure and phase behavior in the LC phases are not

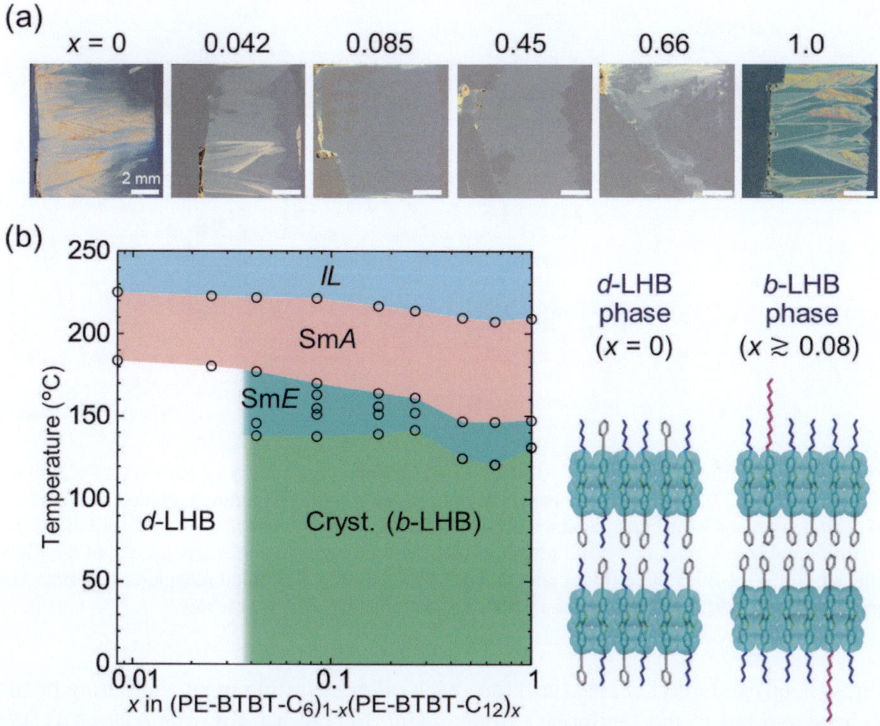

Fig. 7.9 **a** Optical microscope images of crystalline films of $(\text{PE-BTBT-C}_6)_{1-x}(\text{PE-BTBT-C}_{12})_x$, where x is the mole fraction of longer-alkylated molecules. **b** Temperature–composition phase diagram of $(\text{PE-BTBT-C}_6)_{1-x}(\text{PE-BTBT-C}_{12})_x$, with schematics for the layered molecular packing in pristine PE-BTBT-C_6 ($x = 0$) and in mixtures with $x \gtrsim 0.08$ in the right panel. Reprinted (adapted or reprinted in part) with permission from Ref. [36]. Copyright 2022 Wiley

well understood, although many unsymmetric π-core–alkyl-chain linked molecules exhibit liquid crystallinity.

7.3.4 *p*Tol-BTBT-C_n

We developed a series of para-tolyl/alkylated-BTBT (pTol-BTBT-C_n) compounds (n = 9–14), in which a methyl group is added to the terminal phenyl group of Ph-BTBT-C_n, as shown in Fig. 7.10 [29]. The solubility of pTol-BTBT-C_n is approximately twice as high as that of Ph-BTBT-C_n. Notably, pTol-BTBT-C_n exhibits a remarkable phenomenon where the interlayer stacking mode of the monomolecular layer changes dramatically depending on the parity of the alkyl carbon number. For odd n ($n = 9, 11, 13$), the molecules form a b-LHB packing motif in which the π-conjugated

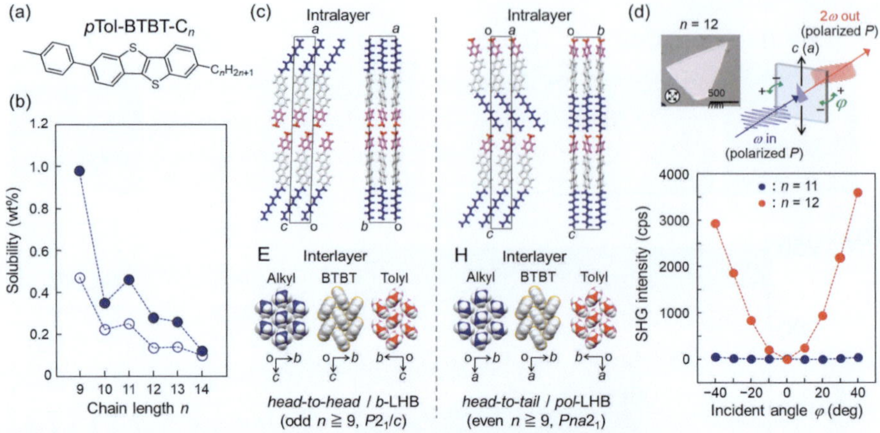

Fig. 7.10 **a** Chemical structures of *p*Tol-BTBT-C_n. **b** Solvent solubility of *p*Tol-BTBT-C_n (filled circles) and Ph-BTBT-C_n (open circles) in chlorobenzene at room temperature, plotted as a function of chain length *n*. **c** Molecular packings of *p*Tol-BTBT-C_n at $n = 9$ and $n = 10$. **d** Schematic of the experimental setup for optical SHG measurements and the incident angle dependence of the SHG intensity for *p*Tol-BTBT-C_{11} (blue) and *p*Tol-BTBT-C_{12} (red). Reprinted (adapted or reprinted in part) from Ref. [29]. Licensed under a Creative Commons CC BY 4.0 License

cores (head) and alkyl chains (tail) face each other, resulting in an alternating head-to-head and tail-to-tail (antipolar) arrangement. In contrast, for even *n* ($n = 10$, 12, 14), a polar LHB (pol-LHB) packing motif is observed, where the molecular long axes in each layer are aligned in the same direction, forming a uniform head-to-tail (polar) arrangement. The single-crystal thin films show excellent OFET characteristics, demonstrating unique polar/antipolar dependence and improved subthreshold swing (SS) in the polar films. Additionally, the polar films exhibit an enhanced second-order nonlinear optical response normal to the film plane. Intermolecular interaction calculations indicate that the observed parity effect originates from the difference in the bonding orientation of the terminal ends of the alkyl chains. It is evident that the remarkable odd–even effect observed in *p*Tol-BTBT-C_n becomes apparent when the molecules involve effectively weakened head-to-head affinity in Ph-BTBT-C_n.

7.4 Solution-Based Thin-Film Processing and Device Characteristics

A class of π-core–alkyl-chain linked molecules provides OSCs with exceptional layered crystallinity, enabling the production of highly homogeneous semiconductor thin films through suitable solution processes. This self-organizing tendency of the molecules allows for the facile formation of ultrathin, uniform crystalline layers with molecular-level thickness over a wide area, as the solution provides an ideal environment for mild solute crystallization. Many of these materials can be easily fabricated into uniaxially oriented polycrystalline thin films, with molecular layers parallel to the substrate surface, using common methods such as spin coating [22, 25, 37, 38] or drop casting [26]. They also enable patterned single-crystal thin-film fabrication via double-shot inkjet printing [12] or large-area single-crystal thin-film fabrication on a wafer scale using blade coating techniques [18].

In the blade coating technique, a meniscus (curved due to surface tension) near the tip of a solution droplet spread on a substrate serves as the site for crystal growth. This technique is effective for obtaining large-area single-crystalline thin films because the growing edge of the thin films on the meniscus can be moved slowly and continuously by sweeping the blade across the substrate. Figure 7.11 shows an optical micrograph of the Ph-BTBT-C_{10} thin film produced by the blade coating technique [17]. By optimizing film-forming conditions, single-crystalline thin films several millimeters in size with a film thickness of approximately 5–40 nm (corresponding to multiple molecular layers) can be achieved. Atomic force microscope (AFM) measurements of the thin-film surface reveal a clear step-terrace structure with a period equal to one bilayer thickness (5 nm). When a top-contact OFET with an SiO_2 layer as a gate insulating layer is constructed using these single-crystal thin films, mobilities exceeding 10 $cm^2\ V^{-1}\ s^{-1}$ are confirmed [17, 19].

Nonetheless, the obtained OSC layers are not ideal for two-dimensional transport of gate-induced carriers when OFETs are manufactured with conventional gate dielectric layers. The issue lies in the cleanliness of the semiconductor–gate dielectric interface, which significantly affects on/off switching characteristics [39, 40]. An effective approach to improve and achieve a sufficiently clean enough semiconductor–insulator interface is to produce OSC layers on top of gate dielectric layers that

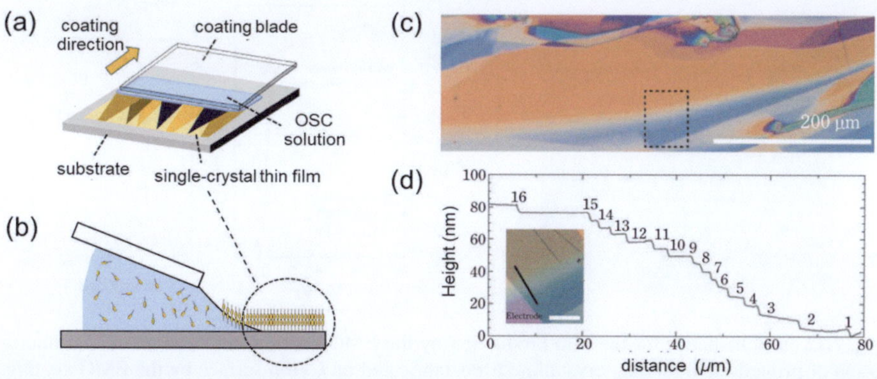

Fig. 7.11 **a**,**b** Schematic illustrations of blade-coating method. **c** Optical micrograph of a single-crystalline thin film of Ph-BTBT-C_{10} grown on an SiO_2/Si substrate. **d** Height profile measured by AFM on the area surrounded by a dashed square shown in (c). Each step corresponds to the bilayer thickness. Reprinted (adapted or reprinted in part) with permission from Ref. [17]. Copyright 2017 by the American Physical Society

have very low surface energy and high flatness [41]. For this purpose, the amorphous perfluoropolymer Cytop is the most promising insulating material for gate dielectric layers. However, the requirement that the surface energy is low also imparts a highly liquid-repellent nature, causing the solution repelled. Thus, the applied solution does not spread wet on the surface, which has been in truth a significant obstacle to realize solution-based uniform thin-film processing on the highly liquid-repellent surfaces.

Recently, we developed an extended meniscus-guided (EMG) coating technique that allows the production of highly uniform single-crystalline OSC films on top of the highly liquid-repellent Cytop surface [42, 43]. The "extended meniscus" refers to the ability of the technique to create a state in which the OSC solution can spread and wet even on highly liquid-repellent surfaces. This is crucial, as the thin OSC layer grown at air–liquid interface can only be transferred to the substrate surface when the solution meniscus is fully extended. In this technique, a U-shaped metal film pattern is first manufactured on the Cytop surfaces, and then the simple blade coating of the OSC solution is carried out over the entire surface, as schematically shown in Fig. 7.12a. Film growth starts from the bottom area of the lyophilic U shape due to the extended meniscus. The extended meniscus and the resulting film growth are maintained even on the liquid-repellent area enclosed by the U-shaped metal film and the blade. This ultimately enables the production of an ultrathin, large-area, and highly uniform semiconductor layer on the enclosed area of the Cytop surface.

Figure 7.12b shows the OFET device characteristics of bottom-gate-bottom-contact (BGBC)-type OFETs composed of Ph-BTNT-C_{10} single-crystalline thin films manufactured by the EMG coating on the Cytop gate dielectric surface. The devices exhibit a subthreshold swing (SS) value close to the theoretical limit, averaging 67 mV dec^{-1}, with hysteresis-free and stable operation at low voltages. These

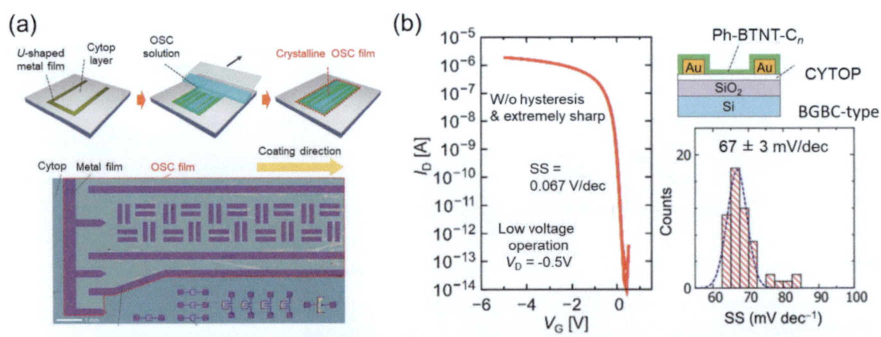

Fig. 7.12 **a** Schematics for the film production by the EMG coating (top) and an optical micrograph of printed Ph-BTNT-C_n crystalline films fabricated on Cytop surface by the EMG coating. **b** Transfer characteristics (left) and distribution of SS values for 56 devices (right) of BGBC-type FETs composed of Ph-BTNT-C_n single-crystal thin films produced by EMG coating. Reprinted (adapted or reprinted in part) from Ref. [42]. Licensed under a Creative Commons CC BY 4.0 License

Table 7.1 Summary of device characteristics in the linear region for various π-core–alkyl-chain linked OSCs

OSCs	Mobility ($cm^2\ V^{-1}\ s^{-1}$)	SS ($mV\ dec^{-1}$)
Ph-BTBT-C_{10}/-C_{14}	9.1	62
Ph-BTBT-C_{11}	8.3	60
*p*Tol-BTBT-C_{11}/-C_{13}	11.8 (11.2)	63
*p*Tol-BTBT-C_{11}	12.0 (13.6)	68
2-C_{12}-BTNT	10.9	86
Ph-BTNT-C_{10}/-C_{12}	6.0 (6.0)	68

Device adopted BGBC-type OFETs composed of single crystal thin films produced on Cytop gate dielectric surfaces by the EMG coating. The values in parentheses show the mobility in the saturation regime

results are attributed to the clean semiconductor–insulator interface with significantly fewer carrier traps. The device characteristics for the types of highly layered-crystalline OSCs based on π-core–alkyl-chain linked molecular systems produced by the EMG coating are summarized in Table 7.1. Excellent device characteristics with significantly suppressed trap formation at the semiconductor–insulator interface were obtained with a simplified process for forming films at room temperature and atmospheric pressure.

7.5 Summary

In summary, we have explored the concept of "layered crystallinity," a crucial attribute for the construction of high-performance OFETs. We detailed the exceptional layered-crystalline characteristics exhibited by π-core–alkyl-chain linked molecules. The high layered crystallinity is attributed to the synergistic effects between the π-electron skeleton and long-chain alkyl groups, both of which inherently favor layered ordering. We also discussed a film formation process that leverages the layered liquid crystallinity occurring at the air–liquid interface. Additionally, we presented molecular design strategies that manipulate the size of the π-electron skeleton and the types and positions of substituents while maintaining layered crystallinity. This approach aims to optimize properties for film-forming processes and field-effect device structures, an area that has not been thoroughly investigated thus far. From this viewpoint, it is likely that the full potential of molecular components, such as the π-electron skeleton and various substituents, has yet to be fully utilized. We hope this chapter will contribute to further material development, fully taking advantage of these intrinsic characteristics.

References

1. T. Hasegawa, S. Inoue, Extremely sharp switching operation of printed transistors using highly layered crystalline organic semiconductors. JSAP Rev. 220206 (2022). https://doi.org/10.11470/jsaprev.220206.
2. X. Zhao, H. Zhang, J. Zhang, J. Liu, M. Lei, L. Jiang, Organic semiconductor single crystal arrays: preparation and applications. Adv. Sci. **10**, 2300483 (2023)
3. S.K. Park, J.H. Kim, S.Y. Park, Organic 2D optoelectronic crystals: charge transport, emerging functions, and their design perspective. Adv. Mater. **30**, 1704759 (2018)
4. V.C. Coropceanu, J. Cornil, D.A. da Silva Filho, Y. Olivier, R. Silbey, J.-L. Brédas, Charge transport in organic semiconductors. Chem. Rev. **107**, 926–952 (2007)
5. H. Sirringhaus, T. Sakanoue, J.-F. Chang, Charge-transport physics of high-mobility molecular semiconductors. Phys. Status Solidi B **249**, 1655–1676 (2012)
6. H. Sato, S.A.A. Rahman, Y. Yamada, H. Ishii, H. Yoshida, Conduction band structure of high-mobility organic semiconductors and partially dressed polaron formation. Nat. Mater. **21**, 910–916 (2022)
7. K. Takimiya, S. Shinamura, I. Osaka, E. Miyazaki, Thienoacene-based organic semiconductors. Adv. Mater. **23**, 4347–4370 (2011)
8. T. Okamoto, C.P. Yu, C. Mitsui, M. Yamagishi, H. Ishii, J. Takeya, Bent-shaped *p*-type small-molecule organic semiconductors: a molecular design strategy for next-generation practical applications. J. Am. Soc. Chem. **142**, 9083–9096 (2020)
9. H. Iino, T. Usui, J. Hanna, Liquid crystals for organic thin-film transistors. Nat. Commun. **6**, 6828 (2015)
10. H. Ebata, T. Izawa, E. Miyazaki, K. Takimiya, M. Ikeda, H. Kuwabara, T. Yui, Highly soluble [1]benzothieno[3,2-*b*]benzothiophene (BTBT) derivatives for high-performance, solution-processed organic field-effect transistors. J. Am. Chem. Soc. **129**, 15732–15733
11. T. Izawa, E. Miyazaki, K. Takimiya, Molecular ordering of high-performance soluble molecular semiconductors and re-evaluation of their field-effect transistor characteristics. Adv. Mater. **20**, 3388–3392 (2008)
12. H. Minemawari, T. Yamada, H. Matsui, J. Tsutsumi, S. Haas, R. Chiba, R. Kumai, T. Hasegawa, Inkjet printing of single-crystal films. Nature **475**, 364–367 (2011)
13. Y. Diao, B.C.-K. Tee, G. Giri, J. Xu, D.H. Kim, H.A. Becerril, R.M. Stoltenberg, T.H. Lee, G. Xue, S.C.B. Mannsfeld, Z. Bao, Solution coating of large-area organic semiconductor thin films with aligned single-crystalline domains, Nat. Mater. **12**, 665–671 (2013)
14. R. Janneck, N. Pilet, S.P. Bommanaboyena, B. Watts, P. Heremans, J. Genoe, C. Rolin, Highly crystalline c8-btbt thin-film transistors by lateral homo-epitaxial growth on printed templates. Adv. Mater. **29**, 1703864 (2017)
15. S. Inoue, H. Minemawari, J. Tsutsumi, M. Chikamatsu, T. Yamada, S. Horiuchi, M. Tanaka, R. Kumai, M. Yoneya, T. Hasegawa, Effects of substituted alkyl chain length on solution-processable layered organic semiconductor crystals. Chem. Mater. **27**, 3809–3812 (2015)
16. H. Minemawari, M. Tanaka, S. Tsuzuki, S. Inoue, T. Yamada, R. Kumai, Y. Shimoi, T. Hasegawa, Enhanced layered-herringbone packing due to long alkyl chain substitution in solution-processable organic semiconductors. Chem. Mater. **29**, 1245–1254 (2017)
17. T. Hamai, S. Arai, H. Minemawari, S. Inoue, R. Kumai, T. Hasegawa, Tunneling and origin of large access resistance in layered-crystal organic transistors. Phys. Rev. Appl. **8**, 054011 (2017)
18. S. Arai, S. Inoue, T. Hamai, R. Kumai, T. Hasegawa, Semiconductive single molecular bilayers realized using geometrical frustration. Adv. Mater. **30**, 1707256 (2018)
19. T. Hamai, S. Inoue, S. Arai, T. Hasegawa, Trap-state suppression and band-like transport in bilayer-type organic semiconductor ultrathin single crystals. Phys. Rev. Mater. **4**, 074601 (2020)
20. C.C. Mattheus, A.B. Dros, J. Baas, A. Meetsma, J.L. de Boer, T.T.M. Palstra, Polymorphism in pentacene. Acta Cryst. **C57**, 939–941 (2001)

21. T. Siegrist, C. Besnard, S. Haas, M. Schiltz, P. Pattison, D. Chernyshov, B. Batlogg, C. Kloc, A polymorph lost and found: the high-temperature crystal structure of pentacene. Adv. Mater. **19**, 2979–2082 (2007)
22. S. Inoue, S. Shinamura, Y. Sadamitsu, S. Arai, S. Horiuchi, M. Yoneya, K. Takimiya, T. Hasegawa, Extended and modulated thienothiophenes for thermally durable and solution-processable organic semiconductors. Chem. Mater. **30**, 5050–5060 (2018)
23. T. Higashino, S. Inoue, Y. Sadamitsu, S. Arai, S. Horiuchi, T. Hasegawa, Bilayer-type layered herringbone packing in 3-*n*-octyl-9-phenyl-benzothieno[3,2-*b*]naphtho[2,3-*b*]thiophene. Chem. Lett. **48**, 453–456 (2019)
24. T. Higashino, S. Arai, S. Inoue, S. Tsuzuki, Y. Shimoi, S. Horiuchi, T. Hasegawa, R. Azumi, Architecting layered molecular packing in substituted benzobisbenzothiophene (BBBT) semiconductor crystals. CrystEngComm **22**, 3618–3626 (2020)
25. S. Inoue, T. Higashino, S. Arai, R. Kumai, H. Matsui, S. Tsuzuki, S. Horiuchi, T. Hasegawa, Regioisomeric control of layered crystallinity in solution-processable organic semiconductors. Chem. Sci. **11**, 12493–12505 (2020)
26. T. Higashino, S. Inoue, S. Arai, H. Matsui, N. Toda, S. Horiuchi, R. Azumi, T. Hasegawa, Architecting layered crystalline organic semiconductors based on unsymmetric π-extended thienoacenes. Chem. Mater. **33**, 7379–7385 (2021)
27. S. Inoue, K. Nikaido, T. Higashino, S. Arai, M. Tanaka, R. Kumai, S. Tsuzuki, S. Horiuchi, H. Sugiyama, Y. Segawa, K. Takaba, S. Maki-Yonekura, K. Yonekura, T. Hasegawa, Emerging disordered layered-herringbone phase in organic semiconductors unveiled by electron crystallography. Chem. Mater. **34**, 72–83 (2022)
28. T. Higashino, S. Inoue, S. Arai, S. Tsuzuki, H. Matsui, R. Kumai, K. Takaba, S. Maki-Yonekura, H. Kurokawa, I. Inoue, K. Tono, K. Yonekura, T. Hasegawa, Effects of thiophene-fused isomer on high-layered crystallinity in π-extended and alkylated organic semiconductors. Chem. Mater. **36**, 848–859 (2024)
29. S. Inoue, T. Higashino, K. Nikaido, R. Miyata, S. Matsuoka, M. Tanaka, S. Tsuzuki, S. Horiuchi, R. Kondo, R. Sagayama, R. Kumai, D. Sekine, T. Koyanagi, M. Matsubara, T. Hasegawa, Control of polar/antipolar layered organic semiconductors by the odd-even effect of alkyl chain. Adv. Sci. **11**, 2308270 (2024)
30. T. Yamamoto, K. Takimiya, Facile synthesis of highly π-extended heteroarenes, dinaphtho[2,3-*b*:2',3'-*f*]chalcogenopheno[3,2-*b*]chalcogenophenes, and their application to field-effect transistors. J. Am. Chem. Soc. **129**, 2224–2225 (2007)
31. K. Nakayama, Y. Hirose, J. Soeda, M. Yoshizumi, T. Uemura, M. Uno, W. Li, M.-J. Kang, M. Yamagishi, Y. Okada, E. Miyazaki, Y. Nakazawa, A. Nakao, K. Takimiya, J. Takeya, Patternable solution-crystallized organic transistors with high charge carrier mobility. Adv. Mater. **23**, 1626–1629 (2011)
32. K. Kawabata, S. Usui, K. Takimiya, Synthesis of soluble dinaphtho[2,3-*b*:2',3'-*f*]thieno[3,2-*b*]thiophene (DNTT) derivatives: one-step functionalization of 2-bromo-DNTT. J. Org. Chem. **85**, 195–206 (2020)
33. T. Gruene, J.T.C. Wennmacher, C. Zaubitzer, J.J. Holstein, J. Heidler, A. Fecteau-Lefebvre, S. De Carlo, E. Müller, K.N. Goldie, I. Regeni, T. Li, G. Santiso-Quinones, G. Steinfeld, S. Handschin, E. van Genderen, J.A. van Bokhoven, G.H. Clever, R. Pantelic, Rapid structure determination of microcrystalline molecular compounds using electron diffraction. Angew. Chem. Int. Ed. **57**, 16313–16317 (2008)
34. K. Kato, K. Takaba, S. Maki-Yonekura, N. Mitoma, Y. Nakanishi, T. Nishihara, T. Hatakeyama, T. Kawada, Y. Hijikata, J. Pirillo, L.T. Scott, K. Yonekura, Y. Segawa, K. Itami, Double-helix supramolecular nanofibers assembled from negatively curved nanographenes. J. Am. Chem. Soc. **143**, 5465–5469 (2021)
35. K. Takaba, S. Maki-Yonekura, S. Inoue, T. Hasegawa, K. Yonekura, Protein and organic-molecular crystallography with 300kv electrons on a direct electron detector. Front. Mo. Biosci. **7**, 61226 (2021)
36. K. Nikaido, S. Inoue, R. Kumai, T. Higashino, S. Matsuoka, S. Arai, T. Hasegawa, Mixing-induced orientational ordering in liquid-crystalline organic semiconductors. Adv. Mater. Interfaces **9**, 2201789 (2022)

37. R. Miyata, S. Inoue, K. Nakajima, T. Hasegawa, Insulating polymer blend organic thin-film transistors based on bilayer-type alkylated benzothieno[3,2-*b*]naphtho[2,3-*b*]thiophene. ACS Appl. Mater. Interfaces **14**, 17719–17726 (2022)
38. R. Miyata, S. Inoue, K. Nikaido, K. Nakajima, T. Hasegawa, Friction force mapping of molecular ordering and mesoscopic phase transformations in layered-crystalline organic semiconductor films. ACS Appl. Mater. Interfaces **16**, 39701–39707 (2024)
39. S. Riera-Galindo, F. Leonardi, R. Pfattner, M. Mas-Torrent, Organic semiconductor/polymer blend films for organic field-effect transistors. Adv. Mater. Technol. **4**, 1900104 (2019)
40. C. Jiang, H.W. Choi, X. Cheng, H. Ma, D. Hasko, A. Nathan, Printed subthreshold organic transistors operating at high gain and ultralow power. Science **363**, 719–723 (2019)
41. B. Blülle, R. Häusermann, B. Batlogg, Approaching the trap-free limit in organic single-crystal field-effect transistors. Phys. Rev. Applied **1**, 034006 (2014)
42. G. Kitahara, S. Inoue, T. Higashino, M. Ikawa, T. Hayashi, S. Matsuoka, S. Arai, T. Hasegawa, Meniscus-controlled printing of single-crystal interfaces showing extremely sharp switching transistor operation. Sci. Adv. **6**, eabc8847 (2020)
43. K. Murata, G. Kitahara, S. Inoue, T. Higashino, S. Matsuoka, S. Arai, R. Kumai, T. Hasegawa, Stability of ternary interfaces and its effects on ideal switching characteristics in inverted coplanar organic transistors. Phys. Rev. Appl. **21**, 024005 (2024)

Chapter 8
Control of Molecular Orientation in Asymmetric Thienoacene-Based Organic Semiconductors

Hiroshi Katagiri

Abstract Polycyclic aromatic hydrocarbons (PAHs) and heterocyclic aromatic hydrocarbons (hetero-PAHs) have been the focus of considerable attention as promising organic semiconductor materials with potential applications in organic field-effect transistors (OFETs). High-performance OFETs depend on well-ordered molecular orientation in the crystalline thin film of the active layer. Research has so far concentrated on the synthesis of fused rings in symmetric and linear structures, as exemplified by pentacene or dinaphthothienothiophene (DNTT), with the objective of expanding the π-electron system and strengthening intermolecular interactions. However, this strategy often reduces solubility, impeding solution-based fabrication, which is a key advantage of organic materials. This study investigates asymmetric molecular structures with the aim of balancing solubility with molecular orientation. It encompasses bent-shaped acenes with thiophene and thieno[2,3-*b*]thiophene as key structural components, as well as extended π-conjugated molecules with asymmetric terminals based on thieno[3,2-*b*]thiophene. To be able to control molecular orientation in asymmetric OFET materials, we also examine how molecular structure and intermolecular interactions affect charge transport properties. Our results reveal that controlling molecular shape and fine-tuning crystal structure by exploiting weak intermolecular forces, such as London dispersion forces and intermolecular iodine–iodine interactions, are important methods of achieving high charge transport properties in the design of high-performance OFET materials.

Keywords π-conjugation · Thienoacene · Asymmetric molecule · Organic field-effect transistors · Herringbone structure · Structure–property relationship

---

H. Katagiri (✉)
Graduate School of Organic Materials Science, Yamagata University, Jonan, Yonezawa, Japan
e-mail: kgri7078@yz.yamagata-u.ac.jp

S. Kobatake and H. Uekusa (eds.), *Advances in Organic Crystal Chemistry*,
https://doi.org/10.1007/978-981-96-4479-7_8

8.1 Introduction

Organic field-effect transistor (OFET) devices, which rely on organic crystalline materials as their active components, have attracted considerable attention for their potential applications in flexible and printable electronics [1]. OFET performance depends on efficient charge transfer in the organic semiconductor layer, primarily determined by the overlap of the highest occupied molecular orbital (HOMO) for p-type and lowest unoccupied molecular orbital (LUMO) for n-type [2–4]. Acenes and thienoacenes, characterized by their planar, linear, and rigid extended π-conjugated systems, have been extensively studied as organic semiconductors for OFETs [5–7]. These polycyclic aromatic hydrocarbons, composed of linearly fused benzene and heteroaromatic rings, exhibit strong intermolecular interactions due to enhanced London dispersion forces. Such interactions facilitate the formation of close-packed structures, including herringbone packing and π-stacking motifs, which are highly favorable for efficient charge transport in OFETs [8–11]. However, π-extension leads to decreased solubility and stability, making it necessary to improve the trade-off between molecular orientation and solubility for applications in rapidly evolving printable electronics (Fig. 8.1).

One solution to improving solubility is to attach solubilizing groups, such as long alkyl chains or triisopropylsilylethynyl (TIPS) moieties [12–15]. More recently, attention has shifted toward the development of asymmetric thienoacenes [16–19]. These structurally asymmetric molecules have demonstrated improved solubility in organic solvents while maintaining highly ordered crystalline structures, potentially offering a solution to the long-standing solubility-crystallinity dilemma (Fig. 8.2).

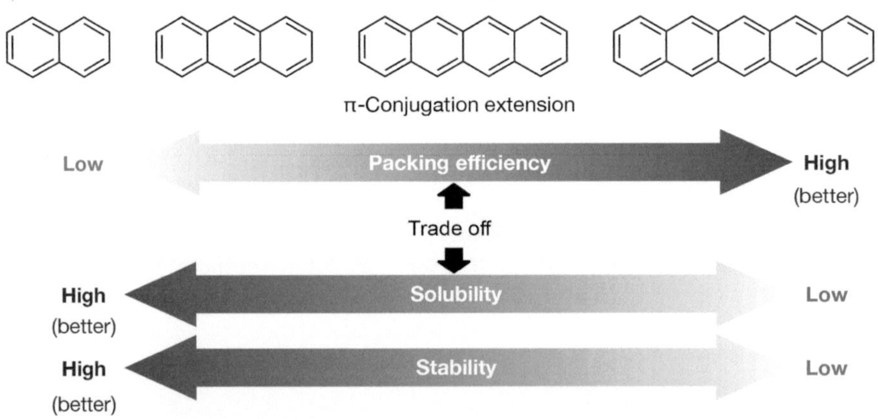

Fig. 8.1 Perceived trade-off between packing efficiency and solubility or stability as a function of extension of π-conjugated systems

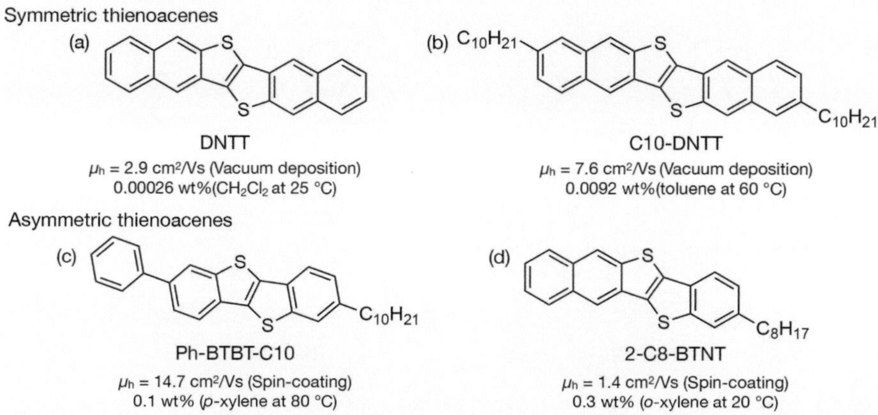

Fig. 8.2 Examples of symmetric and asymmetric thienoacenes with charge mobility and solubility. **a** DNTT [6], **b** C10-DNTT [13], **c** Ph-BTBT-C10 [16], and **d** 2-C8-BTNT [18]

In this chapter, we describe our previous research on the synthesis, structural and physicochemical properties, and organic field-effect transistor (OFET) performance of thienoacenes with bent structures and asymmetric configurations. By elucidating the relationships between molecular structure and functional properties, we provide crucial insights into the design principles that govern these advanced organic semiconductors (Fig. 8.3).

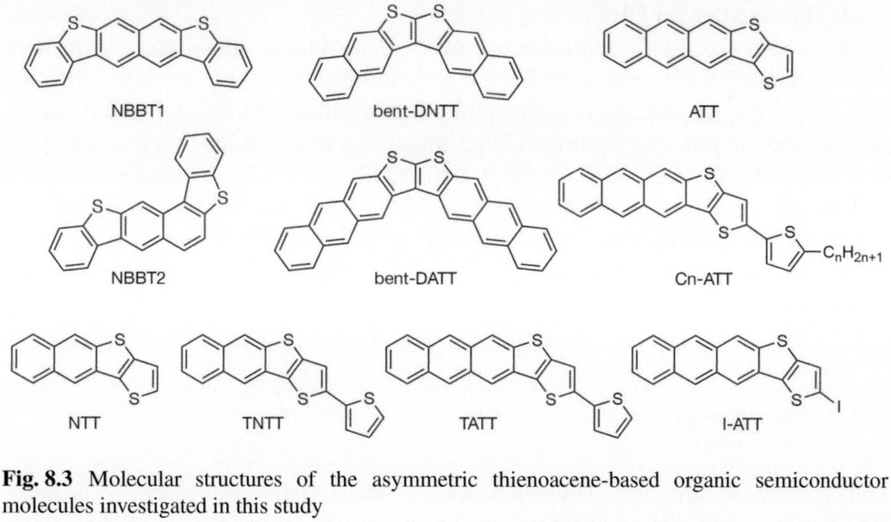

Fig. 8.3 Molecular structures of the asymmetric thienoacene-based organic semiconductor molecules investigated in this study

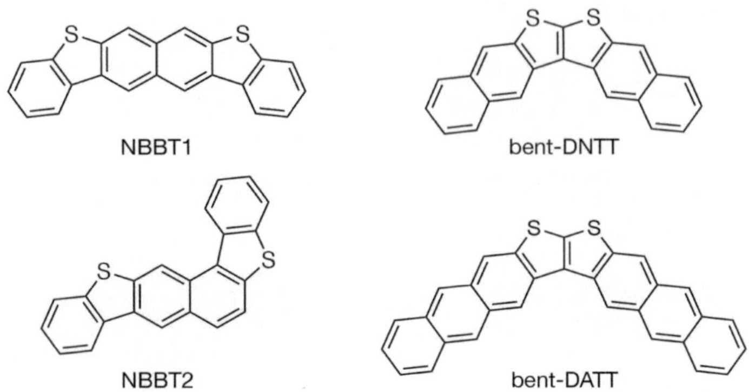

Fig. 8.4 Molecular structures of bent-shaped thienoacenes

8.2 Bent-Shaped Thienoacenes

8.2.1 Objective and Background

Linear expansion of fused ring systems in simple hydrocarbons often results in increased HOMO values and decreased stability and solubility. The incorporation of sulfur atoms into aromatic systems, as exemplified by thienoacenes, has successfully addressed these challenges by reducing HOMO values and improving air stability [7]. Recent research has highlighted the potential of bent thienoacenes to optimize material properties for OFETs [20].

We synthesize and characterize novel bent-shaped thienoacenes, namely naphtho[2,3-*b*:7,6-*b*′]- and naphtho[2,1-*b*: 5,6-*b*′]bisbenzo[*b*]thiophenes (NBBTs) [21], as well as naphtho[2,3-*b*]naphtho[2′,3′:4,5]thieno[3,2-*d*]thiophene (bent-DNTT) and anthra[2,3-*b*]anthra[2′,3′:4,5]thieno[3,2-*d*]thiophene (bent-DATT) [22] (Fig. 8.4). These compounds exhibit remarkable solubility in organic solvents, without the need for additional solubilizing functional groups. They also exhibit a close-packed structure such as herringbone packing, which is known to promote charge transport in the general OFET device structure with electrodes arranged horizontally.

8.2.2 Synthesis

The synthesis of bent-shaped thienoacenes, including NBBTs, bent-DNTT, and bent-DATT, was carried out using the series of synthetic strategies shown in Scheme 8.1. The methods employed, including Friedel–Crafts reactions, acid-promoted cyclizations, and reductive aromatizations provide a versatile approach to the preparation of novel higher thienoacenes.

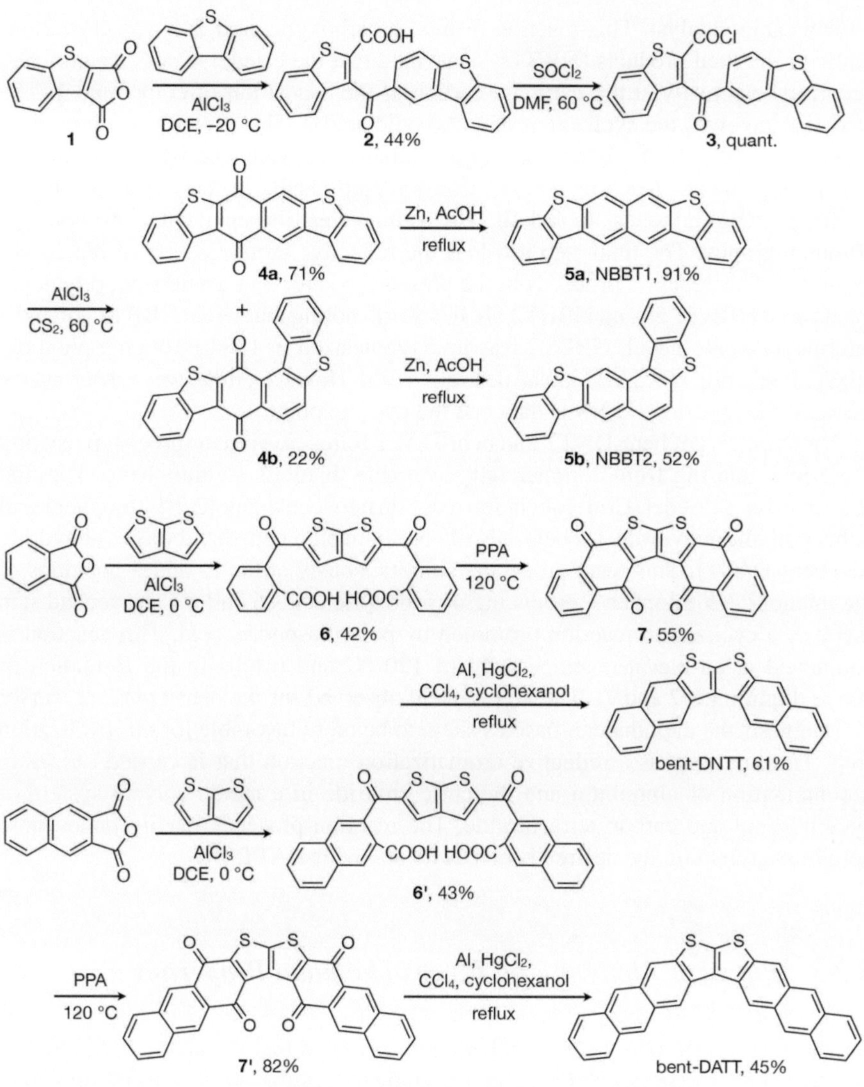

Scheme 8.1 Synthesis of naphtho[2,3-*b*:7,6-*b*′]- and naphtho[2,1-*b*:5,6-*b*′]bisbenzo[*b*]thiophene (NBBT1 and NBBT2), and naphtho[2,3-*b*]naphtho[2′,3′:4,5]thieno[3,2-*d*]thiophene (bent-DNTT) and anthra[2,3-*b*]anthra[2′,3′:4,5]thieno[3,2-*d*]thiophene (bent-DATT). DCE, 1,2-dichloroethane; PPA, polyphosphoric acid

NBBTs are synthesized from benzothiophene-2,3-dicarboxylic anhydride **1** [23, 24] in four steps. This approach builds upon previously reported methods for synthesizing fused quinones containing sulfur atoms, with modifications that enable the creation of larger acene structures. The first step involves a Friedel–Crafts reaction between dibenzothiophene and compound **1**, utilizing aluminum chloride as a Lewis acid catalyst. This reaction results in carboxylic acid **2**. X-ray crystallography of the final products (NBBTs) confirmed that the Friedel–Crafts reaction had occurred selectively at the carbonyl carbon at the 3-position of compound **1**. The next step involves the cyclization of carboxylic acid **2**. This process is achieved by first converting **2** to its corresponding acid chloride **3**, followed by an intramolecular Friedel–Crafts reaction. This cyclization step results in the formation of two isomeric acene-quinones, **4a** and **4b**, which can be readily separated using silica gel chromatography. The final step involves the reductive aromatization of the acene-quinones. This reaction proceeds in the presence of zinc and acetic acid, producing the desired NBBT1 **5a** and NBBT2 **5b**. It is worth noting that while NBBT1 and most intermediates are novel, NBBT2 was first synthesized in 1984 through a photocyclization reaction of a diarylethene derivative [25]. However, there was a discrepancy between the reported ^{1}H NMR data and the melting point.

The synthesis of bent-DNTT and bent-DATT follows a similar three-step reaction procedure, starting from commercially available thieno[2,3-*b*]thiophene. The first step involves a Friedel–Crafts acylation reaction between thieno[2,3-*b*]thiophene and either phthalic anhydride (for bent-DNTT) or 2,3-naphthalenedicarboxylic anhydride (for bent-DATT). This reaction occurs simultaneously at the 2- and 5-positions of the thieno[2,3-*b*]thiophene, producing dicarboxylic acids **6** and **6′**. The second step involves a cyclization reaction promoted by polyphosphoric acid. This reaction is conducted at an elevated temperature of 120 °C and results in the formation of acene-diquinones **7** and **7′**. The higher yield observed for the bent-DATT precursor (**7′**) suggests the naphthalene-based system to be more favorable for this cyclization step. The final step is a reductive aromatization reaction that is carried out using a combination of aluminum and mercuric chloride in a mixed solvent system of cyclohexanol and carbon tetrachloride. The reaction proceeds readily under these conditions, yielding the desired bent-DNTT and bent-DATT.

8.2.3 Physical, Optical, and Electrochemical Properties

Density functional theory (DFT) calculations revealed that the HOMOs of NBBT1 and NBBT2 are spread over the entire molecule, with substantial contributions from the sulfur atoms (Fig. 8.5). This distribution of electrons is favorable for hole transport in OFET devices. The more curved NBBT2 exhibits a lower HOMO level and a larger HOMO–LUMO energy gap than seen in NBBT1, suggesting that the π-conjugation extension is less effective with a highly curved structure. Similar calculations for bent-DNTT and bent-DATT showed that the HOMO–LUMO energy gap of bent-DATT is smaller than that of bent-DNTT, indicating that the expansion of aromatic units at both ends effectively extends the π-conjugation system, even in the bent skeleton.

Physical, optical, and electrochemical properties of bent-shape thienoacenes are summarized in Table 8.1. Despite their extended acene skeletons, the bent-shaped thienoacenes exhibit enhanced solubility than seen in their linear counterparts.

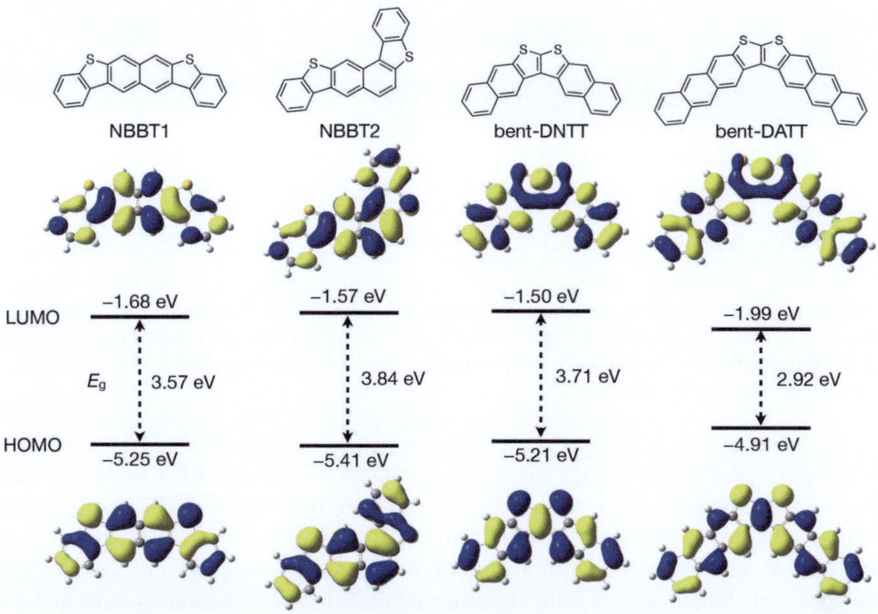

Fig. 8.5 Frontier molecular orbitals (HOMO and LUMO) and orbital energies of bent-shape thienoacenes calculated by DFT//B3LYP/6-31G(d)

NBBT1, consisting of six aromatic rings, demonstrates a solubility of 1.4 g/L in dichloromethane at 25 °C. The more curved isomer, NBBT2, shows an even higher solubility of 16 g/L under the same conditions. Bent-DNTT exhibits a solubility of 2.7 g/L in dichloromethane, surpassing that of its structural isomer NBBTs. This enhanced solubility is attributable to the remarkable structural curvature of bent-DNTT. In contrast, bent-DATT, composed of eight aromatic fused rings, displays a lower solubility of 0.07 g/L in dichloromethane, likely due to its larger π-conjugated system.

The hypochromic shift in solution in NBBT2 relative to NBBT1 reflects the reduced π-conjugation in the more curved structure of NBBT2. Greater red-shifts were observed in thin films than in the solutions of both compounds. NBBT1 showed a 23-nm red-shift in the thin film, while NBBT2 showed a smaller 13-nm red-shift. These bathochromic shifts are indicative of the formation of the J-aggregate-like molecular arrangements seen in herringbone packing. The absorption spectra of bent-DNTT and bent-DATT solutions were similarly red-shifted in the thin-film state. The large difference in λ_{max} (84 nm) between bent-DNTT and bent-DATT in solution highlights the significant influence of the extended π-conjugation in bent-DATT on its electronic properties.

The cyclic voltammetry results showed NBBT1 to exhibit a reversible oxidation peak at +0.74 V, while NBBT2 showed a higher oxidation potential of +0.94 V. This difference indicates a lower HOMO energy level for NBBT2 than for NBBT1,

Table 8.1 Physical, optical, and electrochemical properties of bent-shaped thienoacenes

Compound	Solution[a]			Thin film[b]		
	Solubility (g/L)	λ_{max} (nm)	E_{ox} (V)[c]	λ_{max} (nm)	E_g (eV)[d]	E_{homo} (eV)[e]
NBBT1	1.4	395	+0.68	414	2.8	−5.5
NBBT2	16	377	+0.86	383	3.1	−5.7
bent-DNTT	2.7	377	+0.57	401	2.9	−5.4
bent-DATT	0.07	461	+0.38	489	2.4	−5.2

[a]In CH_2Cl_2. [b]Prepared by spin coating (NBBTs) and vacuum deposition (bent-DNTT and bent-DATT). [c]Measured by cyclic voltammetry versus Fc/Fc^+ in CH_2Cl_2. [d]Estimated from the absorption edge by intersecting the line tangent to the long-wavelength side of the absorption band and the corrected baseline. [e]Estimated by photoelectron spectroscopy in air (PESA) using thin films prepared by vacuum deposition on PS-treated SiO_2 substrates

consistent with the reduced π-conjugation in the more curved NBBT2 structure and the observed hypsochromic shift in its UV–vis spectrum. Bent-DNTT displayed an oxidation potential of +0.57 V, while bent-DATT showed a lower oxidation potential of +0.38 V. The lower oxidation potential of bent-DATT suggests a higher HOMO energy level, which aligns with its more extended π-conjugation system and its absorption at longer wavelengths. Importantly, all compounds exhibited reversible oxidation peaks, indicating the high stability of their respective cation radicals. Overall, the trends observed in these electrochemical measurements correlate closely with the structural features and spectroscopic data of the compounds. The oxidation potentials decrease in the order of NBBT2 > NBBT1 > bent-DNTT > bent-DATT, reflecting the increasing extent of π-conjugation and the consequent rise in HOMO energy levels across this series. This trend is mirrored in the bathochromic shifts observed in the UV–vis spectra, with λ_{max} values increasing in the same order.

8.2.4 Single-Crystal Structures

NBBT1 exhibits a γ-type (flattened herringbone) packing motif, characterized by predominant face-to-face interactions and minor edge-to-face contributions (Fig. 8.6a) [8, 9]. The structure features π-stacking between aromatic planes at a distance of 3.46 Å, as well as S···C van der Waals interactions at 3.52 Å. Notably, NBBT1 crystallizes in the noncentrosymmetric orthorhombic space group $Pca2_1$, with its calculated dipole moment (1.95 D) aligned along the *c*-axis. In contrast, NBBT2 presents two crystallographically independent molecules in face-to-face orientation, forming separate columns with π-stacking distances of 3.44 and 3.51 Å (Fig. 8.6b). These columns, grown along the *c*-axis and oppositely tilted within the

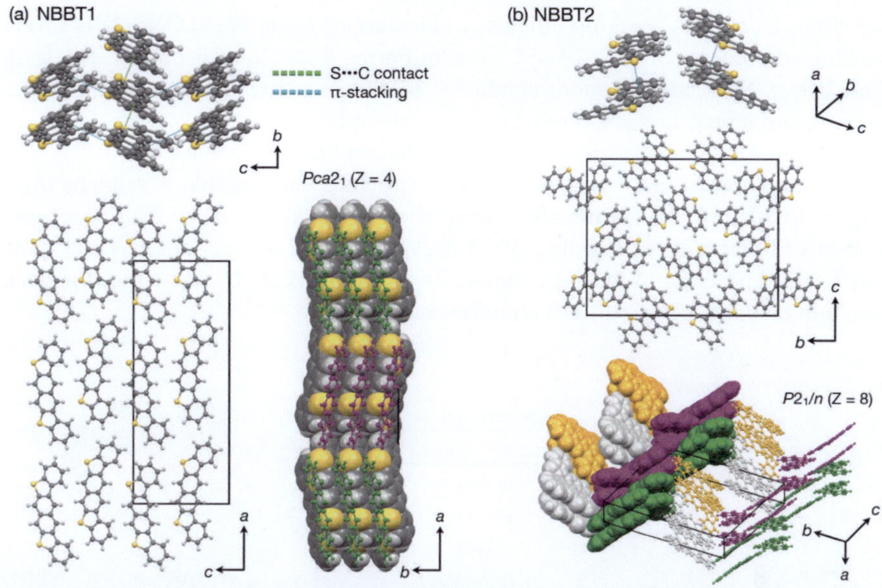

Fig. 8.6 Crystal structures of **a** NBBT1 and **b** NBBT2. In crystal packing, molecules colored gray, green, yellow, and magenta are symmetrically equivalent

a–*b* plane, result in a *γ*-type herringbone configuration in the centrosymmetric monoclinic space group $P2_1/n$. Despite its noncentrosymmetric nature, NBBT1 demonstrates superior packing efficiency, with a 73.4% filled space, in contrast to NBBT2's 72.7% as determined by PLATON/VOID calculations.

For bent-DNTT, two distinct polymorphs were identified: Form A ($P2_1/n$ space group) and Form B ($P2_1/c$ space group), obtained through sublimation and recrystallization, respectively. Form A of bent-DNTT exhibits a *θ*-type herringbone structure characterized by strong edge-to-face interactions among naphthalene moieties (Fig. 8.7a). In contrast, Form B displays a *γ*-type herringbone structure with notable π–π interactions between bent-DNTT molecules (Fig. 8.7b). The transfer integrals for holes show marked differences between the two polymorphs. In Form A, moderate transfer integrals (17 meV) are observed along the *b*-axis in the face-to-face direction (t_1) due to π–π stacking. The edge-to-face orientations exhibit varying transfer integrals, with very small values (1 and 3 meV) between the perpendicular long axes of the naphthalene moieties (t_2, t_3) and moderate values (35 and 13 meV) between the perpendicular short axes (t_4, t_5). These moderate transfer integrals arise from largish intermolecular orbital couplings that include π–π interactions and sulfur–sulfur interatomic contacts. Form B demonstrates significantly larger transfer integrals, particularly in the face-to-face orientation (t_1, 110 meV), attributable to tighter molecular packing and smaller slip-stack π–π interactions than those seen with Form A. The edge-to-face orientations in Form B also show moderate transfer integrals (21 meV for t_4 and t_5, 20 and 4 meV for t_2 and t_3, respectively) due to large sulfur–sulfur interatomic contacts. Packing efficiency calculations using PLATON/VOID reveal packing coefficients of 71.9% and 73.7% for Forms A and B, respectively. The form B of bent-DNTT appears more conducive to charge transport, as it offers tighter packing and larger, less anisotropic transfer integrals.

For bent-DATT, single crystals obtained by sublimation belong to the $P2_12_12_1$ space group and form a γ-type herringbone structure with notable π–π interactions (Fig. 8.7c). However, the transfer integrals are generally smaller than seen with bent-DNTT Form B, with values of 35 meV along π–π interactions (t_1) and quite small values for edge-to-face orientations (t_2–t_5). This is likely due to the slip-stack structure lacking sulfur–sulfur interatomic contacts.

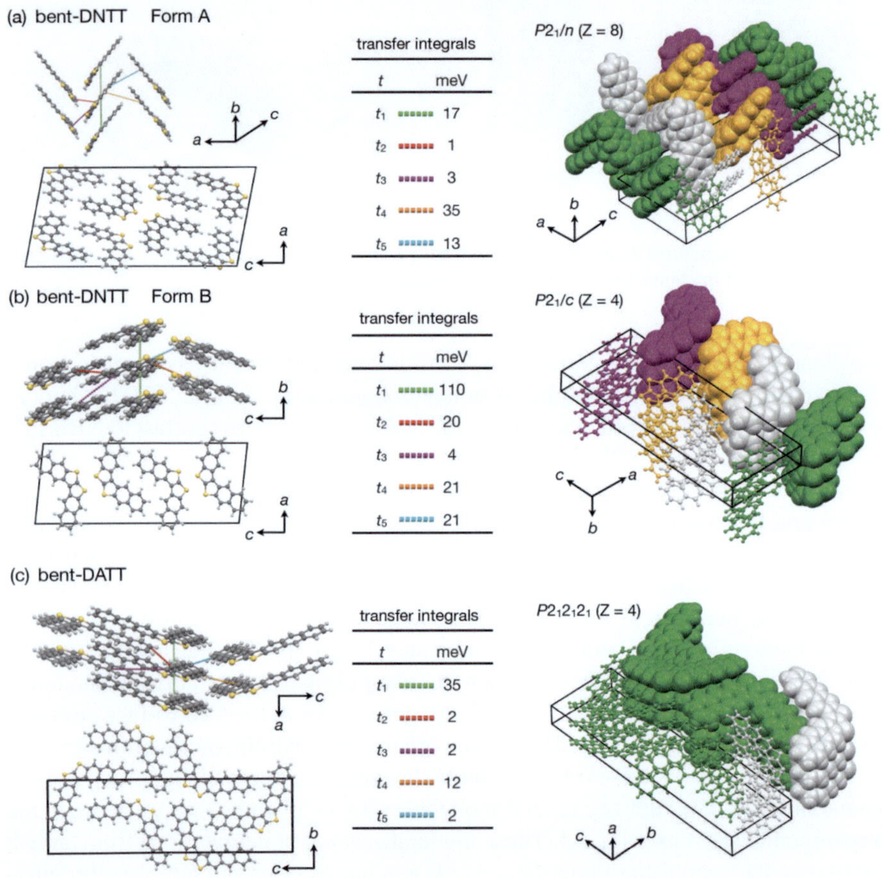

Fig. 8.7 Crystal structures and transfer integrals of **a** Form A of bent-DNTT, **b** Form B of bent-DNTT, and **c** bent-DATT. Transfer integral calculations were conducted utilizing the PW91/TZ2P level. In crystal packing, molecules colored gray, green, yellow, and magenta are symmetrically equivalent

X-ray diffraction (XRD) measurements of powder samples revealed the presence of multiple polymorphs in sublimated bent-DNTT, including Form A crystals. However, recrystallized single crystals of bent-DNTT exhibited only the Form B pattern. Bent-DATT powder samples prepared by sublimation yielded a single-crystal form.

8.2.5 OFET Properties and Thin-Film Structures

The OFET properties of NBBTs, bent-DNTT, and bent-DATT were investigated to evaluate their potential for use as organic semiconductor materials. Despite improved solubility, NBBT1 and NBBT2 presented challenges in solution-processed thin-film formation, requiring vacuum deposition for device fabrication. NBBT1 exhibited p-type characteristics in a top-contact OFET configuration on SiO_2, with a field-effect mobility (μ_{FET}) of 7×10^{-5} cm^2 V^{-1} s^{-1}. Although these performance metrics are modest, they underscore the potential of NBBT1 as an organic semiconductor. In contrast, NBBT2 thin films did not exhibit FET characteristics, possibly due to differences in molecular orientation on the substrate, particularly the presence or absence of end-on orientation. Bent-DNTT showed more promising results in terms of thin-film formation and device performance. In particular, spin-coated devices using bent-DNTT exhibited p-type characteristics with a mobility of 5.1×10^{-5} cm^2 V^{-1} s^{-1}. This solution processability and functional OFET performance highlight the advantages of bent-DNTT over NBBTs in terms of ease of fabrication and device characteristics. Bent-DATT-based OFETs fabricated by vacuum deposition exhibited p-type behavior with a mobility of 2.2×10^{-4} cm^2 V^{-1} s^{-1}. However, the extremely low solubility of bent-DATT hindered the creation of FET-active thin films by spin-coating, limiting its potential for solution-processed devices.

Thin-film X-ray diffraction (XRD) studies provided further insights into the molecular ordering and polymorphism of these materials (Fig. 8.8a, b, e, f). Vacuum-deposited films of bent-DNTT showed a complex mixture of crystal forms, including forms A and B. In contrast, spin-coated films showed well-oriented crystal structures with an edge-on lamellar arrangement, which is generally favorable for charge transport in OFETs. Bent-DATT thin films, on the other hand, showed no obvious peaks in XRD measurements, indicating lower crystallinity than seen in bent-DNTT.

Atomic force microscopy (AFM) and polarized optical microscopy further revealed the morphological differences between the films (Fig. 8.8c, d, g, h). Vacuum-deposited bent-DNTT films displayed large, non-uniformly distributed crystal grains and aggregates, unsuitable for high-performance OFETs. Conversely, spin-coated bent-DNTT films comprised uniformly grown grains in a three-dimensional mode, demonstrating the advantages of solution processing for this compound.

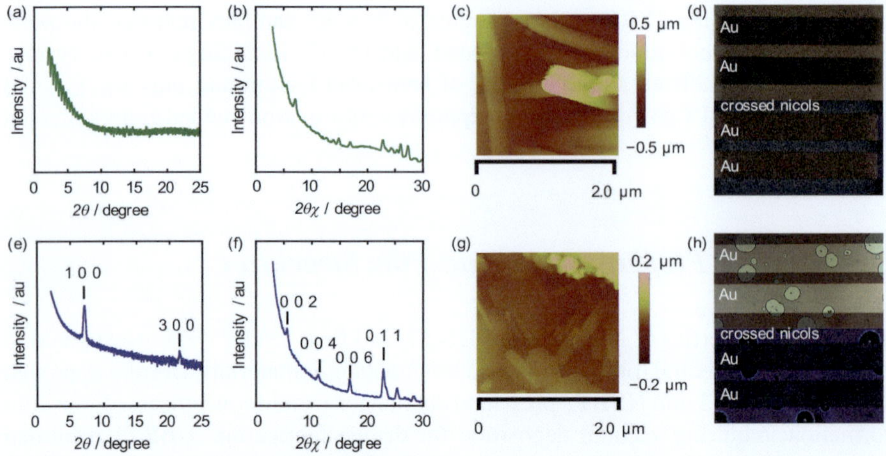

Fig. 8.8 Thin-film structures of bent-DNTT: **a–d** vacuum-deposited film **e–h** spin-coated film. **a**, **e** out-of-plane XRD patterns, **b**, **f** in-plane XRD patterns. The samples for the XRD measurements came from the same fabrication batch as the OFET devices. **c**, **g** AFM images; **d**, **h** polarized optical microscope images. Reproduced from Ref. [22] with permission. Copyright 2017 Elsevier

8.3 Asymmetric Alkyl- and Alkylthienyl-ATT Derivatives

8.3.1 *Objective and Background*

Molecular design strategies for OFET materials have up to now focused on expanding the π-conjugated system to enhance intermolecular interactions. However, extensive π-conjugation often results in reduced solubility in various solvents. Thus, determining the optimal balance between these competing properties is of significant importance. Recent advances in molecular design for OFET materials have focused on asymmetric monoalkylated thienoacenes, which offer improved solubility while maintaining or improving semiconductor performance. However, examples of readily synthesizable asymmetric acenes are still limited, and the design of asymmetric scaffolds that allow regioselective substitution is particularly challenging. This section describes the development of anthra[2,3-*b*]thieno[2,3-*d*]thiophene (ATT) as a readily synthesized thienoacene, as well as its monoalkyl and monoalkylthienyl derivatives derived from selective modification at the terminal thienothiophene on ATT (Fig. 8.9) [26].

8.3.2 *Synthesis*

The synthesis of ATT commences with commercially available thieno[3,2-*b*]thiophene **1** and 2,3-naphthalenedicarboxylic anhydride **2** (Scheme 8.2). The initial

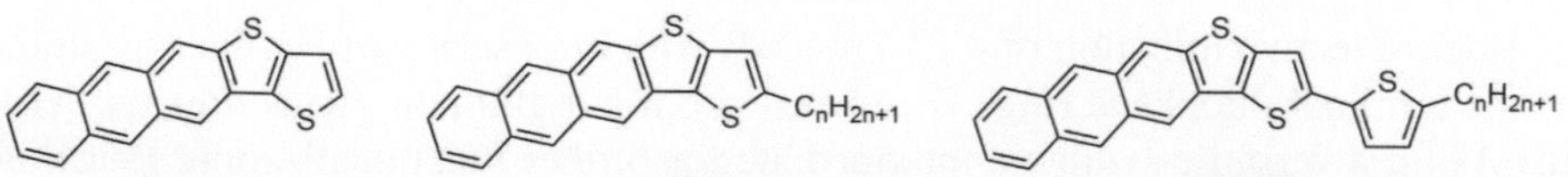

Fig. 8.9 Molecular structures of anthra[2,3-*b*]thieno[2,3-*d*]thiophene (ATT) and its monoalkyl and monoalkylthienyl derivatives (CnATT and CnTATT)

step involves Friedel–Crafts acylation between these two compounds, conducted in the presence of $AlCl_3$ and using 1,2-dichloroethane as the solvent. This reaction yielded monocarboxylic acid **3**. Subsequent to the acylation, the synthesis proceeds via a cyclization reaction promoted by polyphosphoric acid at 80 °C, resulting in the formation of acene-quinone intermediate **4**. The final step in the ATT synthesis involves reductive aromatization, efficiently carried out using $Al/HgCl_2$ in a cyclohexanol/CCl_4 solvent system, providing the desired ATT product at a yield of 73%.

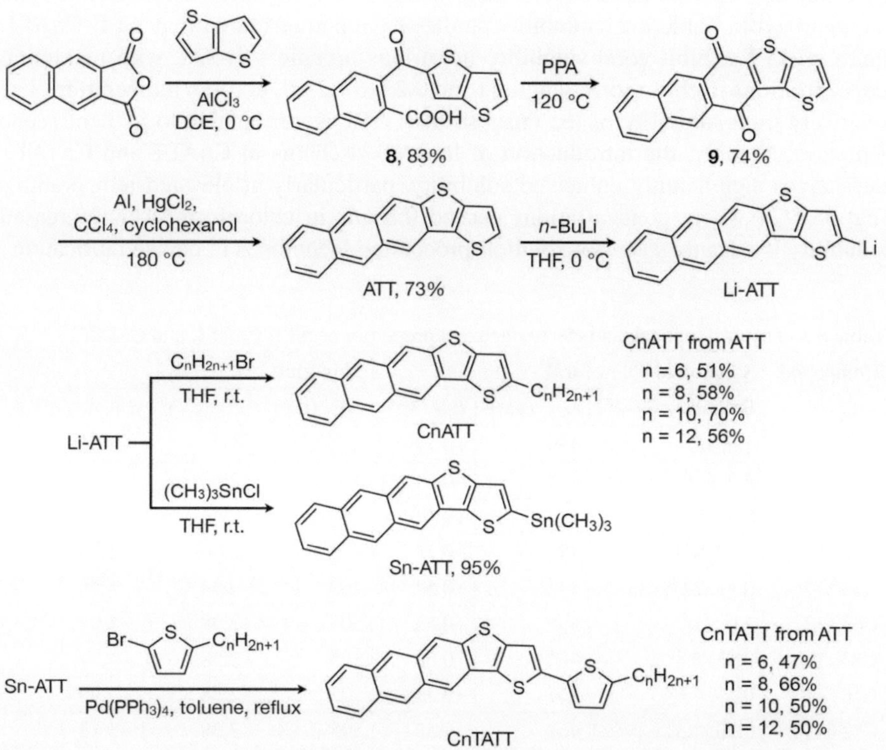

Scheme 8.2 Synthesis of anthra[2,3-*b*]thieno[2,3-*d*]thiophene (ATT), 2-alkyl ATT (CnATT), and 2-alkylthienyl ATT (CnTATT). DCE, 1,2-dichloroethane; PPA, polyphosphoric acid

Regioselective lithiation of ATT proceeds with high selectivity at the α-position of the terminal thiophene ring. Treatment with n-butyllithium yields lithiated ATT (Li-ATT), a versatile synthetic intermediate for further functionalization. Reaction of Li-ATT with various alkyl halides yields 2-alkyl ATT derivatives (CnATT) at a 51–70% yield, demonstrating the efficacy of this approach. Alternatively, stannylation of Li-ATT produces Sn-ATT, another key intermediate. This stannylated derivative, typically used without extensive purification, exhibits > 99% purity as determined by NMR analysis. Sn-ATT serves as a valuable precursor for the synthesis of 2-alkylthienyl ATT derivatives (CnTATT) via Stille cross-coupling reactions with various alkylthienyl bromides. These cross-coupling reactions proceed at 47–66% yield, further exemplifying the utility of this synthetic strategy.

8.3.3 Physical, Optical, and Electrochemical Properties

Physical, optical, and electrochemical properties of ATT, CnATT, and CnTATT are summarized in Table 8.2. Solubility studies have demonstrated that ATT, CnATT, and CnTATT exhibit good solubility in various organic solvents, with maximum concentrations in chloroform ranging from 0.23 to 2.2 g/L at room temperature. The relatively high solubility of the unsubstituted ATT is attributable to its bent acene structure. Notably, the introduction of long alkyl chains in CnATT and CnTATT derivatives significantly enhanced solubility, particularly at elevated temperatures (ca. 55 °C), where concentrations exceed 0.8 g/L in chloroform. This increased solubility is advantageous for solution-processing techniques in device fabrication.

Table 8.2 Physical, optical, and electrochemical properties of ATT, CnATT, and CnTATT

Compound	Solution in chloroform			Thin film[c]		
	Solubility (g/L)[a]	λ_{max} (nm)	E_{ox} (V)[c]	λ_{max} (nm)[d]	E_g (eV)[e]	E_{HOMO} (eV)[f]
ATT	1.6, 1.9[b]	439	+0.59	472	2.52	−5.30
C6ATT	2.2, 6.2[b]	442	+0.54	464	2.60	−5.55
C8ATT	0.41, 1.1[b]	442	+0.54	460	2.64	−5.53
C10ATT	0.30, 0.88[b]	442	+0.54	460	2.64	−5.63
C12ATT	0.43, 2.0[b]	442	+0.54	460	2.64	−5.66
C6TATT	1.0, 2.0[b]	459	+0.51	505	2.38	−5.07
C8TATT	0.53, 3.8[b]	461	+0.50	505	2.38	−4.91
C10TATT	0.35, 4.6[b]	461	+0.51	505	2.38	−4.90
C12TATT	0.23, 4.8[b]	461	+0.52	505	2.38	−5.13

[a]Determined by colorimetry. [b]At 55 °C. [c]Measured by cyclic voltammetry versus Fc/Fc^+ in CH_2Cl_2. [d]Prepared by vacuum deposition onto quartz substrates. [e]Estimated from absorption edge as determined from the intersection of the line tangent with the long-wavelength side of the absorption band and the corrected baseline. [f]Estimated by photoelectron spectroscopy in air (PESA) using thin films prepared by vacuum deposition on PS-treated SiO_2 substrates

The optical properties of these compounds have been investigated using UV–vis spectroscopy in both solution and thin-film form. In chloroform solutions, ATT and C8ATT exhibit similar absorption maxima (λ_{max}) and spectral profiles. However, C8TATT shows a bathochromic shift of approximately 20 nm higher than seen in ATT and C8ATT, which is attributable to the extended π-conjugation that results from the additional thiophene unit. Thin films of these compounds, prepared by vacuum deposition, display red-shifted absorption spectra relative to their solution counterparts, indicating the formation of crystalline structures with J-aggregate characteristics. This phenomenon is particularly pronounced in ATT and C8TATT films, suggesting a more effective increase in HOMO levels than seen in C8ATT.

Cyclic voltammetry measurements in dichloromethane have revealed the oxidation potentials of ATT, C8ATT, and C8TATT to be +0.59, +0.54, and +0.50 V (onset vs. Fc/Fc^+), respectively. These values reflect the electron-donating nature of the alkyl groups and the effective π-expansion that is contributed by the additional thiophene ring. It is important to note that these electrochemical properties are primarily attributable to single-molecule characteristics.

8.3.4 Single-Crystal Structures

Figure 8.10 illustrates the crystal structure of sublimated ATT. The ATT molecules exhibit a slight bend along the *b*-axis, resulting in a noncentrosymmetric space group $P2_1$. The crystal structure of ATT adopts a herringbone packing arrangement, with molecular orientations alternating between edge-to-face and face-to-face configurations. This packing motif is characteristic of high-performance OFET materials, such as pentacene and oligothiophene derivatives. However, a slight slip-stacking of ATT molecules is observed along the *c*-axis, accompanied by an antiparallel arrangement of ATT cores. The thienothiophene moieties do not engage in intermolecular interactions, and consequently, no interatomic contacts occur within the van der Waals radii. As a result, the larger size of the sulfur atom fails to contribute effectively to intermolecular coupling in this particular packing configuration. The distribution of the ATT integral shows significant anisotropy, which is unfavorable for charge transport, with large values of 54 meV in the t_3 and t_5 directions and small values of 4–7 meV in the other directions. This suggests that the slip-stacking structure in the *c*-axis direction is unfavorable for orbital interactions. An analysis of packing efficiency, conducted using PLATON/VOID, reveals that ATT possesses a packing coefficient of 73.3% filled space, which is lower than that of pentacene (76.3%). The single-crystal structures of CnATT and CnTATT remain undetermined due to the lack of availability of suitable single crystals for XRD analysis.

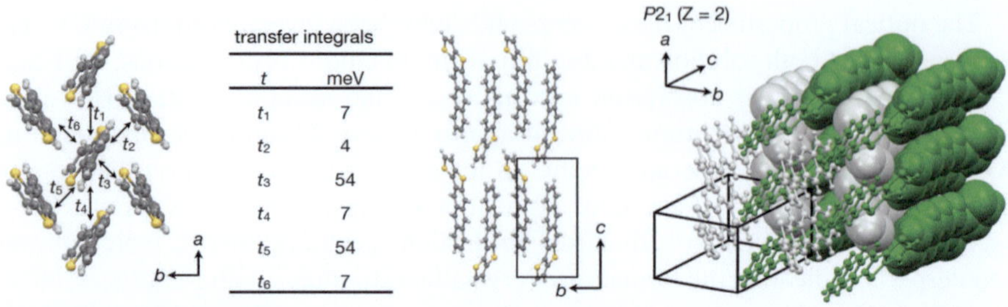

Fig. 8.10 Crystal structures and transfer integrals of ATT. Transfer integral calculations were conducted utilizing the PW91/TZ2P level. In crystal packing, molecules colored gray and green are symmetrically equivalent

8.3.5 *OFET Properties and Thin-Film Structures*

OFET devices using ATT, CnATT, and CnTATT were fabricated by vacuum deposition and drop-casting. The characteristics of these devices are summarized in Table 8.3. Vacuum-deposited devices exhibited typical p-type characteristics with mobilities ranging from 1.2×10^{-2} cm^2 V^{-1} s^{-1} for ATT, 9.0×10^{-4} to 6.6×10^{-2} cm^2 V^{-1} s^{-1} for CnATT, and 0.15–1.9 cm^2 V^{-1} s^{-1} for CnTATT. The highest mobility was observed in C8TATT. The mobility values of ATT- and CnATT-based devices showed only slight differences ($n > 6$), suggesting that the introduction of alkyl groups to ATT does not significantly contribute to FET performance. In contrast, CnTATT-based devices demonstrated significantly higher mobilities, indicating a more substantial contribution to mobility from the additional thiophene ring in the ATT framework.

Table 8.3 Characteristics of ATT-, CnATT-, and CnTATT-based FET devices

Compound	Vacuum deposition[a]			Drop-casting[a]		
	μ_{FET} (cm^2 V^{-1} s^{-1})	I_{on}/I_{off}	V_{th} (V)	μ_{FET} (cm^2 V^{-1} s^{-1})	I_{on}/I_{off}	V_{th} (V)
ATT	1.2×10^{-2}	10^3	−39	4.0×10^{-2}	10^4	−30
C6ATT	9.0×10^{-4}	10^3	−16	–	–	–
C8ATT	2.0×10^{-2}	10^4	−20	–	–	–
C10ATT	2.3×10^{-2}	10^4	−32	–	–	–
C12ATT	6.6×10^{-2}	10^4	−2	–	–	–
C6TATT	0.65	10^4	−2	0.26	10^2	1
C8TATT	1.9	10^5	−26	1.6[b]	10^{5}[b]	−10[b]
C10TATT	0.47	10^5	−10	0.11	10^3	−24
C12TATT	0.15	10^5	−11	0.10	10^3	−5

[a]Top-contact bottom-gate OFET devices. [b]Bottom-contact top-gate OFET devices

While drop-cast ATT- and CnATT-based devices showed no FET response, CnTATT-based devices exhibited mobilities of 0.10–0.26 $cm^2\ V^{-1}\ s^{-1}$ for drop-casting. These results suggest that the introduction of an additional thiophene ring effectively enhances both solution processability and mobility. The highest mobility of 1.6 $cm^2\ V^{-1}\ s^{-1}$ was observed in the bottom-contact top-gate OFET C8TATT-based device fabricated by drop-casting. Furthermore, this device demonstrated an on/off current ratio of 10^5 in an ambient atmosphere. The implementation of a CYTOP dielectric layer with a low dielectric constant and low interface trap sites facilitated the mitigation of interface trap and mobility issues.

X-ray diffraction (XRD) profiles of ATT-, C8ATT-, and C8TATT-based thin films are presented in Fig. 8.11. The XRD profiles of vacuum-deposited films for all molecules exhibit distinct differences between the out-of-plane and in-plane directions, indicating the formation of well-oriented crystalline structures. The out-of-plane XRD profiles of ATT and C8TATT reveal a series of peaks clearly assignable to (00L) reflections, whereas the in-plane XRD profiles show three typical peaks assignable to π-stacking orientations [(11L), (02L), (12L)]. These observations suggest an edge-on lamellar structure, with molecules arranged in a herringbone fashion. In contrast, the XRD profile of the C8ATT vacuum-deposited film differs from those of the ATT and C8TATT films. A single strong peak observed in the in-plane XRD of C8ATT is likely assignable to a planar π stacking distance. The XRD profile from an ATT-based spin-coated film exhibits weak intensity, indicating a less-oriented structure. Solution-processed C8ATT-based films display XRD profiles similar to those of vacuum-deposited films, which demonstrate a field-effect transistor (FET) response. XRD profiles of other CnATT and CnTATT films (i.e., $n \neq 8$) show similarities to those of the C8 derivatives. These results demonstrate the advantages of C8TATT for the fabrication of solution-processed films suitable for FET devices.

8.4 Monoiodinated Asymmetric ATT Derivative: I-ATT

8.4.1 Objective and Background

Asymmetric molecular structures exhibit higher solubility than that seen in their symmetrical linear counterparts. However, asymmetric molecules without substituents often adopt antiparallel molecular orientations that reduce dipole moments and steric hindrance. This arrangement results in less overlap between molecular orbitals and consequently impedes charge transport. This phenomenon has previously been observed in anthra[2,3-*b*]thieno[2,3-*d*]thiophene (ATT) systems. To address this issue, the packing effect of sufficiently long alkyl chains has been widely employed to achieve a parallel orientation conducive to charge transport. With this approach, the long alkyl chains contribute to both solubility and molecular orientation. Nevertheless, these compounds present certain drawbacks, including reduced

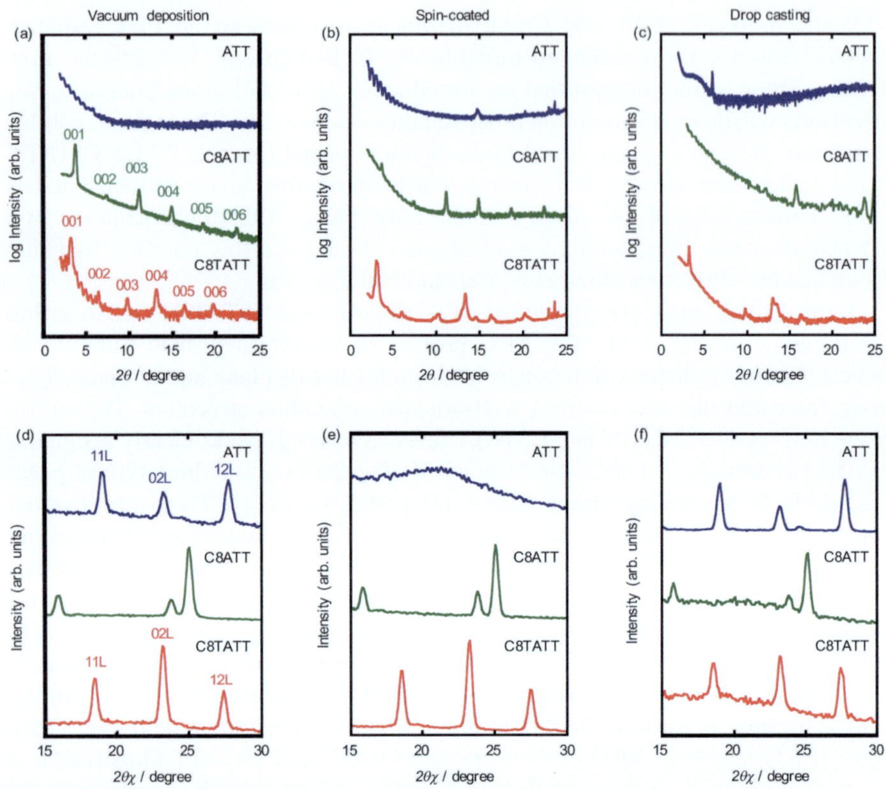

Fig. 8.11 XRD profiles of thin films of ATT, C8ATT, and C8TATT: **a**–**c** out-of-plane patterns, **d**–**f** in-plane patterns. The left, middle, and right columns show results for vacuum-deposited, spin-coated, and drop-cast films, respectively. XRD measurements were made on samples made from the same fabrication batch as the OFET devices. Reproduced from Ref. [26] with permission. Copyright 2017 American Chemical Society

thermal stability due to lowered melting points and increased access resistance resulting from the insulation volume of the alkyl chains.

This section presents an investigation of the structural and physical properties of an iodinated asymmetric acene molecule, 2-iodoanthra[2,3-*b*]thieno[2,3-*d*]thiophene (I-ATT) (Fig. 8.12) [27]. This compound demonstrates solution processability, exhibits FET characteristics, and allows for molecular orientation control through iodine–iodine interactions, resulting in improved carrier mobility. Among the halogens, iodine possesses a particularly large polarizability, generating substantial London dispersion forces that produce attractive interactions between iodine atoms. This strategy represents a novel approach to achieving compatibility between solubility and molecular orientation in organic semiconductors.

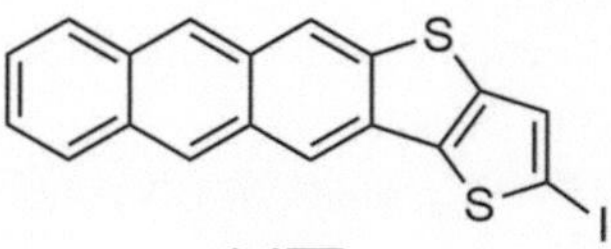

Fig. 8.12 Molecular structure of 2-iodoanthra[2,3-*b*]thieno[2,3-*d*]thiophene (I-ATT)

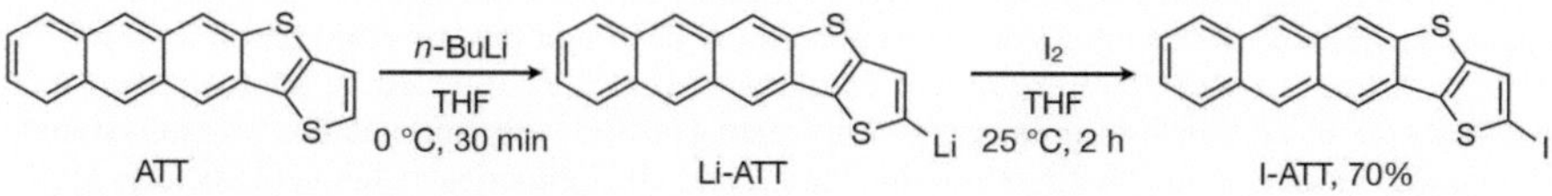

Scheme 8.3 Synthesis of iodinated ATT (I-ATT)

8.4.2 Synthesis

I-ATT was synthesized via selective α-lithiation of ATT, followed by iodination (Scheme 8.3). This reaction allows for the precise introduction of an iodine atom at a specific position within the ATT molecular framework. The incorporation of this halogen atom into the ATT structure presents additional avenues for further functionalization and derivatization, thereby expanding the potential applications of this compound class.

8.4.3 Physical, Optical, and Electrochemical Properties

The thermal stability, solubility, optical, and electrochemical properties of ATT and I-ATT are presented in Table 8.4. Despite the general instability of iodine compounds due to C–I bond cleavage, I-ATT showed high thermal stability. Thermogravimetric analysis (TGA) revealed a 5% decay temperature of 233 °C for I-ATT, which is comparable to the 294 °C observed for ATT. Iodine incorporation only slightly decreased solubility; both ATT and I-ATT remained highly soluble due to their bent acene structure.

In chloroform solutions and drop-cast films, I-ATT exhibited absorption at longer wavelengths than with ATT, a phenomenon attributable to the electron-withdrawing effect of iodine, as corroborated by theoretical calculations. A drastic red-shift in absorption spectra was observed in thin films, unlike in solutions for both compounds, indicative of the J-aggregate-like packing structure commonly observed in linear acenes. The thin-film form of I-ATT displayed absorption at longer wavelengths than ATT, suggesting enhanced intermolecular interactions due to the presence of iodine. Cyclic voltammetry in dichloromethane showed reversible oxidation peaks at +0.57 V (ATT) and +0.61 V (I-ATT) vs Fc/Fc^+, indicating stable cation radicals. The ionization energy (E_{IP}) of I-ATT (5.51 eV) is lower than that of ATT (5.30 eV),

Table 8.4 Physical, optical, and electrochemical properties of ATT and I-ATT

Compound	5% weight loss (°C)[a]	Solubility ($g\ L^{-1}$)[b]	Solution			Thin film	
			λ_{max} (nm)[c]	E_{ox} (V)[d]	E_g (eV)[e]	λ_{max} (nm)[f]	E_{IP} (eV)[g]
ATT	294	1.6	439	+0.57	2.53	464	5.30
I-ATT	233	0.94	444	+0.61	2.45	477	5.51

[a]Determined by TGA. [b]Determined by colorimetry as chloroform solutions. [c]In chloroform. [d]Versus Fc/Fc^+ couple. [e]Estimated from absorption edge as determined from the intersection of the line tangent to the long-wavelength side of the absorption band and the corrected baseline. [f]Prepared by drop-casting 0.1 wt% tetralin solution on glass substrates. [g]Estimated by photoelectron spectroscopy in air using thin films prepared by drop-casting on SiO_2 substrates treated with polystyrene. Reproduced from Ref. [27] with permission. Copyright 2021 American Chemical Society

but it remains sufficiently high to allow efficient hole injection from the Au electrode (work function: 5.1 eV).

8.4.4 OFET Properties and Crystal Structures

Top-contact bottom-gate OFET devices were fabricated by drop-casting ATT and I-ATT tetralin solutions. Table 8.5 summarizes the mobility, on/off ratio, and threshold voltage of these devices, along with the transfer integrals and reorganization energies, which are discussed in subsequent paragraphs. The I-ATT device exhibited a hole mobility of 2.2 $cm^2\ V^{-1}\ s^{-1}$, which represents a two-order-of-magnitude increase over that of the ATT device (0.04 $cm^2\ V^{-1}\ s^{-1}$). Furthermore, the superiority of I-ATT was corroborated by its enhanced on/off ratio and threshold voltage characteristics.

Single-crystal X-ray analysis shows that ATT and I-ATT form layered-herringbone structures typical of linear acene molecules (Fig. 8.13a–d). However, a notable distinction lies in their molecular orientations: ATT adopts an antiparallel arrangement, while I-ATT exhibits a parallel configuration. The parallel orientation adopted by I-ATT is stabilized by significant iodine–iodine interactions, with I···I contacts observed at distances of 3.9 and 4.0 Å, which fall within the sum of the van der Waals radii of iodine (4.0 Å). The parallel orientation in I-ATT, facilitated by

Table 8.5 FET characteristics and transfer integrals of ATT and I-ATT

Compound	Mobility ($cm^2\ V^{-1}\ s^{-1}$)		I_{on}/I_{off}	V_{th} (V)	Transfer integrals[b] (meV)						
	μ_{ave}[a]	μ_{max}			t_1	t_2	t_3	t_4	t_5	t_6	t_7
ATT	0.03	0.04	10^4	−30	8	73	10	8	10	73	1
I-ATT	1.1	2.2	10^5	−25	42	8	85	42	48	82	5

[a]Average of six devices. [b]Performed at the M06-2X/TZ2P level

these iodine–iodine interactions, leads to advantageous transfer integrals as calculated from single-crystal structures. I-ATT displays smaller anisotropy and larger transfer integral values than ATT. Notably, the transfer integral between layers along the molecular long axis in I-ATT reaches up to 5 meV, likely due to the HOMO distribution on the iodine atoms. These enhanced transfer integrals, resulting from effective orbital overlap in the parallel orientation, are the primary contributors to the high carrier mobility observed in I-ATT.

Thin-film X-ray diffraction studies on drop-cast films of ATT and I-ATT reveal molecularly oriented structures conducive to two-dimensional charge transport (Fig. 8.13e, f). Both compounds exhibit end-on and layered-herringbone orientations, as evidenced respectively by their out-of-plane and in-plane diffraction profiles. For ATT, the out-of-plane periodicity corresponds to the antiparallel arrangement of one molecule, while I-ATT shows a head-to-head (or tail-to-tail) bilayer unit formation. The in-plane diffraction profiles further confirm the herringbone structure in both cases, demonstrating a high degree of structural similarity between the drop-cast films and their respective single-crystal structures.

AFM analysis of the thin films in the device active layer reveals step-terrace morphology with periodic steps for both compounds (Fig. 8.13g, h). ATT exhibits a periodicity of approximately 1.54 nm, consistent with its single molecular length, while I-ATT shows a periodicity of about 3.06 nm, corresponding to a bilayer unit. These observations support an end-on orientation of molecules perpendicular to the substrate, which is favorable for charge transport.

The iodine–iodine interactions can be classified into two types based on their dominant intermolecular forces and geometries (Fig. 8.13i, j). Type I interactions, characterized by weak directionality, are primarily governed by London dispersion forces. In contrast, Type II interactions exhibit strong directionality due to orientation forces via σ-holes, which are regions of low electron density formed on the opposite side of the σ-bond of the iodine atom. The strength of these interactions follows the order I > Br > Cl, making iodine atom interactions the strongest for both types. In I-ATT, the observed interactions are geometrically classified as Type I, generated by London dispersion forces with strengths ranging from 1 to 4 kJ mol^{-1}. This classification suggests that the I-ATT crystal can be described as a van der Waals crystal, formed by a delicate balance of weak iodine–iodine interactions and interactions between the π-conjugated acene skeleton. The resulting structure can be conceptualized as a cooperative supramolecular aggregate held together by these subtle London dispersion forces.

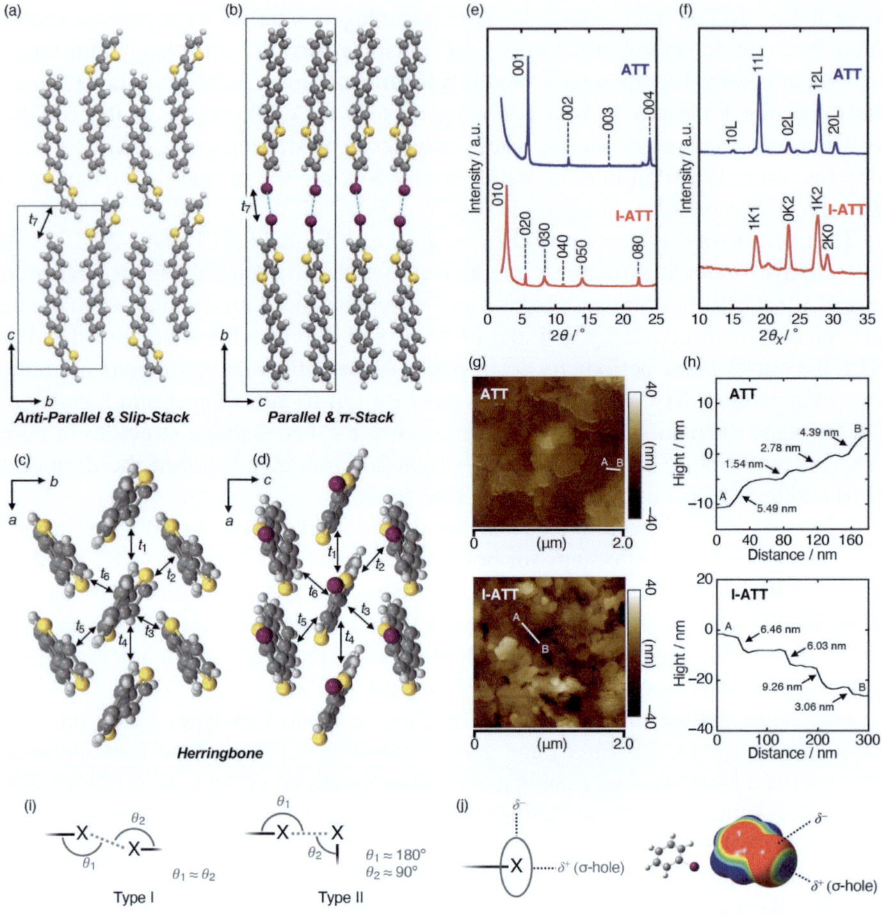

Fig. 8.13 Crystal structures of ATT (**a**, **c**) and I-ATT (**b**, **d**) with positions of the transfer integrals. Overviews showing indications of antiparallel and parallel layered structures (**a**, **b**) and herringbone packing (**c**, **d**). The blue dashed lines in **b** indicate the iodine–iodine contacts. Out-of-plane (**e**) and in-plane (**f**) X-ray diffraction (XRD) profiles. AFM images (**g**) and associated cross-sectional profiles (**h**) of ATT and I-ATT thin films. Thin-film samples acquired from the same fabrication batch as the one used for the OFET devices were subjected to XRD and AFM analysis. Schematic representation of iodine–iodine interactions: London dispersion-force-dominant interaction Type I and orientation-force-dominant interaction Type II (**i**), and σ-hole of iodobenzene (**j**). Reproduced from Ref. [27] with permission. Copyright 2021 American Chemical Society

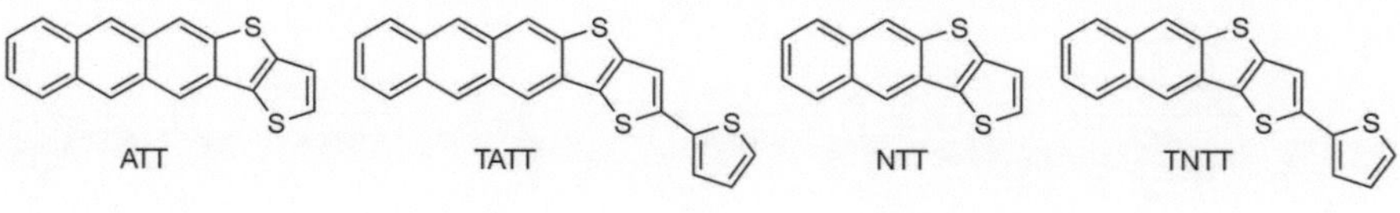

Fig. 8.14 Molecular structures of anthra[2,3-*b*]thieno[2,3-*d*]thiophene (ATT), its thienyl derivative (TATT), naphtho[2,3-*b*]thieno[2,3-*d*]thiophene (NTT), and its thienyl derivative (TNTT)

8.5 Asymmetric Naphtho- and Anthra-Thienothiophene and Their Thienyl Derivatives

8.5.1 Objective and Background

The correlation between molecular shape and crystal structure in systems dominated by London dispersion forces is a fundamental aspect of materials science that has significant implications for organic semiconductor design. This section aims to elucidate the structure–property relationships in asymmetric thienoacene derivatives, focusing on the effects of molecular extension on crystal packing and subsequent electronic properties. Our previous work on anthra[2,3-*b*]thieno[2,3-*d*]thiophene (ATT) and its derivatives has provided valuable insights into this phenomenon. ATT, in its unmodified form, adopts an antiparallel slip-stack structure that is suboptimal for charge transport. However, the introduction of an alkylthienyl group induces a significant change in crystal structure, leading to improved semiconductor properties. This observation underscores the profound impact that even minor structural modifications, such as the addition of a single thiophene ring, can have on the overall crystal structure and carrier transport characteristics. Building on these findings, the present study seeks to further explore the structure-optical property relationships in a series of related compounds. Specifically, we investigate ATT and its thienyl derivative (TATT), as well as naphtho[2,3-*b*]thieno[2,3-*d*]thiophene (NTT) and its corresponding thienyl derivative (TNTT) (Fig. 8.14) [28]. By comparing these pairs of molecules, we aim to elucidate the effects, on both structural and physical properties, of extending thienoacenes with an α-thiophene ring.

8.5.2 Synthesis

The synthesis of NTT, TNTT, and TATT was carried out using a series of strategic reactions (Scheme 8.4). Although NTT is a known compound, we prepared it using a synthetic route similar to that previously reported for ATT. The synthesis involved Friedel–Crafts acylation, followed by cyclization using polyphosphoric acid, and subsequent reduction with zinc in acetic acid. TATT and TNTT were obtained via a two-step process: initial stannylation of ATT and NTT, respectively, using butyllithium, which selectively lithiated the α-position of thienothiophene, followed by

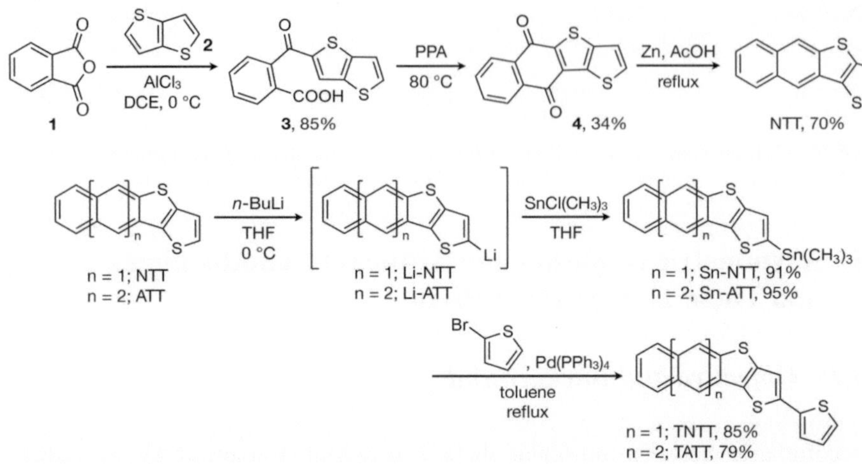

Scheme 8.4 Synthesis of naphtho[2,3-*b*]thieno[2,3-*d*]thiophene (NTT), the thienyl derivatives 2-(thienyl)naphtho[2,3-*b*]thieno[2,3-*d*]thiophene (TNTT), and 2-(thienyl)anthra[2,3-*b*]thieno[2,3-*d*]thiophene (TATT)

Stille cross-coupling with 2-bromothiophene. This synthetic strategy demonstrates the versatility of organometallic chemistry for constructing complex heterocyclic systems, showcasing the selective functionalization of thienothiophene scaffolds and the utility of palladium-catalyzed coupling reactions in forming new carbon–carbon bonds.

8.5.3 Crystal Structures

The crystal structures of NTT, TNTT, ATT, and TATT were analyzed and compared, revealing that all four compounds exhibited herringbone structures primarily characterized by edge-to-face orientations (Fig. 8.15a–d). However, distinct features were observed in each compound's crystal structure, reflecting the influence of molecular design on intermolecular interactions and packing motifs.

NTT and TATT displayed disordered structures along their major and minor components and molecular long axes (Fig. 8.15e, h). The occupancies of major and minor components were found to be 0.71:0.29 for NTT and 0.56:0.44 for TATT, indicating a non-random distribution of molecular orientations. In contrast, TNTT showed a more ordered arrangement (Fig. 8.15g). The additional thiophene ring in TNTT and TATT was observed to be coplanar with the acene core, effectively extending the π-conjugation and promoting a close-packed herringbone structure. The packing arrangements were further classified into full-stack and slip-stack orientations. NTT adopted an antiparallel slip-stack orientation with disorder along the molecular short axis. TNTT exhibited a mixed packing motif, combining slip-stack

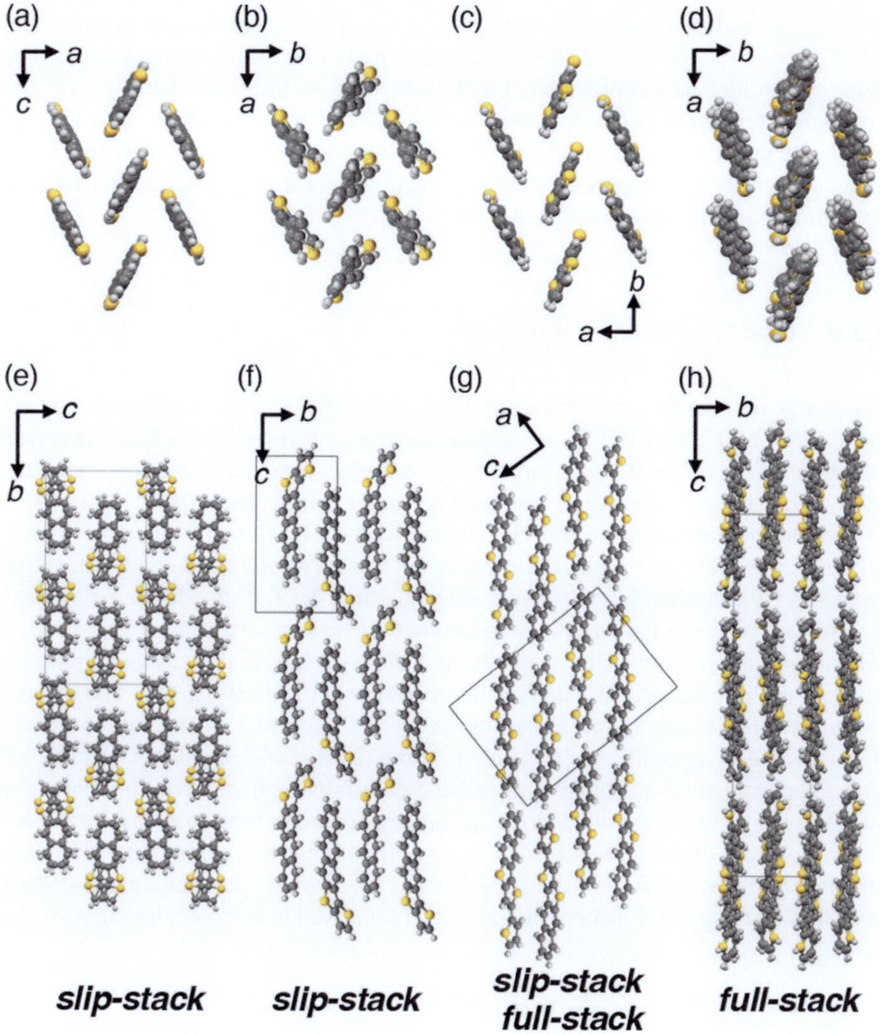

Fig. 8.15 Crystal structures of **a**, **e** naphtho[2,3-*b*]thieno[2,3-*d*]thiophene (NTT), **b**, **f** anthra[2,3-*b*]thieno[2,3-*d*]thiophene (ATT), and their corresponding thienyl derivatives, **c**, **g** TNTT and **d**, **h** TATT. Overviews with indications of **a–d** herringbone packing and **e–h** slip-stack and/or full-stack structures. Reproduced from Ref. [28] with permission. Copyright 2021 Wiley–VCH

and full-stack structures. This arrangement is likely attributable to the tendency of aromatic rings containing hydrocarbon acene moieties and heteroatoms such as thiophene to stack individually.

TATT demonstrated a more complex packing behavior, featuring a disordered structure composed of both parallel and antiparallel forms. Of interest is that the slip-stack property observed in ATT was resolved in TATT, giving rise to a full-stack structure (Fig. 8.15f, h). This transformation was attributable to enhanced dispersion forces resulting from the extended π-conjugated system. The high molecular symmetry of TATT, despite its asymmetric structure, contributed to the difficulty in distinguishing between parallel and antiparallel orientations. The well-defined herringbone and disordered full-stack structures observed in TATT were both ascribed to its high molecular symmetry. The positioning of the curvature at the center of the TATT molecule and the increased symmetry along the long axis played crucial roles in determining its unique packing behavior.

8.5.4 Charge Transfer Integrals

The charge transfer integrals and reorganization energies of three thienoacene derivatives (NTT, TNTT, and TATT) are summarized in Fig. 8.16 and Table 8.6. The crystal structures of the molecules were analyzed to calculate charge transfer integrals in the in-layer herringbone direction, assuming common lateral electrodes and an end-on orientation perpendicular to the FET substrate. TNTT exhibited moderate molecular orbital overlap and transfer integral despite its slip-stack structure. On the other hand, TNTT exhibited larger transfer integrals than NTT, both adopting slip-stack structures. The thienyl group appears to promote close-packed structures and enhance orbital overlap. The study shows that the molecular packing and electronic properties of TATT are significantly affected by incorporation of the thienyl group. The transfer integral of TATT was found to be approximately one order of magnitude greater than that of ATT, regardless of whether the configuration was parallel of antiparallel. This enhancement is attributable to the thienyl group's influence on the molecular structure, placing the bending point at the center of the molecule and increasing its symmetry. Consequently, TATT forms well-defined herringbone and full-stack structures. Moreover, the additional thiophene ring extends the π-conjugated system, contributing to stronger dispersion forces and improved intermolecular interactions.

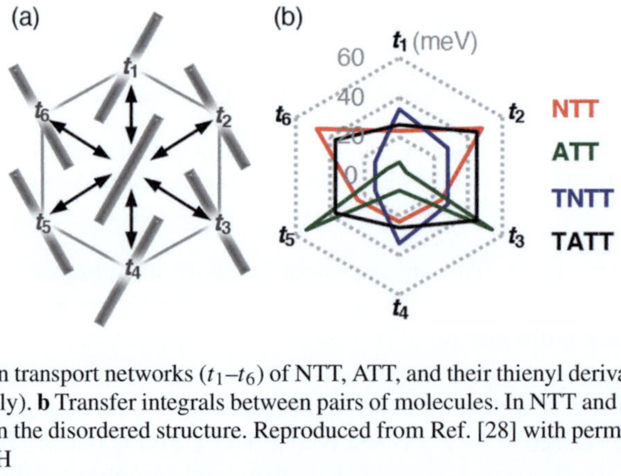

Fig. 8.16 **a** Main transport networks (t_1–t_6) of NTT, ATT, and their thienyl derivatives (TNTT and TATT, respectively). **b** Transfer integrals between pairs of molecules. In NTT and TATT, average of two geometries in the disordered structure. Reproduced from Ref. [28] with permission. Copyright 2021 Wiley–VCH

Table 8.6 Transfer integrals and reorganization energies of naphtho[2,3-*b*]thieno[2,3-*d*]thiophene (NTT), anthra[2,3-*b*]thieno[2,3-*d*]thiophene (ATT), and their thienyl derivatives (TNTT and TATT, respectively)

Compound	Transfer integrals (meV)[a]						Reorganization energy (meV)[c]
	t_1	t_2	t_3	t_4	t_5	t_6	
NTT[b]	23	48	24	23	24	48	139
ATT	7	4	54	7	54	7	110
TNTT	34	28	28	34	14	14	228
TATT[b]	26	45	45	26	37	37	158

[a]Calculated at the PW91/TZ2P level. [b]NTT and TATT have two herringbone structures, each centered on two independent molecules. Average of two geometries in the disordered structure. [c]Calculated at the B3LYP/6-31 + G(d,p) level

Reorganization energies were calculated to assess the structural relaxation associated with charge transfer. ATT and TATT exhibited smaller reorganization energies than did NTT and TNTT, which is likely attributable to the expansion of the conjugated system through larger fused rings. However, the incorporation of additional thiophene moieties into TNTT and TATT resulted in higher reorganization energies of 89 and 48 meV, respectively, than seen in their counterparts without the thiophene ring. This observation suggests that the flexible thiophene ring contributes to higher reorganization energies. Interestingly, the reorganization energy of NTT was found to be 79 meV lower than that of [1]benzothieno[3,2-*b*]benzothiophene (BTBT), a widely used core in OFET materials. This lower reorganization energy is favorable for charge transport, as it indicates reduced energy loss during the charge hopping process.

The study demonstrates that the introduction of thienyl groups can significantly enhance charge transfer integrals through improved molecular packing and extended π-conjugation. However, this modification also leads to increased reorganization energies due to the flexibility of the thiophene ring. These results highlight the delicate balance between molecular design, crystal packing, and electronic properties in organic semiconductors, offering guidance for the rational design of high-performance OFET materials.

8.6 Conclusions and Scope

Our comprehensive study as described in this chapter has demonstrated the effectiveness of various structural design strategies for controlling the molecular orientation and enhancing the performance of thienoacene-based organic semiconductors. The introduction of bent shapes in molecules like NBBT and bent-DNTT proved to simultaneously improve solubility and achieve higher-order orientation in crystalline states without the need for additional functional groups. The development of a simple and efficient synthetic method for ATT, which allows selective chemical modification at the terminal thiophenes, will also assist further developments in asymmetric thienoacenes. Although there is still room for improvement in the formation of oriented thin films on substrates in order to utilize the unsubstituted framework itself as a material, the introduction of alkylthienyl groups to ATT not only improved solubility, but also facilitated the formation of ordered molecular orientations conducive to charge transport. In fact, the additional thiophene unit in CnTATT contributed significantly to the observed high-performance FET characteristics. The incorporation of iodine, as seen in I-ATT, proved to be a novel and effective method of controlling molecular packing. This approach transformed the molecular arrangement from an antiparallel slip-stack to a head-to-head layered-herringbone structure, resulting in larger transfer integrals and improved hole mobility. This achievement highlights the potential of halogen incorporation for fine-tuning the electronic properties of organic semiconductors. Furthermore, our study of asymmetric naphtho- and anthra-thienothiophene derivatives and their thienyl-extended counterparts reveals the importance of molecular symmetry and dispersion forces in determining crystal packing.

While this study has primarily focused on the contributions of molecular structural asymmetry to solubility and molecular orientation, it is important to note that such asymmetry also influences molecular orbitals. Our current understanding of molecular packing and solubilities opens a path to future studies of electronic structure, orbital interactions, and charge transport mechanisms. This broader perspective is likely to reveal new strategies for exploiting asymmetry to significantly improve the performance and versatility of organic electronic devices.

Acknowledgements This research was partially supported by Grant-in-Aid for Young Scientist (B) from the Ministry of Education, Culture, Sports, Science and Technology (MEXT) (25870076), Adaptable and Seamless Technology transfer Program through Target-driven R&D (A-STEP) from Japan Science and Technology Agency (JST) (AS232Z00230D), and Core Research for Evolutional Science and Technology (CREST) from JST (JPMJCR18J2).

References

1. H. Klauk, *Organic Electronics: Materials, Manufacturing, and Applications* (John Wiley & Sons, 2006)
2. J.L. Bredas, J.P. Calbert, D.A. da Silva Filho, J. Cornil, Organic semiconductors: a theoretical characterization of the basic parameters governing charge transport. Proc. Natl. Acad. Sci. U.S.A. **99**, 5804–5809 (2002)
3. J. Cornil, D. Beljonne, J.P. Calbert, J.L. Brédas, Interchain interactions in organic π-conjugated materials: Impact on electronic structure, optical response, and charge transport. Adv. Mater. **13**, 1053–1067 (2001)
4. V. Coropceanu, J. Cornil, D.A. da Silva Filho, Y. Olivier, R. Silbey, J.L. Bredas, Charge transport in organic semiconductors. Chem. Rev. **107**, 926–952 (2007)
5. J.E. Anthony, Functionalized acenes and heteroacenes for organic electronics. Chem. Rev. **106**, 5028–5048 (2006)

6. T. Yamamoto, K. Takimiya, Facile synthesis of highly pi-extended heteroarenes, dinaphtho[2,3-b:2',3'-f]chalcogenopheno[3,2-b]chalcogenophenes, and their application to field-effect transistors. J. Am. Chem. Soc. **129**, 2224–2225 (2007)
7. K. Takimiya, S. Shinamura, I. Osaka, E. Miyazaki, Thienoacene-based organic semiconductors. Adv. Mater. **23**, 4347–4370 (2011)
8. A. Gavezzotti, G.R. Desiraju, A systematic analysis of packing energies and other packing parameters for fused-ring aromatic-hydrocarbons. Acta Crystallogr. Sect. B: Struct. Sci. **44**, 427–434 (1988)
9. G.R. Desiraju, A. Gavezzotti, Crystal-structures of polynuclear aromatic-hydrocarbons - classification, rationalization and prediction from molecular-structure. Acta Crystallogr. Sect. B: Struct. Sci. **45**, 473–482 (1989)
10. M.D. Curtis, J. Cao, J.W. Kampf, Solid-state packing of conjugated oligomers: from p-stacks to the herringbone structure. J. Am. Chem. Soc. **126**, 4318–4328 (2004)
11. C. Wang, H. Dong, W. Hu, Y. Liu, D. Zhu, Semiconducting p-conjugated systems in field-effect transistors: a material odyssey of organic electronics. Chem. Rev. **112**, 2208–2267 (2012)
12. S. Allard, M. Forster, B. Souharce, H. Thiem, U. Scherf, Organic semiconductors for solution-processable field-effect transistors (OFETs). Angew. Chem. Int. Ed. Engl. **47**, 4070–4098 (2008)
13. M.J. Kang, I. Doi, H. Mori, E. Miyazaki, K. Takimiya, M. Ikeda, H. Kuwabara, Alkylated dinaphtho[2,3-b:2′,3′-f]thieno[3,2-b]thiophenes (C(n)-DNTTs): organic semiconductors for high-performance thin-film transistors. Adv. Mater. **23**, 1222–1225 (2011)
14. J.E. Anthony, J.S. Brooks, D.L. Eaton, S.R. Parkin, Functionalized pentacene: improved electronic properties from control of solid-state order. J. Am. Chem. Soc. **123**, 9482–9483 (2001)
15. C.D. Sheraw, T.N. Jackson, D.L. Eaton, J.E. Anthony, Functionalized pentacene active layer organic thin-film transistors. Adv. Mater. **15**, 2009–2011 (2003)
16. H. Iino, T. Usui, J. Hanna, Liquid crystals for organic thin-film transistors. Nat. Commun. **6**, 6828 (2015)
17. S. Inoue, S. Shinamura, Y. Sadamitsu, S. Arai, S. Horiuchi, M. Yoneya, K. Takimiya, T. Hasegawa, Extended and modulated thienothiophenes for thermally durable and solution-processable organic semiconductors. Chem. Mater. **30**, 5050–5060 (2018)
18. S. Inoue, T. Higashino, S. Arai, R. Kumai, H. Matsui, S. Tsuzuki, S. Horiuchi, T. Hasegawa, Regioisomeric control of layered crystallinity in solution-processable organic semiconductors. Chem. Sci. **11**, 12493–12505 (2020)
19. T. Higashino, S. Inoue, S. Arai, S. Tsuzuki, H. Matsui, R. Kumai, K. Takaba, S. Maki-Yonekura, H. Kurokawa, I. Inoue, K. Tono, K. Yonekura, T. Hasegawa, Effects of thiophene-fused isomer on high-layered crystallinity in π-extended and alkylated organic semiconductors. Chem. Mater. **36**, 848–859 (2024)
20. T. Okamoto, C. Mitsui, M. Yamagishi, K. Nakahara, J. Soeda, Y. Hirose, K. Miwa, H. Sato, A. Yamano, T. Matsushita, T. Uemura, J. Takeya, V-shaped organic semiconductors with solution processability, high mobility, and high thermal durability. Adv. Mater. **25**, 6392–6397 (2013)
21. K. Yamamoto, H. Katagiri, H. Tairabune, Y. Yamaguchi, Y.-J. Pu, K.-I. Nakayama, Y. Ohba, Synthesis and properties of naphthobisbenzo[*b*]thiophenes: structural curvature of higher acene frameworks for solubility enhancement and high-order orientation in crystalline states. Tetrahedron Lett. **53**, 1786–1789 (2012)
22. Y. Ogawa, E. Takiguchi, M. Mamada, D. Kumaki, S. Tokito, H. Katagiri, Synthesis, crystal structure, and FET characteristics of thieno[2,3-b] thiophene-based bent-thienoacenes. Tetrahedron Lett. **58**, 963–967 (2017)
23. M.G. Reinecke, J.G. Newsom, K.A. Almqvist, An improved synthesis of thiophene-2,3-dicarboxylic acid by sequential carboxylation. Synthesis **1980**, 327–329 (1980)
24. M.G. Reinecke, J.G. Newsom, L.-J. Chen, Thermolysis of thiophenedicarboxylic acid anhydrides as a route to five-membered hetarynes. J. Am. Chem. Soc. **103**, 2760–2769 (1981)
25. M.L. Tedjamulia, H. Kudo, R.N. Castle, M.L. Lee, Angular polycyclic thiophenes containing two thiophene rings. Part II. J. Heterocycl. Chem. **21**, 321–325 (1984)

26. Y. Ogawa, K. Yamamoto, C. Miura, S. Tamura, M. Saito, M. Mamada, D. Kumaki, S. Tokito, H. Katagiri, Asymmetric alkylthienyl thienoacenes derived from anthra[2,3-*b*]thieno[2,3-*d*]thiophene for solution-processable organic semiconductors. ACS Appl. Mater. Interfaces **9**, 9902–9909 (2017)
27. A. Matsunaga, Y. Ogawa, D. Kumaki, S. Tokito, H. Katagiri, Control of molecular orientation in organic semiconductors using weak iodine–iodine interactions. J. Phys. Chem. Lett. **12**, 111–116 (2021)
28. A. Matsunaga, Y. Ogawa, S. Tamura, K. Yamamoto, H. Katagiri, Molecular structure–property relationships of the asymmetric thienoacenes: naphtho[2,3-*b*]thieno[2,3-*d*]thiophene, anthra[2,3-*b*]thieno[2,3-*d*]thiophene, and their thienyl derivatives. ChemistrySelect **6**, 4506–4510 (2021)

Chapter 9
Polarization-Induced Phenomena in Ferroelectric Liquid Crystals Comprising of Extended π-Conjugated Units

Masahiro Funahashi

Abstract Liquid crystals comprising extended π-conjugated units work as semiconductors. On the other hand, ferroelectricity in liquid crystal systems is induced by molecular chirality in the classical concept. Moreover, ferroelectric columnar phases based on one-dimensional hydrogen-bonding chains and stacking of umbrella-shaped molecules have recently been studied to provide new insights into polarization in liquid crystal media. By incorporating electroactive π-conjugated units into the ferroelectric liquid crystal molecules, ferroelectricity can be coupled with electronic charge carrier transport in the one- or two-dimensional assemblies of π-conjugated units to create new functional materials and devices. In this chapter, π-conjugated ferroelectric liquid crystals with layer and columnar ordered structures are focused from a viewpoint of new phenomena based on electronic carrier transport coupled by spontaneous polarization. This coupling provides new electronic functions such as bulk photovoltaic effect, switched electronic transport, and polarization-induced charge carrier injection leading to polarized electroluminescence.

Keywords Liquid crystalline semiconductor · Ferroelectric liquid crystal · Extended π-conjugated unit · Bulk photovoltaic effect · Polarization-induced electroluminescence · Polarization-induced switching

---

M. Funahashi (✉)
Department of Chemical Science and Engineering, Graduate School of Engineering, Kobe University, Kobe, Hyogo, Japan
e-mail: funahashi.masahiro@phoenix.kobe-u.ac.jp

Health and Medical Research Institute, National Institute of Advanced Industrial Science and Technology, Takamatsu, Kagawa, Japan

Research Center for Membrane and Film Technology, Kobe University, Kobe, Hyogo, Japan

S. Kobatake and H. Uekusa (eds.), *Advances in Organic Crystal Chemistry*,
https://doi.org/10.1007/978-981-96-4479-7_9

9.1 Carrier Transport in Liquid Crystalline Semiconductors

Electronic charge carrier transport in organic semiconductors has been studied extensively since 1960s. In crystals, amorphous solids, and polymers comprising extended π-conjugated units, efficient electronic charge carrier transport has been studied by time-of-flight (TOF), microwave conductivity, space charge limited current, and field-effect mobility measurements [1]. Applications of organic semiconductors exhibiting high carrier mobilities to light-emitting diodes, field-effect transistors, and solar cells have been reported extensively [2].

Even liquid crystals comprising extended π-conjugated units work as organic semiconductors [3]. In columnar and ordered smectic phases, π-conjugated chromophores are self-assembled into columnar or layer orders in which transfer integrals between the chromophores are maximized, resulting in efficient electronic charge carrier transport. Liquid crystalline (LC) semiconductors are characterized by the solution-processabilities and anisotropic carrier transport properties. Polarized light-emitting diodes (LEDs), field-effect transistors (FETs), and solar cells have been studied so far [4]. In particular, LEDs based on LC semiconductors can emit linearly polarized light when π-conjugated chromophores are aligned uniaxially in the light-emission layers [5]. In solar cells using LC semiconductors, heterojunction between electron-donating and withdrawing components is indispensable to enhance power conversion efficiency [6].

Carrier transport properties of liquid crystalline semiconductors have been studied by the TOF method because this technique provides drift mobility including information on the energetic disorder, traps, and impurities [7]. Liquid crystalline semiconductors which exhibit wide mesomorphic temperature ranges are required for the extraction of disorder parameters and there have not been so many studies on carrier transport mechanism in the liquid crystal phases. A few liquid crystals of which carrier transport mechanisms were studied are indicated in Fig. 9.1.

In the ordered columnar phases of triphenylene dimers **1** and perylene bisimide derivatives **2** bearing oligosiloxane chains, electronic charge carrier transport mechanisms were analyzed by the one-dimensional Gaussian disorder model [8, 9]. The studies indicated that the electronic charge carrier transport was described as a charge carrier hopping process in disordered systems, unlike organic molecular crystals. The distribution width of energy levels of π-molecular orbitals was around 50 meV, which is around half as those in organic amorphous semiconductors. The hole and electron mobilities are typically on the order of 10^{-2} to 10^{-1} cm^2 V^{-1} s^{-1}. A few columnar liquid crystals exhibit high carrier mobility exceeding 1 cm^2 V^{-1} s^{-1} [10].

In the ordered smectic phases of alkynylterthiophene **3**, carrier transport mechanism is also electronic carrier hopping process in disordered systems, which is described by the two-dimensional Gaussian disorder model [11]. The distribution width of the energy levels is around 50 meV. The carrier mobilities are typically on the order of 10^{-2} to 10^{-1} cm^2 V^{-1} s^{-1}. Phenylterthiophene derivative **4** exhibits temperature-independent hole mobility around 0.1 cm^2 V^{-1} s^{-1} between 200 °C

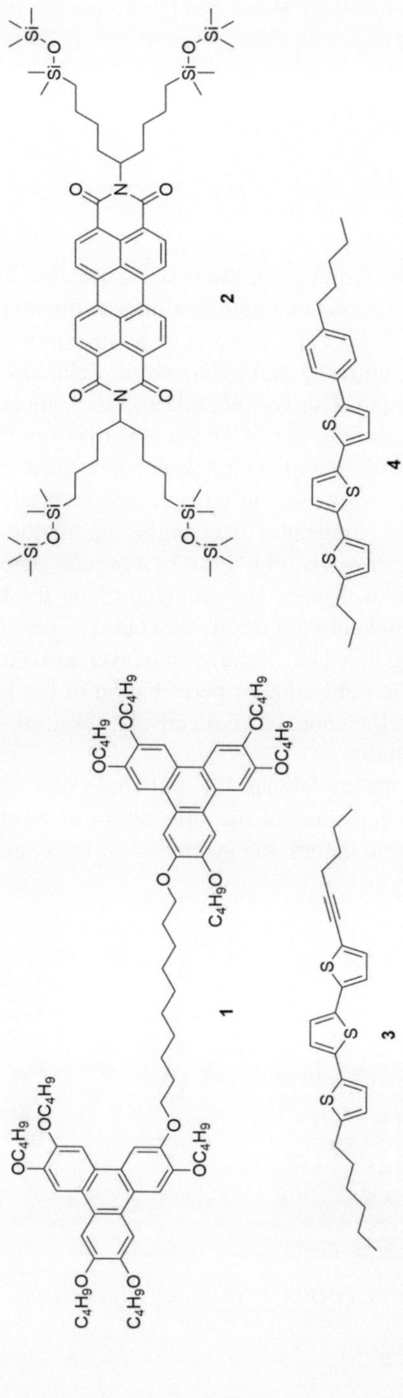

Fig. 9.1 Molecular structures of liquid crystals of which carrier transport mechanisms were studied by the TOF technique

and −100 °C. In this case, the hole transport is regarded as a band-like transport influenced from trapping in localized states and phonon scattering [12].

9.2 Ferroelectric Liquid Crystals

9.2.1 Chiral Systems

Figure 9.2 shows a molecular structure of the first ferroelectric liquid crystal (FLC) and an illustration of the supramolecular aggregation structure of the FLC phase. The first FLC **5** was reported by Meyer and coworkers, as shown in Fig. 9.2a [13]. They paid attention to molecular chirality and synthesized a cinnamoyl ester derivative bearing a chiral alkyl side chain. This compound exhibits a smectic phase which has a layer structure. Within the layers, the LC molecules tilt from the layer normal. In a smectic C (SmC) phase consisting of achiral molecules, the direction of molecular tilt is the same for all layers. However, in a ferroelectric chiral smectic C (SmC*) phase, the director of the LC molecules is changing continuously along the layer normal, drawing a helix, as shown in Fig. 9.2b. In a ground state, total polarization is canceled because molecular dipoles are averaged along the helical axis. Under a DC bias application, the helical structure is unwound to generate a macroscopic polarization, as shown in Fig. 9.2c [14]. Polarization inversion is induced by a change in the polarity of the electric field without perturbation of the layer structure. The LC molecules rotate around the cone axis collectively, resulting in the switching of the optical axis of the LC phase.

The coupling between the molecular tilt and molecular chirality causes the symmetry break to induce deviation of the orientation of electrical dipoles [15]. From a viewpoint of the group theory, the symmetry of the system is reduced from

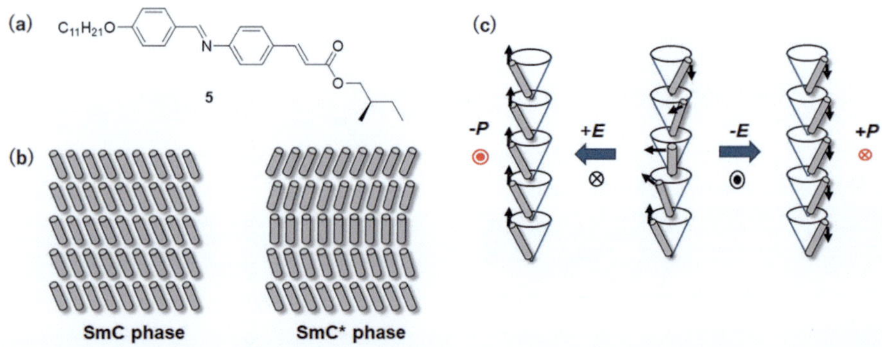

Fig. 9.2 **a** Molecular structure of the first FLC **5**. **b** Schematics for molecular aggregation structures in the SmC and SmC* phase. **c** Illustration for polarization switching in the SmC* phase with arrows indicating molecular dipole moments

C_{2h} in the SmC phase to C_2 in the SmC* phase by introduction of chirality into the system. Based on this concept, many FLCs have been synthesized because this kind of FLCs could be applied to high-speed displays.

Basically, the FLCs exhibiting the SmC* phase consist of a rigid core, polar functional groups, and chiral side chains. A spontaneous polarization generally increases when a polar functional group is located near the chiral center and the core moiety.

9.2.2 *Achiral Systems*

9.2.2.1 Bent-Core Type FLCs

Macroscopically polarized systems can also be constructed by bent-core type molecules with a strong dipole moment perpendicular to the molecular axis [16]. Figure 9.3 shows molecular structures of typical bent-core ferroelectric liquid crystals **6** and **7** and the supramolecular aggregation structure of the FLC phase. Bent-core molecules are self-organized into layer structures to form smectic phases. In the case of rod-like molecules with a dipole perpendicular to the molecular axis, the dipole moments are canceled by molecular rotation, degrading the symmetry of the system to C_∞. Due to steric interaction between the bent-core molecules besides dipole–dipole interaction, macroscopic polarization along the symmetry axis is induced in the smectic phases consisting of the polar bent-core molecules. Polarization inversion occurs by rotation of the bent-core molecules around the molecular axis without destruction of the layer structures. The polar smectic phases are named B1 ~ B8 phases.

Most of bent-core FLCs are characterized as an oligoester structure including a resorcinol unit which has a *m*-substituted structure. The plural ester carbonyl groups induce strong polarizations along the symmetry axis.

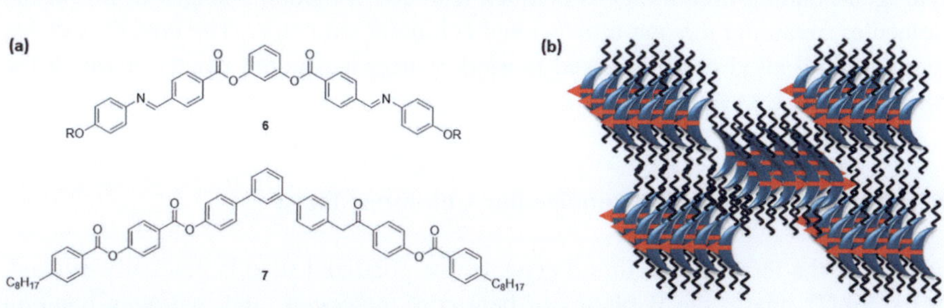

Fig. 9.3 **a** Molecular structures of typical bent-core FLCs **6** and **7**. **b** Illustration for supramolecular aggregation structure in a ferroelectric B1 phase. Red arrows indicate molecular diopole moments

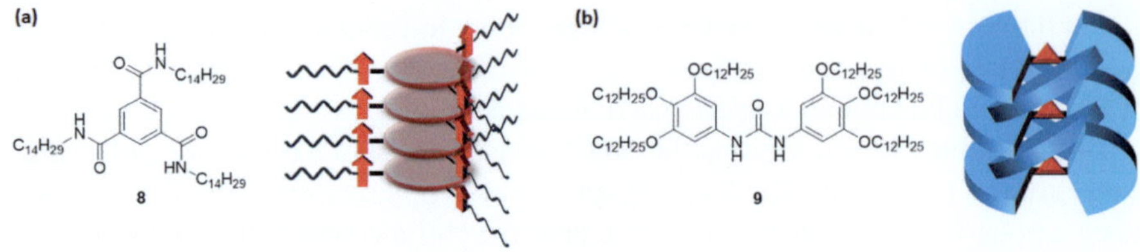

Fig. 9.4 Molecular structures and schematics of their supramolecular aggregation structures for hydrogen-bonding ferroelectric columnar liquid crystals consisting of **a** benzenetricarboxylic amide derivative **8** with red arrows indicating molecular dipole moments and **b** urea derivative **9**

9.2.2.2 Ferroelectric Hydrogen-Bonding Columnar Phase

The symmetry of columnar phases consisting of disk-like molecules is C_∞, indicating no macroscopic polarization. However, disk-like molecules bearing a strong dipole parallel to the disk normal can form a polarized columnar aggregate. In particular, polarization in columnar aggregates can be reinforced by hydrogen-bonding interaction. The examples of hydrogen-bonding ferroelectric columnar phases are indicated in Fig. 9.4.

Matsunaga and coworkers reported a columnar phase of disk-like molecule **8** bearing amide moieties, exhibiting wide mesomorphic temperature ranges. In the columnar phases, π–π stacking interaction as well as hydrogen bonding between the amide groups stabilized the columnar assemblies of the amide molecules (Fig. 9.4a) [17]. Akutagawa and coworkers confirmed ferroelectricity in a few materials of the amide columnar liquid crystals later [18]. Polarization inversion occurred through rotation of amide moieties, retaining the columnar structures.

Kishikawa and coworkers synthesized urea derivative **9** bearing two fan-like hydrophobic units, which formed one-dimensional hydrogen-bonding chains (Fig. 9.4b) [19]. During the polarization inversion, the collective recombination of hydrogen-bonding urea moieties couples with conformation changes of the whole molecules, retaining the one-dimensional columnar structures. The urea derivatives bearing chiral alkyl chains formed twisted hydrogen-bonding chains in which the polarized state was stabilized [20].

9.2.2.3 Ferroelectric Supramolecular Columnar Phase

In most of the ferroelectric liquid crystals, the polarized state is gradually relaxed. In the SmC* phase, the B phases of bent-core molecules, and hydrogen-bonding columnar phases, polarization inversion can proceed without perturbing layer or columnar structures. On the other hand, the polarized states can be relaxed to retain the layer or columnar structures.

In contrast, the columnar phases consisting of bowl-shaped molecules can be polarized because of asymmetric molecular shapes. However, polarization inversion should require rearrangement of the columnar aggregates.

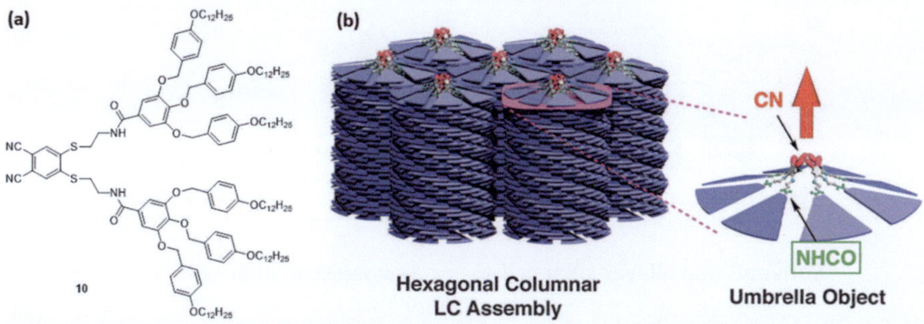

Fig. 9.5 **a** Molecular structure and **b** its schematic of self-assembling structure for ferroelectric supramolecular columnar liquid crystal **10**. Red arrows indicate electrical polarization of the supramolecular aggregates. Reprinted with the permission from Ref. [21]. Copyright 2011 AAAS

Miyajima and coworkers synthesized fan-shaped molecule **10** consisting of six alkyl chains, two amide moieties and two cyano groups (Fig. 9.5a) [21]. Umbrella-like aggregates consisting of four compound **10** are assembled into a columnar aggregate. The four fan-shaped units bind weakly through hydrogen-bonding interaction and the umbrella-like aggregates can be inverted inducing macroscopic polarization inversion, as shown in Fig. 9.5b.

9.3 Ferroelectric Liquid Crystals Comprising Extended π-Conjugated Units

9.3.1 Bulk Photovoltaic Effect

The conventional photovoltaic effects are based on p-n or Schottky junctions. The driving force is a local electric field formed at the interface between p-type and n-type semiconductor layers or between a semiconductor layer and a metallic electrode, as shown in Fig. 9.6a. Therefore, open-circuit voltage is limited by the bandgaps of semiconductors used for active layers of the photovoltaic devices [1]. There is a theoretical limit for the energy conversion efficiency (Shockley–Queisser limit) around 30% because a decrease in the bandgap results in a decrease in the open-circuit voltage as well as an increase in the spectral range which can be utilized for the energy conversion [22].

In contrast, the bulk photovoltaic effect has been observed in ferroelectrics [23]. The driving force is an electric field produced by spontaneous polarization of ferroelectrics, as shown in Fig. 9.6b. Because of a strong polarization field in a ferroelectric material, a higher open-circuit voltage can be realized in the bulk photovoltaic effect. The carrier generation proceeds in the bulk of the ferroelectrics and excitons formed whole in the bulk can be utilized for energy production. Theoretically, the open-circuit voltage is proportional to the thickness of the active layer and fabrication of

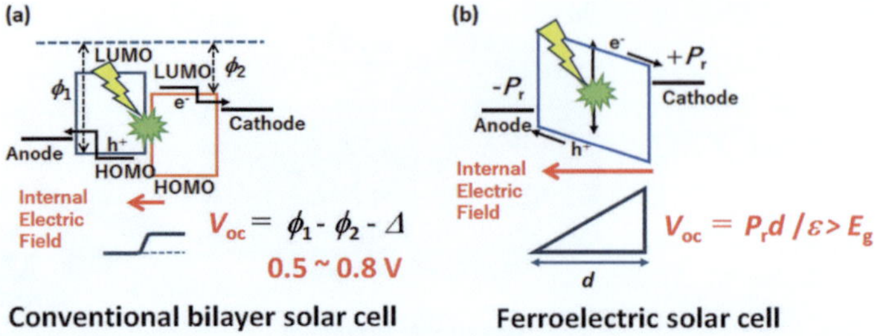

Fig. 9.6 Illustration of the band diagrams for **a** a conventional pn-junction-based photovoltaic effect and **b** a bulk photovoltaic effect in ferroelectrics. Reproduced from Ref. [39] by permission of The Royal Society of Chemistry

thin films is not required. Anodes with low work functions, such as Ca and Mg, are not required. The polarity of the cell can be inverted.

The bulk photovoltaic effect was first observed in $LiNiO_3$ and $BaTiO_3$ which have non-centrosymmetric crystal structures [24, 25]. These materials could generate higher open-circuit voltages than their bandgaps although the photocurrents were very low because of the high band gaps and low conductivities. Bismuth ferrite exhibits spectral sensitivity for visible light illumination at room temperature because of the low band gap [26]. However, power conversion efficiency has still been low although a few two-dimensional ferroelectrics such as $CuInP_2S_6$ induce large photocurrent [27].

9.3.1.1 Bulk Photovoltaic Effect in Organic Ferroelectrics

There have been very limited studies on the bulk photovoltaic effect in organic ferroelectrics. In 1981, Sasabe and coworkers studied the photovoltaic effect of dye-doped poly(vinylidene fluoride) film, which was known as a typical ferroelectric polymer (Fig. 9.7a) [28]. High open-circuit voltages on the order of 10^4 V were generated while very weak photocurrent response resulted in a low power conversion efficiency around 0.25% for white light illumination. In 2004, Tasaka and coworkers studied the bulk photovoltaic effect in crystalline thin films of triphenylene hexaester (Fig. 9.7a) [29]. In this material, polarization inversion was induced by rotation of ester carbonyl groups connected to an extended π-conjugated triphenylene unit. The triphenylene unit absorbed UV light and form paths for electronic charge carrier transport. The polarization field should be on the order of 1 MV cm^{-2}. However, low photocurrent for UV light illumination resulted in a low power conversion efficiency around 0.5%.

In 2017, Nakamura and coworkers reported a shift current in ferroelectric crystals of a charge transfer complex between tetrathiafulvalene and chloranil (Fig. 9.7c) [30]. The charge transfer complex formed a non-centrosymmetric crystal with a spontaneous polarization of 5 mC cm^{-2} below 80 K. For visible light illumination,

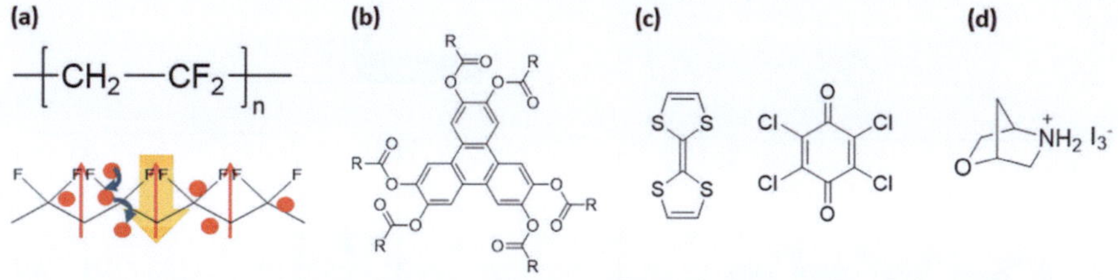

Fig. 9.7 Examples of the bulk photovoltaic effect in organic materials. **a** Dye-doped poly (vinylidene fluoride), **b** triphenylene hexaester, and **c** charge transfer complex of tetrathiafulvalene, chloranil, and chiral ammonium iodide

unidirectional photocurrent was induced based on the asymmetric crystal structure. However, the ferroelectric phase was limited only below 80 K and the photovoltaic effect disappeared at room temperature.

The bulk photovoltaic effect was recently reported in 2-oxa-5-azabicyclo [2.2.1]heptonium triodide which forms a crystal structure with a chiral symmetry (Fig. 9.7d). The power conversion efficiency (0.48%) with the open-circuit voltage of 0.58 V and the short-circuit current of 1.73 mA cm^{-2} for 100 mW cm^{-2} solar light illumination was obtained in a polarized state [31].

9.3.1.2 Bulk Photovoltaic Effect in FLCs

FLCs consisting of extended π-conjugated units and chiral alkyl chains have been synthesized and their application to photorefractive effect has been studied by Sasaki and coworkers [32]. However, carrier mobilities in the ferroelectric liquid crystal phase were not reported.

Funahashi and coworkers reported a photovoltaic effect in the SmC* phase of phenylterthiophene derivative **11** bearing a chiral alkyl chain in 2015 [33]. Compound **11** exhibited the SmC* phase, below which more ordered SmG* phase appeared (Fig. 9.8a). In the non-polarized SmC* phase, clear photocurrent response was not observed while a strong photocurrent response was obtained for a UV light illumination in a polarized SmC* phase formed by an application of a DC bias. The polarity of the photocurrent response was opposite to that of the poling bias and inversion of the poling bias polarity caused the reverse of the photocurrent polarity (Fig. 9.8b). These characteristics were attributed to the bulk photovoltaic effect driven by ferroelectric polarization of the liquid crystal [34, 35].

Seki and Funahashi synthesized (*R*)- and (*S*)-forms of the ferroelectric liquid crystal **11** and studied the dependence of the photocurrent response in the SmC* phase on the enantiomer purity [36]. The photocurrent response and spontaneous polarization in the SmC* phase decreased by decreasing the enantiomer purity and they disappeared in the racemic mixture (Fig. 9.8c). This result indicated the bulk photovoltaic effect should be attributed to the symmetry break by the molecular chirality but not to polarization of ionic impurity.

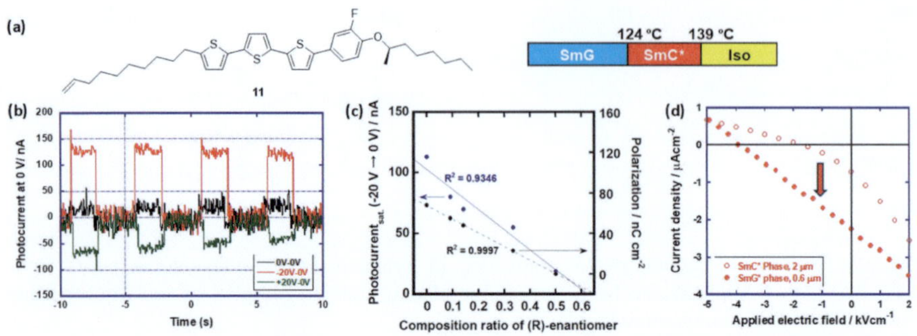

Fig. 9.8 **a** Molecular structure of π-conjugated FLC **11** with a phase transition diagram, **b** photocurrent response in the SmC* phase at 127 °C for UV light illumination (3 mW cm^{-2}), **c** photocurrent at 127 °C as a function of enantiomer purity, and **d** enhancement of the bulk photovoltaic effect for UV light (6 mW cm^{-2}) illumination in the ordered chiral smectic phase at 100 °C. Reproduced from Ref. [33, 36] by permission of The Royal Society of Chemistry

Seki and Funahashi also verified the bulk photovoltaic effect in the ordered chiral smectic phases appearing below the SmC* phase, as shown in Fig. 9.8d [37]. Polarization in the SmC* phase could be immobilized in the ordered chiral smectic phase by cooling the SmC* phase to the ordered smectic phase under an application of a DC bias. The carrier mobility increased in the chiral ordered smectic phase, resulting in the enhancement of the photovoltaic effect.

Construction of bulk heterojunctions in the ferroelectric liquid crystals is effective to enhance the photovoltaic effect [38]. A mixture of ferroelectric LC **11** and fullerene derivative **12** formed a segregated ferroelectric phase cooled from the SmC* phase. In the LC phase, fullerene-rich and LC-rich domains were formed, as shown in Fig. 9.9a. Electrons and holes were transported through the fullerene- and LC-rich domains, respectively. The hole and electron mobilities were on the order of 10^{-2} cm^2 V^{-1} s^{-1}. The spectral sensitivity was extended to the visible area. The polarity of the photovoltaic effect was reversible (Fig. 9.9b).

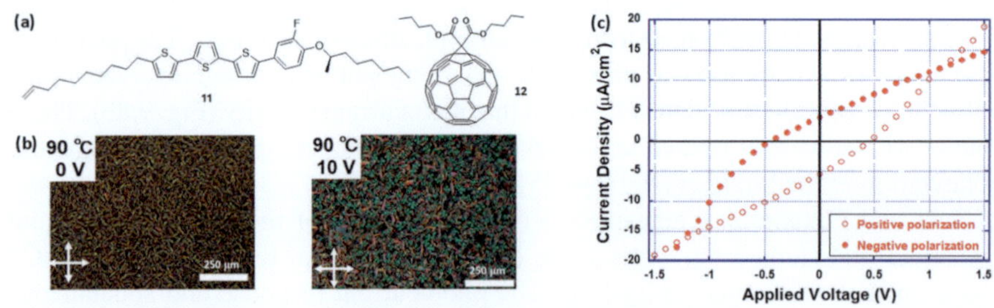

Fig. 9.9 **a** Molecular structure of FLC **11** and fullerene derivative **12**, **b** polarizing optical micrographs of a mixture of compounds **11** and **12**, and **c** current–voltage characteristics of a mixture of compounds **11** and **12**. Reproduced from Ref. [38] by permission of Elsevier

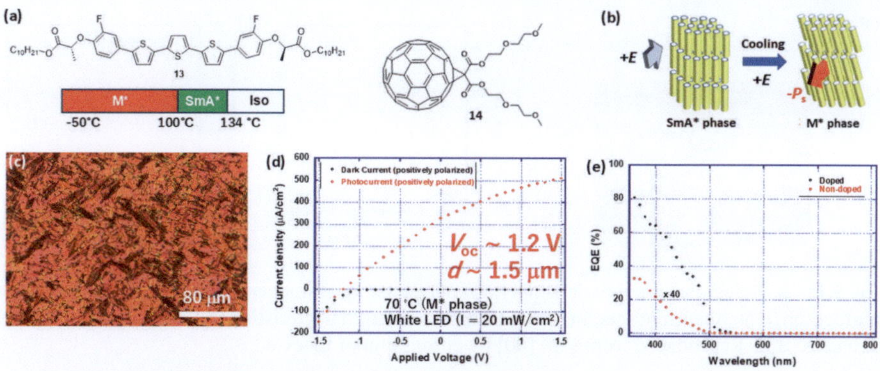

Fig. 9.10 **a** Molecular structures of FLC **13** and fullerene derivative **14**. **b** Molecular aggregation structure in the SmA* and M* phases. **c** Polarizing optical micrograph, **d** current–voltage characteristic, and **e** photocurrent spectra in the M* phase of a mixture of compounds **13** and **14**. Reproduced from Ref. [39] by permission of The Royal Society of Chemistry

The enhancement of the photovoltaic effect by the construction of bulk heterojunctions was more remarkable for the oligothiophene derivative **13** bearing lactic ester moieties (Fig. 9.10a) [39]. The oligothiophene derivatives exhibit the ordered ferroelectric smectic (M*) phase below the SmA phase. In the M* phase, the molecules tilted from the layer normal, as shown in Fig. 9.10b, to induce a large polarization by the orientation of polar lactic moieties. In the mixtures of the FLC **13** and fullerene derivative **14**, crystallites of compound **14** with sizes of several µm were produced to form bulk heterojunctions in the M* phase, as shown in Fig. 9.10c. The photovoltaic effect was enhanced in the M* phase, resulting in the open-circuit voltage exceeding 1.2 V and the quantum efficiency reaching 70% in the near UV and blue area (Fig. 9.10d, e).

Funahashi also synthesized lactate-introduced quinquethiophene derivative **15** which had a more extended π-conjugated unit (Fig. 9.11a) [40]. Compound **15** exhibited the SmC* phase and a more ordered chiral ferroelectric smectic phase. In a mixture of compound **15** and PCBM, crystallites of PCBM were segregated in the ordered smectic phase, as shown in Fig. 9.11b. The absorption edge was extended to 580 nm, resulting in an extension of the spectral sensitivity to 600 nm, as shown in Fig. 9.11c. The fill factor was also improved to 0.68 and the power conversion efficiency was increased to 0.75%.

The origin of the polarization in the ordered chiral smectic phases is the molecular tilt from the layer normal coupled with the molecular chirality of the LC molecules. Funahashi and coworkers synthesized diastereomers **16** and **17** which had the same π-conjugated core and lactate side chains (Fig. 9.12a). The LC structures and the photovoltaic effect in the ordered smectic phases were compared [41]. As shown in Fig. 9.12b, diastereomer **16** exhibited an ordered smectic phase in which the π-conjugated chromophores tilted in 45 degrees from the layer normal while the chromophores were parallel to the layer normal in the high-temperature phase. In

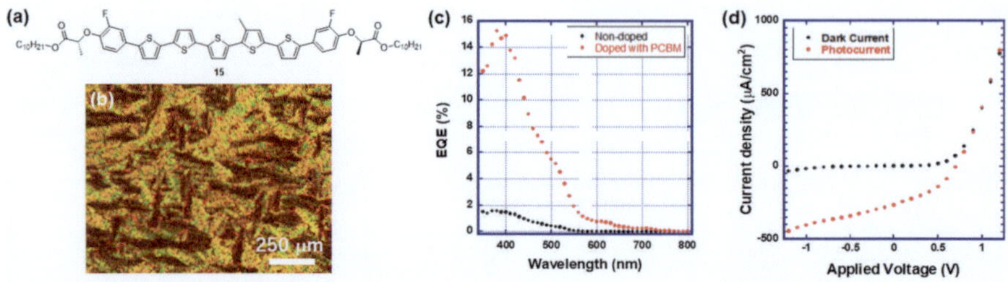

Fig. 9.11 **a** Molecular structure of lactate-introduced quinquethiophene derivative **15**. **b** Polarizing micrograph, **c** photocurrent spectra, and **d** current-voltage characteristics of a mixture of compound **15** and PCBM. Reproduced from Ref. [40] by permission of Elsevier

contrast, the π-conjugated chromophores of diastereomer **17** were parallel to the layer normal in the ordered smectic phase as in the high-temperature phase. As shown in Fig. 9.12c, a photovoltaic effect was observed in the ordered smectic phase of diastereomer **16**. However, an ambiguous photocurrent response based on a temporal polarization was observed in diastereomer **17** (Fig. 9.12d). An atomic force microscopic observation of a spin-coated film of diastereomer **16** revealed a characteristic surface morphology where flat domains on a molecular level and bulging boundaries, as shown in Fig. 9.12e.

It should be noted that the current–voltage characteristics in the polarized smectic phases of the FLCs bearing lactic ester moieties were diode-type in spite of the symmetrical structure of the cells. These results suggested the formation of inhomogeneous electric fields by charge carrier injection.

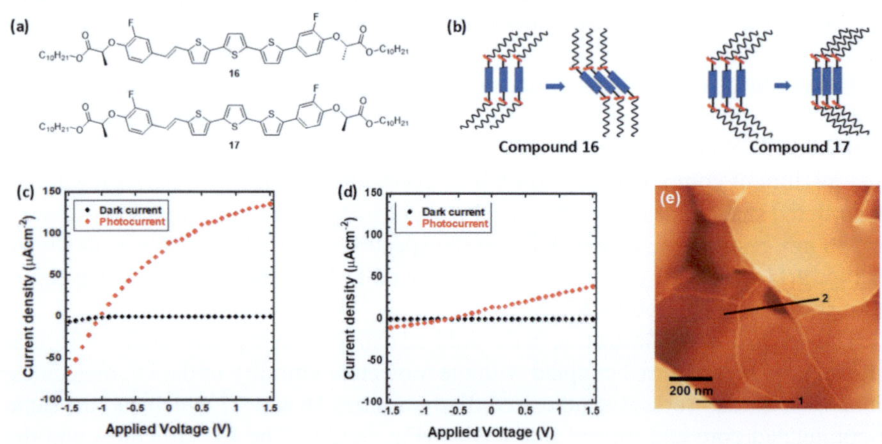

Fig. 9.12 **a** Molecular structures of diastereomer **16** and **17**. **b** Schematics for the changes of the supramolecular aggregations from the high-temperature to the low-temperature phase of compounds **16** and **17**. Current voltage characteristics in the low temperature phases of **c** compound **16** and **d** **17** doped with fullerene derivative **14**. **e** AFM height image of a spin-coated film of compound **16**. Reproduced from Ref. [41] by permission of The Chemical Society of Japan

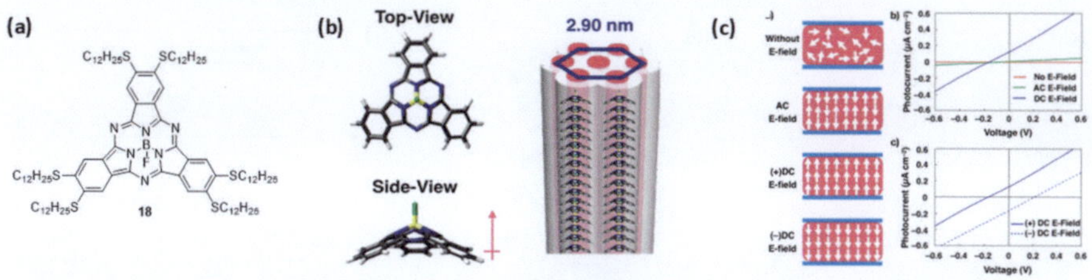

Fig. 9.13 **a** Molecular structure of liquid crystalline subphthalocyanine **18**, **b** illustration for polar columnar aggregates of compound **18**, and **c** current–voltage characteristics of the polarized columnar phase of compound **18** under a solar simulator (0.1 W cm^{-2}). Reprinted with permission from Ref. [42]. Copyright 2011 American Chemical Society

Miyajima and Araoka synthesized ferroelectric columnar liquid crystal **18** based on a subphthalocyanine core which was non-planar and had a strong dipole perpendicular to the π-conjugated unit (Fig. 9.13a) [42]. The non-planar π-conjugated unit stacked to form polar columnar aggregates, as shown in Fig. 9.13b. Polarization was induced in the columnar phase under the DC bias application while no polarization was induced without a DC bias. The open-circuit voltage was 0.2 V for white light illumination. The current–voltage characteristic was linear and shifted upward or downward depending upon the polarization, as shown in Fig. 9.13c. The subphthalocyanine core has an absorption band around 550 nm and the spectral sensitivity red-shifted to 650 nm.

Lehmann and Eremin also synthesized subphthalocyanine derivatives bearing three shape-persistent arms (Fig. 9.14a). Compound **19** formed polar dimers which assembled into polar columnar aggregates, as shown in Fig. 9.14b [43]. Three arms worked as antenna harvesting visible light, resulting in the extension of absorption edge to 650 nm. In the columnar phase, photocurrent response for white light illumination was observed based on the bulk photovoltaic effect. The open-circuit voltage was 0.7 V. As shown in Fig. 9.14c, the current–voltage characteristics were linear in the polar columnar phases of the subphthalocyanine derivatives, unlike the diode character in the ordered chiral smectic phases of the oligothiophene derivatives bearing lactic ester moieties. The power conversion efficiencies were still low and less than 0.1% [44].

Yoshio and coworkers reported the bulk photovoltaic effect in a polar columnar phase of polycatenary liquid crystal **20** based on a bis(oxadiazolylphenyl)bithiophene core (Fig. 9.15a) [45]. Figure 9.15b shows an electron density map determined by an X-ray diffraction pattern for the columnar phase of compound **20**. Compound **20** formed a dimeric columnar structure at room temperature. The open-circuit voltage was 1.0 V and the short-circuit current was 0.35 μA cm^{-2} for AM1.5 illumination. It is noted that the current–voltage characteristic was diode-type in the polarized columnar phase, as shown in Fig. 9.15c.

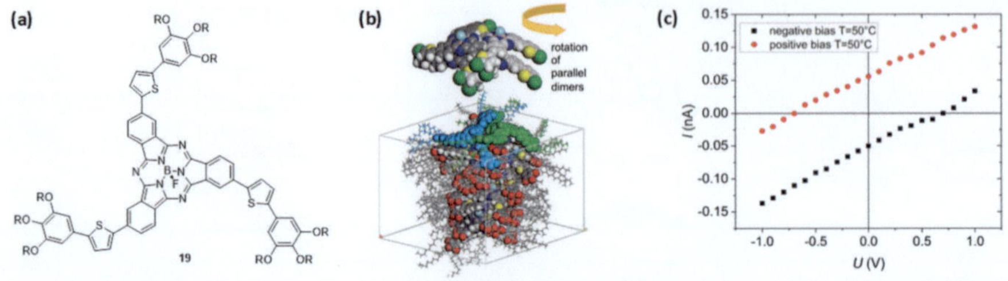

Fig. 9.14 **a** Molecular structure of subphthalocyanine derivative **19**, **b** illustration for supramolecular self-assembling structure of compound **19**, and **c** current–voltage characteristics in the polarized columnar phase of compound **19** for tungsten lamp illumination (5 mW cm^{-2}). Reproduced from Ref. [43, 44] by permission of Wiley–VCH

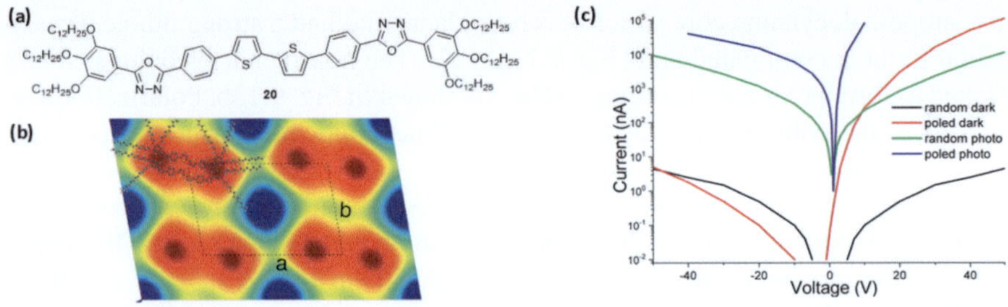

Fig. 9.15 **a** Molecular structure of polycatenary liquid crystal **20**, **b** the electron density map calculated from the X-ray diffraction pattern for the columnar phase of compound **20**, and **c** current–voltage curve in the columnar phase of compound **20** for AM 1.5 illumination. Reproduced from Ref. [45] by permission of The Royal Society of Chemistry

9.3.2 Polarization-Induced Electroluminescence

The electrical polarization of the FLCs influences not only the electric field of the bulk but also the interfacial electronic structures. In a sandwich type device, there are hole and electron injection barriers between an ITO electrode and the FLCs based on terthiophene or quinquethiophene. Under a DC bias application, the current is injection-limited, resulting in no light emission.

In the polarized ferroelectric phase of terthiophene derivative **13**, electroluminescence was observed under a DC bias application [46]. The electroluminescence device in a polarized state indicated a rectification character in spite of the symmetrical structure consisting of two ITO electrodes (Fig. 9.16a). The light-emission was observed in a forward bias for 2 μm-thick cell under the application of 20 V, corresponding to the electric field of 10^5 V cm^{-1}. This value is one order of magnitude lower than that of typical organic LEDs. No light emission was obtained in a reverse bias. The polarity could be inverted by the change of the polarity in the poling operation.

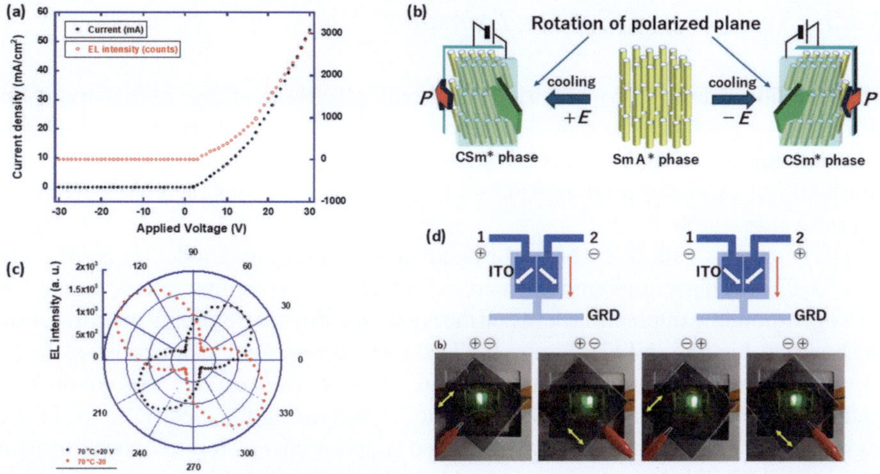

Fig. 9.16 **a** Current- and luminescence-voltage characteristics in the polarized ordered smectic phase of compound **13** in an ITO/ITO cell with a thickness of 2 μm. **b** Schematic for rotation of polarization plane of the electroluminescence. **c** Polar plot of the electroluminescence from the polarized smectic phase of compound **13**. **d** Switchable orthogonally polarized electroluminescence from a single device. Reproduced from Ref. [46] by permission of The Royal Society of Chemistry

In the non-polarized state, the current in the forward bias was described by injection-limited mechanism while it was a space charge-limited current in the polarized phase. This result indicated that the hole and electron injection barriers were remarkably reduced by the polarization at the interface between the electrodes and the LC layer.

Moreover, the polarization axis of the electroluminescence can be rotated in the light-emitting device using the FLCs. Figure 9.16b shows a schematic for the rotation of the polarization plane of the electroluminescence. In these ferroelectric phases, chromophores of the FLC molecules tilted from the layer normal. Accompanied by polarization switching, the FLC chromophores rotate to the symmetrical position around the layer normal. The change of the tilt angles of the FLC molecules was 90° in the case of compound **13**.

The surface of the ITO electrode was treated by a friction transfer method and the LC compounds were capillary-filled in the LC cells. Uniaxial alignment of the LC molecules was achieved by cooling the cell from the isotropic phase to the ferroelectric phase under a DC bias application. The orientation of the LC molecules could be changed by the inversion of the polarity of the DC bias in the poling process, as shown in Fig. 9.16c. Two orthogonally polarized lights could be produced from a single device and the polarization plane could be switched (Fig. 9.16d). This is the first example of the organic LEDs which can rotate the polarization axis of the light emission. This polarization switching was also observed in compound **16** [41].

9.3.3 Switching of Carrier Transport

The coupling between ferroelectricity and bulk electronic charge carrier transport produces a switchable current depending upon the polarization direction of the material. Kemerink and coworkers studied ferroelectric columnar liquid crystals consisting of extended π-conjugated units and amide moieties or oligo(vinylidene fluoride) side chains.

They synthesized a carbonyl-bridged triarylamine trisamide derivative **21** (Fig. 9.17a). This compound formed a ferroelectric columnar phase based on hydrogen-bonding interaction between the amide moieties surrounding the core units, as shown in Fig. 9.17b [47]. Figure 9.17c shows current-field characteristics in the columnar phase of compound **21**. When an Au anode was used, a low conductive state transitioned to a high conductive state by increasing the electric field. In the high conductive state, the current indicated a space-charge limited characteristic because interfacial dipole orientation above the coercive field to reduce the hole injection barrier. In the low conductive state, an ohmic injection-limited current was observed. It should be noted that this switching occurred by the local electric field produced by the orientation of interfacial dipoles. Such a resistive switching was also observed in a ferroelectric phase of C_3-symmetric benzotrithiophene tricarboxamide derivative [48].

In contrast, a conductivity switching based on a bulk polarization effect was observed in ferroelectric columnar phases of phthalocyanine and perylene bisimide derivatives **22** and **23** bearing oligo(vinilydene fluoride) side chains as well as subphthalocyanine derivative **24** bearing amide moieties, as shown in Fig. 9.18a [49, 50]. In these materials, switching from a high conductive to a low conductive state occurred by an increase in the electric field (Fig. 9.18b). The bulk polarization induced an asymmetric electrostatic potential for charge carrier hopping, as shown in Fig. 9.18c.

Akutagawa and coworkers synthesized pyrene derivative **25** bearing four alkylamide groups which exhibited ferroelectric columnar phases (Fig. 9.19a) [51]. In the columnar phases, hydrogen-bonding chains along the columnar aggregates were formed and collective rotation of alkylamide groups induced polarization inversion. The remanent polarization was around 1 μC cm^{-2} and the macroscopic polarization

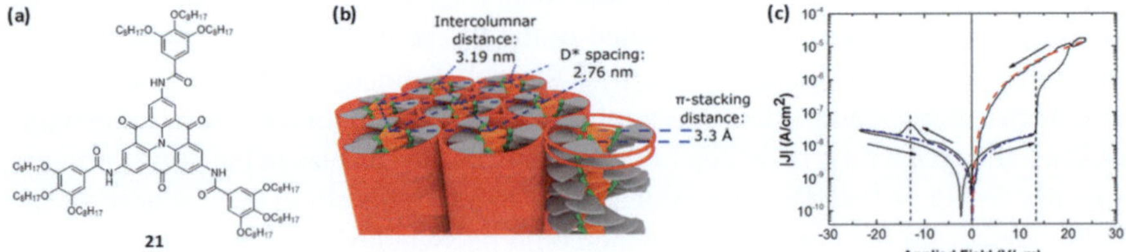

Fig. 9.17 **a** Molecular structure of carbonyl-bridged triarylamine trisamide derivative **21**. **b** Illustration for a supramolecular aggregation structure of the ferroelectric columnar phase of compound **21**. **c** Resistivity switching in the columnar phase of compound **21** in an Al/compound **21**/Au device. Reproduced from Ref. [47] by permission of American Chemical Society

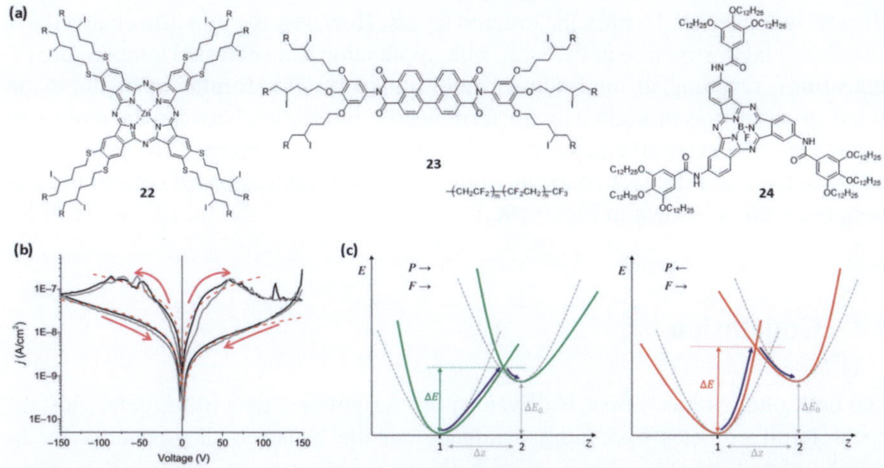

Fig. 9.18 **a** Molecular structures of columnar FLCs **22–24**. **b** Current switching in the ferroelectric columnar phase of compound **23**. **c** Schematics for current switching based on asymmetric electrostatic potential. Arrows labeled P and F indicate direction of the fields. Reproduced from Ref. [50] by permission of AAAS

affected the electronic charge carrier transport through π-stacks of pyrene cores. The linear I-V curve shifted upward and downward depending upon the polarization, as shown in Fig. 9.19b.

They also synthesized benzothienobenzothiophene (BTBT) derivatives bearing an alkylamide moiety [52, 53]. These compounds exhibited a smectic E phase where the π-conjugated units were organized within a bilayer structure and hydrogen-bonding networks comprising amide moieties, inducing ferroelectricity. In particular, compound **26** (Fig. 9.19c) formed a thin film with an ordered bilayer structure in which dipole moments are canceled, resulting in a non-polarized state. Holes are

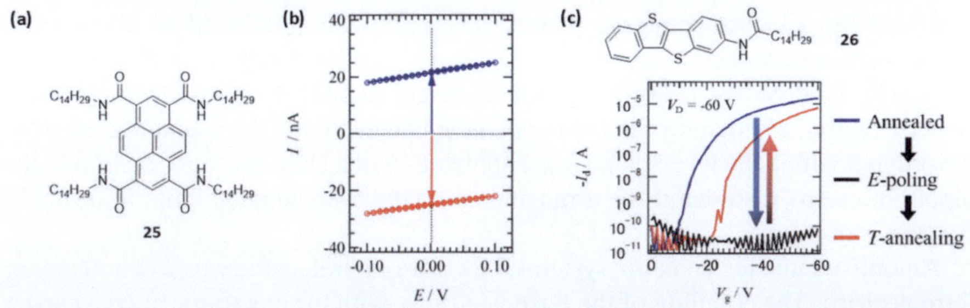

Fig. 9.19 **a** Molecular structure of pyrene derivative **25** and **b** current switching in the ferroelectric columnar phase. **c** Molecular structure of BTBT derivative **26** and switching of a transfer characteristic of a FET device. Reproduced from Ref. [51, 53] by permission of The American Chemical Society

efficiently transported within the ordered layers. However, the thin film changed to a disordered layer structure under a DC bias application at an elevated temperature (T-annealing), resulting in inhibition of hole transport. This formation of the disordered structure was promoted by the ferroelectric interaction between the molecular diploes and the external electric field (E-poling). The switching of the polarization was observed as a transfer characteristics of the FET devices based on thin films of compound **26**, as shown in Fig. 9.19c.

9.4 Conclusion

The bulk photovoltaic effect is driven by strong polarizations in ferroelectrics and open-circuit voltages exceeding bandgaps can be achieved theoretically. In the chiral ordered smectic phases of oligothiophene derivatives, photovoltaic effect with reversible polarity was observed in symmetrical cells consisting of two ITO electrodes. In particular, π-conjugated FLCs bearing lactic ester moieties exhibit large open-circuit voltage exceeding 1 V. Hybridization with fullerene derivatives enhanced the photovoltaic effect remarkably.

The bulk photovoltaic effect was confirmed in achiral ferroelectric columnar phases. Ferroelectric columnar phases of subphthalocyanine derivatives bearing alkyl side chains or shape-persistent units also indicated bulk photovoltaic effect for visible light illumination. Polycatenary liquid crystals based on bis(phenyloxadiazolyl)bithiophene unit also exhibited the bulk photovoltaic effect.

The polarization also affects the electronic state at the interface. Electroluminescence was observed in the FLC phases of oligothiophene derivatives bearing lactic ester moieties. In a symmetrical cell consisting of two ITO electrodes, current–voltage characteristic was diode-like to emit light in a forward bias. The hole and electron injection barriers were reduced by the polarization at the electrodes.

Phthalocyanine and perylene bisimide derivatives bearing oligo(vinylidene fluoride) side chains as well as subphthalocyanine derivatives bearing amide moieties exhibited ferroelectric columnar phases. The polarization induced an asymmetric charge carrier transport.

A BTBT derivative bearing an amide moiety exhibited a ferroelectric bilayer smectic phase in which dipole moments were canceled and the π-conjugated units are ordered within the layers. Under the application of a DC bias, reorientation of the dipole moments disordered the arrangement of the π-conjugated units to decrease the hole mobility.

Among organic ferroelectric systems, FLCs are extending the world of soft matter ferroelectrics. The coupling of the ferroelectricity with the electronic charge carrier transport in the liquid crystal phases will create new categories of functional materials and devices. Development of new ferroelectric materials will extend this concept.

Acknowledgements The author thanks Grant-in-Aid for Scientific Research on Innovative Areas (Element-Block Polymers, no. 15H00753) from the Ministry of Education, Culture, Sports, Science and Technology (MEXT), a Grant-in-Aid for Scientific Research (B) (no. 15H03797, 21H01904) from the Japan Society for the Promotion of Science (JSPS), the Iketani Foundation, the Ogasawara Toshiaki Memorial Foundation and the Fukuoka Naohiko Memorial Foundation for financial supports.

References

1. A. Kohler, H. Bassler, *Electronic Processes in Organic Semiconductors* (Wiley-VCH Verlag GmbH & Co. KGaA, Weinheim, 2015)
2. H. Naito (ed.), *Organic Semiconductors for Optoelectronics* (Wiley Interscience, Hoboken, 2021)
3. T. Kato, M. Yoshio, T. Ichikawa, Ba. Soberats, H. Ohno, M. Funahashi, Transport of ions and electrons in nanostructured liquid crystals. Nat. Rev. Mater., **2**, 17001 (2017)
4. M. Funahashi, Solution-processable electronic and redox-active liquid crystals based on the design of side chains. Flex. Print. Electron. **5**, 043001 (2020)
5. M.P. Aldred, A.E.A. Contoret, S.R. Farrar, S.M. Kelly, D. Mathieson, M. O'Neill, W.C. Tsoi, P. Vlachos, A full-color electroluminescent device and patterned photoalignment using light-emitting liquid crystals. Adv. Mater. **17**, 1368–1372 (2005)
6. T. Hori, Y. Miyake, N. Yamasaki, H. Yoshida, A. Fujii, Y. Shimizu, M. Ozaki, Solution processable organic solar cell based on bulk heterojunction utilizing phthalocyanine derivative. Appl. Phys. Exp. **3**, 101602 (2010)
7. M. Funahashi, Time-of-flight method for determining the drift mobility in organic semiconductors, ed. by H. Naito. *Organic Semiconductors for Optoelectronics* (Wiley interscience, 2021)
8. I. Bleyl, C. Erdelen, H.-W. Schmidt, D. Haarer, One-dimensional hopping transport in a columnar discotic liquid-crystalline glass. Phil. Mag. B **79**, 463–475 (1999)
9. M. Funahashi, A. Sonoda, Electron transport characteristics in nano-segregated columnar phases of perylene tetracarboxylic bisimide derivatives bearing oligosiloxane chains. Phys. Chem. Chem. Phys. **16**, 7754–7763 (2014)
10. Y. Miyake, Y. Shiraiwa, K. Okada, H. Monobe, T. Hori, N. Yamasaki, H. Yoshida, M. J. Cook, A. Fujii, M. Ozaki, Y. Shimizu, High carrier mobility up to 1.4 $cm^2 \cdot V^{-1} \cdot s^{-1}$ in non-peripheral octahexyl phthalocyanine. Appl. Phys. Exp., **4**, 021604 (2011).
11. M. Funahashi, F. Zhang, N. Tamaoki, J. Hanna, Ambipolar transport in the highly ordered smectic phase of 2-alkyl-5"-alkynyl- terthiophene derivative in a wide temperature range. ChemPhysChem **9**, 1465–1473 (2008)
12. M. Funahashi, T. Ishii, A. Sonoda, Temperature-independent hole mobility of a smectic liquid-crystalline semiconductor based on band-like conduction. ChemPhysChem **14**, 2750–2758 (2013)
13. R.B. Meyer, L. Liebert, L. Strzelecki, P. Keller, Ferroelectric liquid crystals. J. Phys. Lett. **36**, L69-71 (1975)
14. N.A. Clark, S.T. Lagerwall, Submicrosecond bistable electro-optic switching in liquid crystals. Appl. Phys. Lett. **36**, 899–901 (1980)
15. J.P.F. Lagerwall, F. Giesselmann, Current topics in smectic liquid crystal research. ChemPhysChem **7**, 20–45 (2006)
16. H. Takezoe, Y. Takanishi, Bent-core liquid crystals: their mysterious and attractive world. Jpn. J. Appl. Phys. **45**, 597–625 (2006)
17. Y. Matsunaga, N. Miyajima, Y. Nakayasu, S. Sakai, M. Yonenaga, Design of novel mesomorphic compounds: *N*, N', N"-trialkyl-1,3,5-benzenetricarbox-amides. Bull. Chem. Soc. Jpn **61**, 207–210 (1988)

18. Y. Shishido, H. Anetai, T. Takeda, N. Hoshino, S. Noro, T. Nakamura, T. Akutagawa, Molecular assembly and ferroelectric response of benzenecarboxamides bearing multiple $-CONHC_{14}H_{29}$ chains. J. Phys. Chem. C **118**, 21204–21214 (2014)
19. K. Kishikawa, S. Nakahara, Y. Nishikawa, S. Kohmoto, M. Yamamoto, A ferroelectrically switchable columnar liquid crystal phase with achiral molecules: superstructures and properties of liquid crystalline ureas. J. Am. Chem. Soc. **127**, 2565–2571 (2005)
20. H. Takahashi, M. Kohri, K. Kishikawa, Axially polar-ferroelectric columnar liquid crystalline system that maintains polarization upon switching to the crystalline phase: implications for maintaining long-term polarization information. ACS Appl. Nano Mater. **6**, 10531–10538 (2023)
21. D. Miyajima, F. Araoka, H. Takezoe, J. Kim, K. Kato, M. Takata, T. Aida, Ferroelectric columnar liquid crystal featuring confined polar groups within core–shell architecture. Science **336**, 209–213 (2012)
22. W. Shockley, H.J. Queisser, Detailed balance limit of efficiency of pn junction solar cells. J. Appl. Phys. **32**, 510–519 (1961)
23. Y. Dang, X. Tao, Recent progress of bulk photovoltaic effect in acentric single crystals and optoelectronic devices. Matter **5**, 2659–2684 (2011)
24. A.M. Glass, D. von der Linde, T.J. Negran, High-voltage bulk photovoltaic effect and the photorefractive process in $LiNbO_3$. Appl. Phys. Lett. **25**, 233–235 (1974)
25. W.T.H. Koch, R. Munser, W. Ruppel, P. Würfel, Bulk photovoltaic effect in $BaTiO_3$. Solid State Commun. **17**, 847–850 (1975)
26. T. Choi, S. Lee, Y.J. Choi, V. Kiryukhin, S.-W. Cheong, Switchable ferroelectric diode and photovoltaic effect in $BiFeO_3$. Science **324**, 63–66 (2009)
27. Y. Li, J. Fu, X. Mao, C. Chen, H. Liu, M. Gong, H. Zeng, Enhanced bulk photovoltaic effect in two-dimensional ferroelectric $CuInP_2S_6$. Nature Commun. **12**, 5896 (2021)
28. H. Sasabe, T. Nakayama, K. Kumazawa, S. Miyata, E. Fukada, Photovoltaic effect in poly(vinilydene fluoride). Polym. J. **13**, 967–973 (1981)
29. A. Sugita, K. Suzuki, S. Tasaka, Ferroelectric properties of a triphenylene derivative with polar functional groups in the crystalline state. Phys. Rev. B **69**, 212201 (2004)
30. M. Nakamura, S. Horiuchi, F. Kagawa, N. Ogawa, T. Kurumaji, Y. Tokura, M. Kawasaki, Shift current photovoltaic effect in a ferroelectric charge-transfer complex. Nat. Commun. **8**, 1–6 (2017)
31. S. Jiao, H. Jiang, C. Fan, C. Xu, J. Jiang, Y. Xu, Z. Tang, X. Sun, P. Ji, X. Yang, K. Ye, L. Xu, Q. You, S. Chen, H.-L. Cai, X. Wu, Metal-free chiral molecular ferroelectric photovoltaics. Chem. Eng. J. **477**, 146805 (2023)
32. T. Sasaki, M. Ikegami, T. Abe, D. Miyazaki, S. Kajikawa, Y. Naka, Real-time dynamic hologram in photorefractive ferroelectric liquid crystal with two-beam coupling gain coefficient of over 800 cm^{-1} and response time of 8 ms. Appl. Phys. Lett. **102**, 063306 (2013)
33. Y. Funatsu, A. Sonoda, M. Funahashi, Ferroelectric liquid-crystalline semiconductors based on a phenylterthiophene skeleton: effect of the introduction of oligosiloxane moieties and photovoltaic effect. J. Mater. Chem. **3**, 1982–1993 (2015)
34. A. Seki, M. Funahashi, Photovoltaic effects in ferroelectric liquid crystals based on phenylterthiophene derivatives. Chem. Lett. **45**, 616–618 (2016)
35. A. Seki, M. Yoshio, Y. Mori, M. Funahashi, Ferroelectric liquid-crystalline binary mixtures based on achiral and chiral trifluoromethylphenylterthiophenes. ACS Appl. Mater. Interfaces **12**, 53029–53038 (2020)
36. A. Seki, Y. Funatsu, M. Funahashi, A. Seki, Anomalous photovoltaic effect based on molecular chirality: Influence of enantiomeric purity on the photocurrent response in π-conjugated ferroelectric liquid crystals. Phys. Chem. Chem. Phys. **19**, 16446–16455 (2017)
37. A. Seki, M. Funahashi, Chiral photovoltaic effect in an ordered smectic phase of a phenylterthiophene derivative. Org. Electron. **62**, 311–319 (2018)
38. Y. Mori, M. Funahashi, Bulk photovoltaic effect in organic binary systems consisting of a ferroelectric liquid crystalline semiconductor and fullerene derivatives. Org. Electron. **87**, 105962 (2020)

39. M. Funahashi, High open-circuit voltage in the bulk photovoltaic effect for the chiral smectic crystal phase of a double chiral ferroelectric liquid crystal doped with a fullerene derivative. Mater. Chem. Front. **5**, 8265–8274 (2021)
40. M. Funahashi, Bulk photovoltaic effect in ferroelectric liquid crystals comprising of quinquethiophene and lactic ester units. Org. Electron. **122**, 106911 (2023)
41. Y. Matoba, S. Uemura, M. Funahashi, Diastereomeric effect on bulk photovoltaic property and polarized electroluminescence in ferroelectric liquid crystals containing an extended π-conjugated unit. Bull. Chem. Soc. Jpn **96**, 247–256 (2023)
42. C. Zhang, K. Nakano, M. Nakamura, F. Araoka, K. Tajima, D. Miyajima, Noncentrosymmetric columnar liquid crystals with the bulk photovoltaic effect for organic photodetectors. J. Am. Chem. Soc. **142**, 3326–3330 (2020)
43. M. Lehmann, M. Baumann, M. Lambov, A. eremin, parallel polar dimers in the columnar self-assembly of umbrella-shaped subphthalocyanine mesogens. Adv. Funct. Mater. **31**, 2104217 (2021)
44. A. Murad, E. Baron, M. Feneberg, M. Baumann, M. Lehmann, A. Eremin, Polarity in liquid crystals formed by self-assembled umbrella-shaped subphthalocyanine mesogens. ACS Appl. Mater. Interfaces **16**, 25025–25032 (2024)
45. K. Iakoubovskii, M. Yoshio, Highly luminescent and photoconductive columnar liquid crystals with a thiophene-oxadiazole backbone. Chem. Commun. **59**, 7443–7446 (2023)
46. M. Funahashi, Y. Mori, Linearly polarized electroluminescence device in which the polarized plane can be rotated electrically using a chiral liquid crystalline semiconductor. Mater. Chem. Front. **4**, 2137–2148 (2020)
47. A.V. Gorbunov, A.T. Haedler, T. Putzeys, R.H. Zha, H.-W. Schmidt, M. Kivala, I. Urbanaviciute M. Wübbenhorst, E.W. Meijer, M. Kemerink, Switchable charge injection barrier in an organic supramolecular semiconductor. ACS Appl. Mater. Interface., **8**, 15535–15542 (2016)
48. N.M. Casellas, I. Urbanaviciute, T. Cornelissen, J.A. Berrocal, T. Torres, M. Kemerink, M. Garcia-Iglesias, Resistive switching in an organic supramolecular semiconducting ferroelectric. Chem. Commun. **55**, 8828–8831 (2019)
49. M. García-Iglesias, B.F.M. de Waal, A.V. Gorbunov, A.R.A. Palmans, M. Kemerink, E.W. Meijer, Versatile method for the preparation of ferroelectric supramolecular materials via radical end-functionalization of vinylidene fluoride oligomers. J. Am. Chem. Soc. **138**, 6217–6223 (2016)
50. A.V. Gorbunov, M.G. Iglesias, J. Guilleme, T.D. Cornelissen, W.S.C. Roelofs, T. Torres, D. González-Rodríguez, E.W. Meijer, M. Kemerink, Ferroelectric self-assembled molecular materials showing both rectifying and switchable conductivity. Sci. Adv. **3**, e1701017 (2017)
51. H. Anetai, Y. Wada, T. Takeda, N. Hoshino, S. Yamamoto, M. Mitsuishi, T. Takenobu, T. Akutagawa, Fluorescent ferro-electrics of hydrogen-bonded pyrene derivatives. J. Phys. Chem. Lett. **6**, 1813–1818 (2015)
52. K. Sambe, T. Takeda, N. Hoshino, W. Matsuda, R. Miura, K. Tsujita, S. Maruyama, S. Yamamoto, S. Seki, Y. Matsumoto, T. Akutagawa, Ferroelectric organic semiconductor: [1]benzothieno[3,2-b][1]benzothiophene-bearing hydrogen-bonding $-CONHC_{14}H_{29}$ Chain. ACS Appl. Mater. Interfaces **15**, 58711–58722 (2023)
53. K. Sambe, T. Takeda, N. Hoshino, W. Matsuda, K. Shimada, K. Tsujita, S. Maruyama, S. Yamamoto, S. Seki, Y. Matsumoto, T. Akutagawa, Carrier transport switching of ferroelectric BTBT derivative. J. Am. Chem. Soc. **146**, 8557–8566 (2024)

Chapter 10
Magneto-Structural Correlation in Organic Radical Crystals Containing Nitroxyl and Triazinyl Spin Centers

Naoki Yoshioka, Masaomi Takii, and Youhei Miura

Abstract Stable organic radicals have drawn attention as promising building blocks for multifunctional materials. In order to realize the tailored magnets based on molecular materials, it is necessary to investigate magneto-structural correlation of open-shell molecules. In the present chapter, we describe the magnetic property of dihydroquinoline *N*-oxyl-based radicals and benztriazynyl-based radicals to discuss the substituent effect on molecular packing and intermolecular magnetic interaction. Relatively large magnetic interactions were observed through π-π stacking for dihydroquinoline *N*-oxyl-based radicals. Spin transition behavior was demonstrated in benztriazynyl-based radical.

Keywords Organic radical crystal · Magneto-structural correlation · Intermolecular magnetic interaction · Molecule-based magnetism

10.1 Introduction

The magnetism of substances is based on the magnetic moment associated with electron spin and their orbital motion. Therefore, atoms, ions, and molecules with imperfect shells exhibit the magnetic property. When electron spin is assembled, the exchange interactions between the spins determine the bulk magnetism of the material. Organic compounds usually exhibit diamagnetism because all electrons of the constituent atoms are paired. From the viewpoint of magnetism, organic radicals and transition metal complexes with unpaired electrons are the target of molecule-based magnetic materials. Magnetism is broadly classified into paramagnetism, ferromagnetism, and antiferromagnetism, depending on the relative magnitude of the exchange interaction between unpaired electrons and the thermal energy of the system. When the exchange interaction is negligible compared to the energy

N. Yoshioka (✉) · M. Takii · Y. Miura
Department of Applied Chemistry, Keio University, Minato, Japan
e-mail: yoshioka@applc.keio.ac.jp

S. Kobatake and H. Uekusa (eds.), *Advances in Organic Crystal Chemistry*,
https://doi.org/10.1007/978-981-96-4479-7_10

of thermal motion, the magnetic moment statistically orients in all directions. Therefore, if no external magnetic field is applied, the magnetic moment is zero. When an external magnetic field is applied, the magnetic moments of the electron spins are aligned in the direction of the magnetic field, and a slight magnetic moment is generated in the direction of the magnetic field. Such a material is called paramagnetic substance, and its properties can be analyzed from the temperature dependence of the molar magnetic susceptibility. The temperature dependence of the molar magnetic susceptibility of pure paramagnets follows the Curie rule, but when the spin of an unpaired electron is oriented in the same or opposite direction as the spin of a neighboring electron, the Curie rule is violated, and the Curie–Weiss rule is followed. In the former case, ferromagnetic interactions between the unpaired electrons are involved, while in the latter, antiferromagnetic interactions are involved. The sign of the Weiss constant, θ can be determined from the temperature dependence of the molar magnetic susceptibility. When θ is positive, ferromagnetic interaction between the spins is present, and when θ is negative, antiferromagnetic interaction between the spins is present. When the exchange interaction between spins exceeds thermal motion below a certain temperature, spontaneous orientation of magnetic moments is observed below a certain temperature. The critical temperature is called the Curie temperature for ferromagnetic interactions and the Néel temperature for antiferromagnetic interactions. Materials used as magnetic materials have Curie points above room temperature. In organic radicals, exchange interactions between molecules are antiferromagnetic in most cases, and examples of ferromagnetic behavior are limited.

While the possibility of an organic ferromagnet has been theoretically proposed, a magneto-structural correlation of a few organic radical crystals has been reported during the 1960-1970s due to the inaccessibility to crystallographic data. Since the discovery of *p*-NPNN as the first organic ferromagnet in 1991 [1], magnetic properties of various organic radical crystals have been reported. In organic radical molecules, evaluation of SOMO corresponding to the delocalization of unpaired electron and distribution of spin density related to spin polarized effect is important because they affect intra- and intermolecular magnetic interactions. Therefore, to construct molecule-based magnetic materials using organic radical crystals, it is necessary to discuss magneto-structural correlations considering the SOMO and spin density distributions.

10.2 Dihydroquinoline *N*-Oxyl-Based Radicals

Fremy's salt, an inorganic nitroxyl radical salt, was reported in the mid-nineteenth century, while porphyrexide, reported by Piloty and Schwerin in 1901, was the first nitroxyl radical and the first organic radical to be isolated (Fig. 10.1) [2]. Subsequently, Wieland et al. reported the synthesis of diaryl nitroxyl radicals, but the stability of these aromatic nitroxyl radicals varied depending on the substituents on the aromatic ring [3, 4]. Similarly, Forrester et al. reported phenyl *tert*-butyl nitroxyl radical revealed that the radical molecules disproportionate to *N*-*tert*-butylaniline and benzoquinone imine *N*-oxide [5]. This structural flexibility often leads to unexpectedly strong antiferromagnetic interaction induced by mutual close contact between spin centers. To overcome these problems, we have focused on a kind of aromatic *N*-oxyl based radical, 2,2-diphenyl-1,2-dihydro-quinoline-*N*-oxyl (**1**) originally synthesized by Colonna et al. [6]. Because **1** has a characteristic fused structure, the N–O spin center is involved in conjugation with the adjacent aryl group. **1** having cardo structure derived from the two phenyl rings at the 2-position is stable both in solution and solid states. The cardo structures have been often used for improving solvent solubility of rigid aromatic polymers. The cardo and fused coplanar structures could synergistically contribute to the prevention of the direct close contact between the N–O moieties. In addition, the magnetically effective multipoint close contacts between the spin sites are also expected for the analogues of **1** (Fig. 10.2).

The magnetic data indicates that the cardo structure of **1** contributes to the prevention of direct close contact between the N–O moieties, which usually causes a

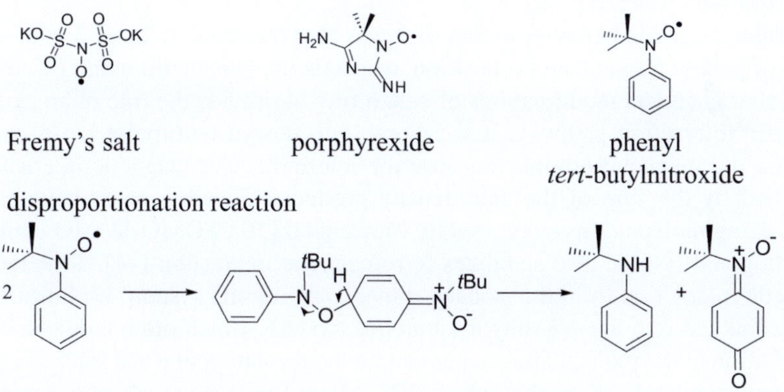

Fig. 10.1 Chemical structures of selected nitroxyl radical and disproportionation reaction for aromatic nitroxyl radical

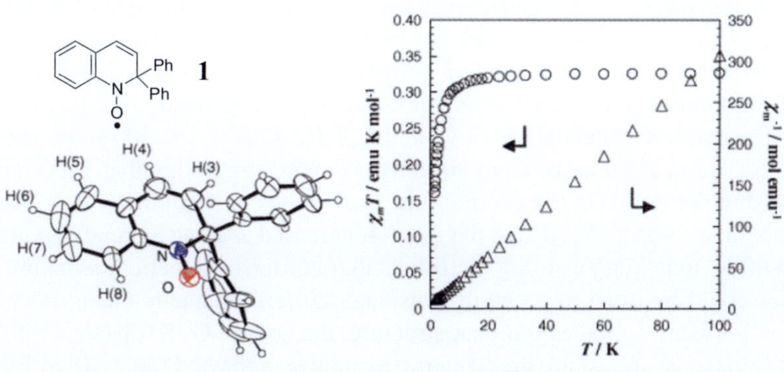

Fig. 10.2 Chemical structure and ORTEP drawing of **1** and temperature dependence of $\chi_m{}^{-1}$ and $\chi_m T$ of **1**. Reprinted from Ref. [7]. Copyright 2005, with permission from Elsevier

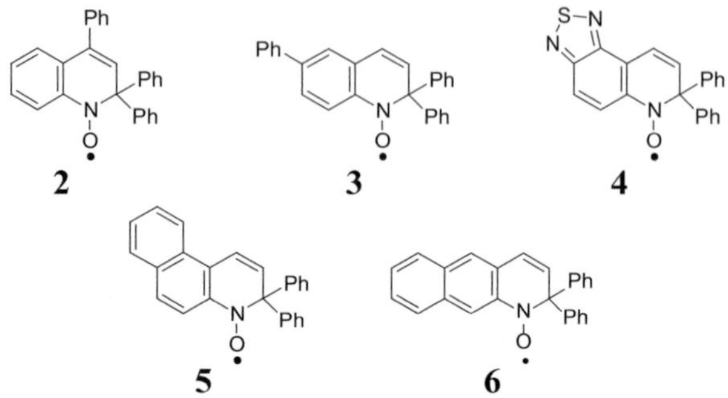

Fig. 10.3 Chemical structures of nitroxyl radical **2–6** having a cardo structure

strong antiferromagnetic interaction often observed in less rigid and hindered *N-tert*-butylphenyl nitroxyl radicals. To evaluate detailed magneto-structural correlation, **2–6** in Fig. 10.3 were synthesized, and their crystal structures and magnetic properties were studied [7–9].

Similar magnetic behavior to that of **1** was also observed in **2** and **3** with substitution of phenyl ring at 4 and 6-position, respectively. Our attention has centered on the further chemical modifications of benzo ring to provide the role of an extended magnetic interaction pathway. It is favorable to design multipoint close contacts between the magnetic orbitals, because the intermolecular magnetic interaction is described by the sum of the spin-density product of nearby atoms belonging to neighboring molecules as suggested by McConnell [10]. SOMO-SOMO contact in an orthogonal fashion also stabilizes ferromagnetic interaction [11]. To extend the magnetic interaction from the isolated dimer to an infinite system, we introduce an -NSN- linkage into the 1,2-dihydroquinoline system, which often causes a drastic perturbation in the molecular arrangement by the formation of a π-π stacked column or intermolecular N···S close contact [12]. Along the strategy, novel π-conjugated nitroxyl radical having a thiadiazole ring (**4**) and its iso π-electronic radical (**5**) to elucidate the effect of the -NSN- linkage.

The observed coplanarity could lead to the large delocalization of unpaired electron over the whole π-conjugated systems (Fig. 10.4). In the crystal, **4** formed a one-dimensional columnar structure consisting of slipped π-stacks along the *c*-axis in the crystal. On the other hand, **5** formed a centrosymmetric dimer, in which benzo[*f*]quinoline units locate an offset π-stacked arrangement.

The temperature dependences of χ_m and $\chi_m T$ for **4** and **5** 1.8–300 K are shown in Fig. 10.5. The $\chi_m T$ values of were virtually constant, suggesting that **5** has a nearly isolated doublet state. On the contrary, the $\chi_m T$ values of **4** gradually de-creased as the temperature was lowered, and the χ_m of **4** increased and had a broad maximum at around 40 K, indicating a strong intermolecular antiferromagnetic interaction. This behavior could be fitted to a one-dimensional antiferromagnetic chain model [13] with $J = -47$ cm^{-1}. In the columnar structure, the nearest O···O distance is 9.37 Å. There seems to be almost no direct interaction between the N–O sites. Therefore, the interaction was understood to be due to the close contact of the delocalized spins. The observed close contacts are longer than the sum of the van der Waals radii; however,

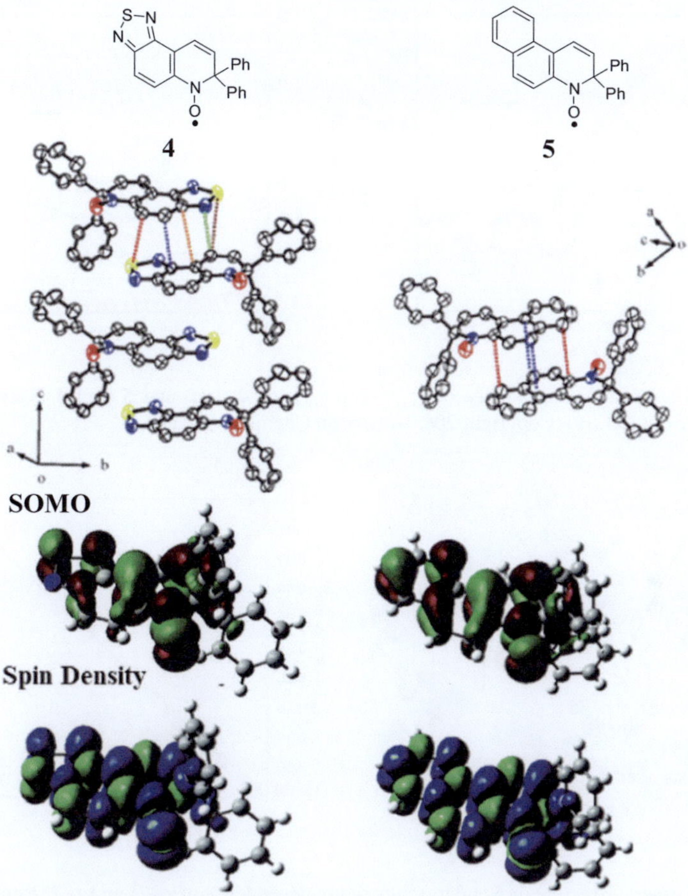

Fig. 10.4 Crystal structure, SOMOs and spin density distribution of **4** and **5**. Reprinted with permission from Ref. [9]. Copyright 2005 American Chemical Society

there are some close contacts between the atoms carrying an α-spin. These contacts are thought to be the main path for the observed intermolecular magnetic interaction. DFT calculations for **4** reproduced a negative J value, supporting the antiferromagnetic interaction through the π-π stacking. Within the column, the delocalized unpaired electron plays an important role in the strong antiferromagnetic interaction. In contrast to the formation of π -stacked dimer in **5**, **6** formed a centrosymmetric edge to edge dimeric structure in its crystal without π-stacking (Fig. 10.6). Similar edge–edge association for the aromatic *N*-oxide derivatives has been reported for the aromatic *N*-oxide derivatives [14]. The temperature dependence of $\chi_m T$ and $\chi_m{}^{-1}$ for **6** are shown in Fig. 10.6. $\chi_m T$ value of **6** increased as the temperature decreased and reached a maximum at 2.2 K. The observed intermolecular interaction was analyzed using the Bleaney–Bowers equation for a dimer model [15], The best-fit parameters were a singlet–triplet energy gap of $2J = +4.6\ \text{cm}^{-1}$.

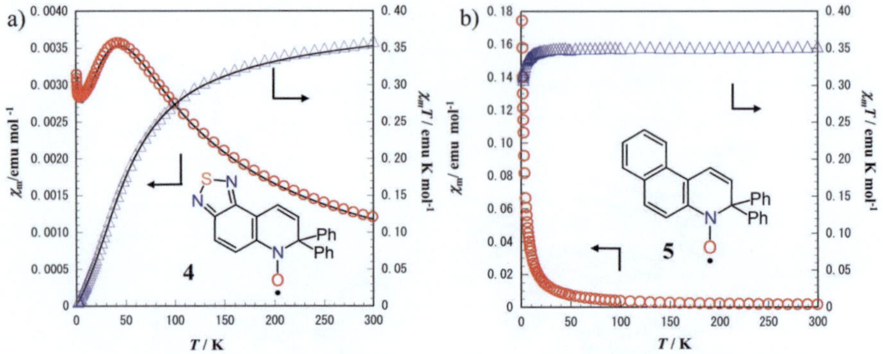

Fig. 10.5 Temperature dependence of $\chi_m{}^{-1}$ and $\chi_m T$ of **4** (left) and **5** (right). Reprinted with permission from Ref. [9]. Copyright 2005 American Chemical Society

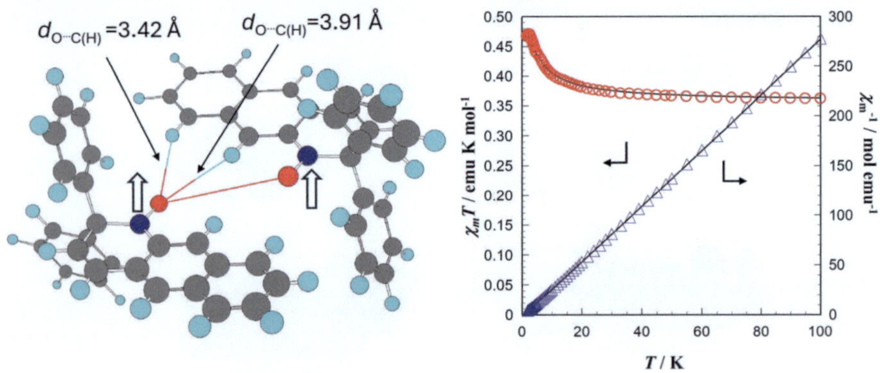

Fig. 10.6 Dimeric structure of **6** and its temperature dependence of $\chi_m{}^{-1}$ and $\chi_m T$. Reprinted with permission from Ref. [7]. Copyright 2005, with permission from Elsevier

Within the dimer of **6**, the nearest O···O distance is 4.94 Å. There seems to be almost no direct interaction between the N–O sites. However, the short distances between the N–O site and the aromatic hydrogen atoms at the 9- and 10-positions of an adjacent molecule are observed. The DFT calculation for **6** gave relatively large negative spin densities for these hydrogen atoms. The observed ferromagnetic coupling is qualitatively interpreted by the McConnell's theorem [10]. The calculation for the dimer found a triplet ground state with $2J = +3.9\ \text{cm}^{-1}$, which was in good agreement with the experimental result ($2J = +4.6\ \text{cm}^{-1}$). Therefore, negatively spin-polarized hydrogen atoms located near the N–O site play an important role in the ferromagnetic interaction.

10.3 Benztriazynyl-Based Radicals

In 1968, Blatter and Lukaszewski reported 1,3-diphenyl-1,4-dihydro-1,2,4-benzotriazin-4-yl (Fig. 10.7), which is known as Blatter's radical [16], however, its derivatives received little attention. Since Koutentis's group reported simple and various synthetic methods of Blatter's radical and its derivatives [17], the situation has dramatically changed. The physicochemical properties and the computational study of the benzotriazinyl radical derivatives have been reported and the enhanced stability of the radicals makes it possible to transform the functional group, i.e., the organometallic reaction of Suzuki–Miyaura, Stille, Sonogashira, Negishi and Heck cross-coupling reactions, debenzylation, acylation and alkylation without protection of the radical moiety. This suggests that highly functionalized 1,2,4-benzotriazin-4-yl derivatives can be synthesized and become important species for advanced organic materials [18]. Regarding the magnetic properties, the magnetic interactions of the benzotriazinyl radicals are typically antiferromagnetic, however, a few derivatives form a 1D column with ferromagnetic interactions [19, 20]. The trifluoromethyl-substituted benzotriazinyl radicals show a 1D ferromagnetic interaction. It implies that the substituent, which causes a dipole moment, can influence the physicochemical properties of the benzotriazinyl derivatives. In this section, we report the crystal structure, the magnetic and study of the 3-*tert*-butyl benzotriazinyl radical derivatives (**7**, **8**) (Fig. 10.7) [21]. It demonstrates an intermolecular ferromagnetic interaction, which follows the Heisenberg 1D chain model, and the substituent effect on the magneto-structural correlation is discussed.

In the molecular packing of radical **7**, there is a head-to-head structure, which forms a 1D column structure along the *b* axis without a π–π interaction between the benzotriazinyl moieties (Fig. 10.8). The heterocyclic ring appears not to have any contacts with other rings, which suggests a weak intermolecular magnetic interaction. Radical **8** also formed a slipped 1D column structure along the *a* axis and there is a head-to-tail structure antiparallel along the b axis (Fig. 10.8). In the 1D column, the distance between the benzotriazinyl rings is 4.35 Å because the bulky *tert*-butyl group prevents close contact of the π conjugation system. The moment of **8** decreased due to the trifluoromethyl group. In the crystal of **7**, the molecules cancel the moment in the column, however, **8** is aligned in the same direction in the column and cancelled the moment between the columns along the *b* axis. The temperature dependence of χ_m^{-1} and $\chi_m T$ for **8** is shown in Fig. 10.9.

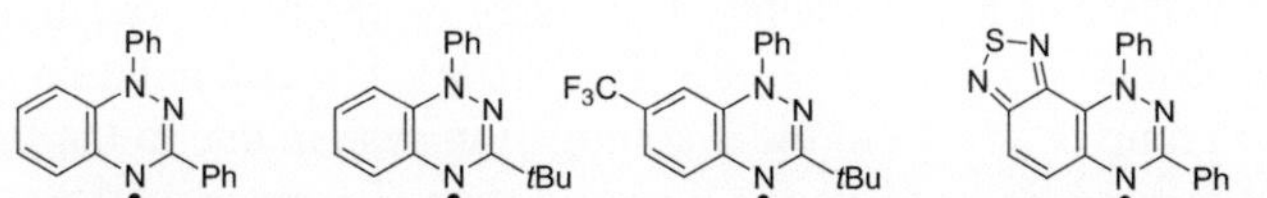

Blatter's radical **7** **8** **9**

Fig. 10.7 Chemical structures of benzotriazinyl radical derivative

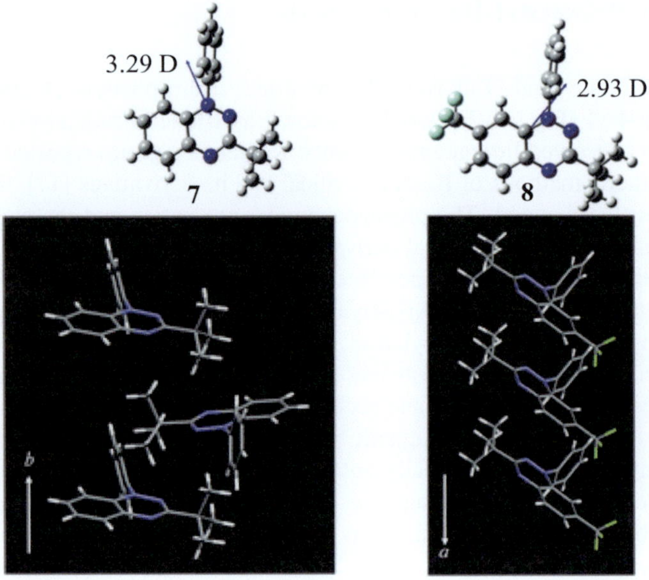

Fig. 10.8 Crystal structures of **7** and **8**. Dipole moments were calculated at the UB3LYP/6-31G(d) level. Reproduced from Ref. [21] with permission from the Centre National de la Rechercher Scientifique (CNRS) and the Royal Society of Chemistry

The behavior of $\chi_m{}^{-1}$ followed the Curie–Weiss law, with $\theta = +0.7$ K above 50 K. The presence of a ferromagnetic interaction was predicted due to the positive Weiss constant θ. This $\chi_m T$ value was almost constant above 40 K and gradually increased with the decreasing temperature. This suggests the existence of a ferromagnetic interaction between the benzotriazinyl radicals. The thermal behavior of the magnetic susceptibility can be analyzed using the Heisenberg 1D chain model. The behavior can be well described using empirical expression reported by Baker [22] and the best-fit result is shown in Fig. 10.9a as a solid line. The estimated magnitude of coupling with $J = +0.63$ cm^{-1} indicates that radical **8** formed a 1D ferromagnetic chain. To clarify the magnetic coupling, the magnetic field dependence of the magnetization was measured at 1.9 K (Fig. 10.9b). The magnetization plots exceed the theoretical Brillouin curve for the $S = 1/2$ spin system and follows the $S = 1$ system. This indicates that a ferromagnetic interaction with $S = 1$ on average was dominant in the column. The SOMO of **8** was mainly distributed on the benzotriazinyl moiety and slightly delocalized over the CF_3 group (Fig. 10.9c). The computation suggests that the N-phenyl group is free from the π conjugation system due to the large dihedral angle against the amidrazonyl ring in the solid state. To understand the contribution of the ferromagnetic interaction in the 1D-column, the magnetic exchange parameter J was estimated by DFT calculation. The calculated value $J = +0.4$ cm^{-1} showed that the ferromagnetic interaction was dominant in the column. The long distance between the benzotriazinyl moieties leads to a reduced overlap of the SOMO and

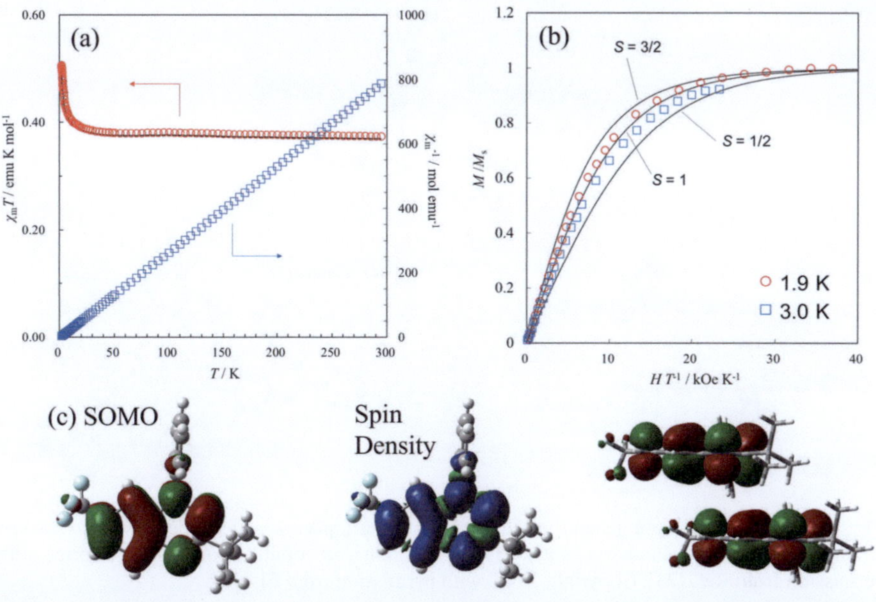

Fig. 10.9 Temperature dependence of $\chi_m{}^{-1}$ and $\chi_m T$ (**a**), magnetization isotherm at 1.9 K (**b**) of **8**, SOMO, and spin density, and overlap of SOMO in the column (**c**). Reproduced from Ref. [21] with permission from the Centre National de la Rechercher Scientifique (CNRS) and the Royal Society of Chemistry

reduction of the overlap integral and ferromagnetic interaction was exhibited. The calculation result was consistent with the magnetic susceptibility measurement.

Benzotriazinyl radical derivative fused by thiadiazole ring (**9**) formed face-to-face stacked arrangement with two types of contact (Fig. 10.10) [23]. One type involves overlapping between thiadiazolobenzotriazinyl moieties, without involving the phenyl ring at 1-position, with an average interplanar distance of 3.54 Å (contact A). This is an overlap between an electron-deficient thiadiazole moiety and an electron-donating triazinyl moiety. The other type of overlap covers all planar moieties, including the phenyl group, at an average interplanar distance of 3.30 Å (contact B). This interplanar distance is shorter than the sum of the van der Waals radii of the carbon atoms (3.4 Å). The short distance π–π inter-action induces the large overlap of spin densities, and it indicates exhibiting the strong magnetic inter-action large overlap of spin densities, and it indicates exhibiting the strong magnetic interaction. In the case of **9**, it is expected to exhibit the strong magnetic interaction derived from the dimer structure with a short distance. Intercolumnar contact was also found to exist between the phenyl rings on N atom, due to π–π interactions of the phenyl rings, with an average distance of 3.55 Å (contact C). In addition, each stacked column was found to present contact through the sulfurs of the thiadiazole ring at a distance of 3.63 Å (contact D), which is close to the sum of the van der

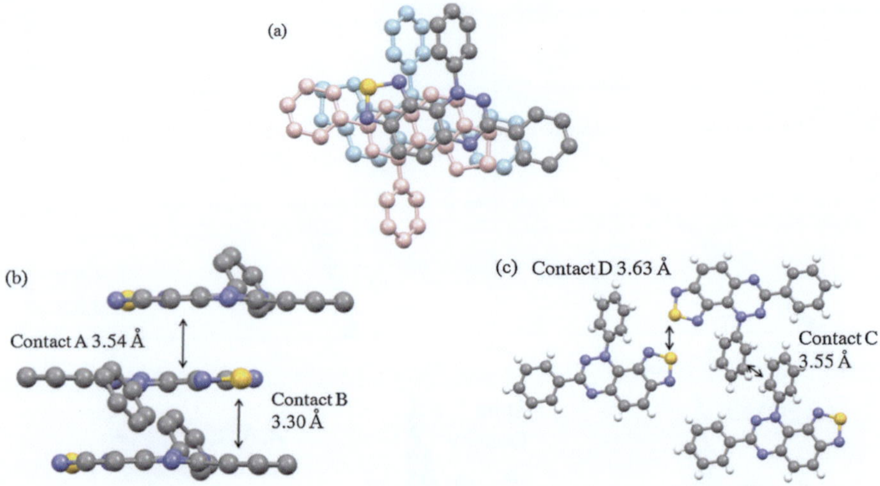

Fig. 10.10 Molecular arrangement in the crystal of **9**: **a** top-view and **b** side-view of *p*-stacked columns and **c** intercolumnar contact. Hydrogen atoms are omitted for clarity. Reprint with permission from Ref. [23]. Copyright 2015, with permission from Elsevier

Waals radii of sulfur (3.6 Å), indicating the presence of intercolumnar sulfur–sulfur interactions.

The temperature dependence of the magnetic susceptibility of **9** is measured in the temperature range of 1.8–300 K by using a SQUID magnetometer. The $\chi_m T$ value at 300 K is about 0.237 emu K mol^{-1}, which is smaller than the theoretical value of 0.375 emu K mol^{-1} for an $S = 1/2$ spin-only system. The temperature dependence of $\chi_m T$ follows the Bleaney–Bowers equation for a dimer model [15] with *ca.* $2J = -302$ cm^{-1}.

10.4 Spin Transition Behavior in Radical Crystal

Many organic radicals have been investigated for developing molecule based multi-functional materials for electronics, spintronics, as well as molecule-based magnets. If their physicochemical properties are modulated in response to external stimuli, such as light, heat, pressure, electric and magnetic fields, the compounds can be used as a switching device by controlling their spin state and molecular arrangement. In the field of coordination chemistry, many metal complexes exhibit a spin crossover (SCO) behavior induced by a thermal or light stimulus to control the spin state of the central metal ion [24]. On the other hand, in the chemistry of organic radical crystals, the magnetic characteristics of the organic radicals significantly depend on their molecular arrangement, and their structural phase transition is one of the triggers to exhibit the spin transition behavior like SCO phenomena. Several radical derivatives showing a phase transition by changing their molecular assembly in the solid state have been reported, whereas almost all the radicals do not demonstrate a spin transition [25]. However, these radical solid have been fortuitously discovered in the study of the magnetic properties of organic radicals. In general, it is difficult to design a group of radical substances exhibiting a structural phase transition involving a spin transition like the SCO. Therefore, in the investigation of physicochemical properties, such as the magneto-structural correlation for the analogues of external stimuli-response organic radical systems, is important to explore molecules displaying those kinds of interesting properties.

Focusing on the benzotriazinyl family, the planar radical derivatives, which have a greater spin delocalization, exhibited a ferromagnetic interaction in the 1D columnar assembly, providing a new basic skeleton for supramolecular chemistry. Therefore, the benzotriazinyl derivatives are promising spin centers for multifunctional materials. Notably, showed a reversible spin transition at *ca.* 58 K fully completed within 5 K caused by the structural phase transition [26]. In the examples where an organic radical exhibited spin transition behavior caused by a structural phase transition, this was accompanied by packing motif changes between the dimer and columnar structure. Therefore, the construction of a columnar structure using a dimer interacted by weak π–π interaction may be one of the methodologies to design an organic radical exhibiting spin transition behavior. Keeping the size of molecules small may also be beneficial. Based on this viewpoint, it is a good approach to exhibit a dipole moment by introducing electron donating or withdrawing group for the formation of the π-dimer. For cancelling the dipole moment between the molecules, dimerization of the π-radicals can be promoted.

In addition, the magnitude of the π–π interaction can be controlled by chemical modification of the bulky group into the radical for formation of the dimer and the columnar structure. Recently, our group investigated the magneto-structural correlation of the 3-*tert*-butylbenzotriazinyl carrying an electron-withdrawing group and reported that **8** forms an ideal 1D ferromagnetic chain. As part of our investigation, in the present section, magneto-structural correlation in connection with the computational results of the 3-*tert*-butylbenzotriazinyls carrying a cyano group at the 7- or 6-position is investigated, and a reversible spin transition of **11** at ambient temperature is demonstrated (Fig. 10.11) [27].

The temperature dependence of the magnetic susceptibility for the polycrystalline sample of **10** in the temperature range of 1.8–300 K is shown in Fig. 10.12. The $\chi_m T$ value gradually decreased with the decreasing temperature, which suggested that the intermolecular antiferromagnetic interaction was dominant in the solid state.

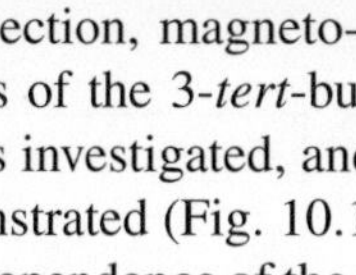

10

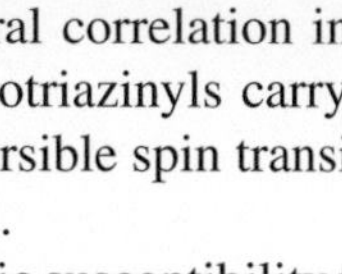

11

Fig. 10.11 Chemical structures of cyano-substituted benzotriazinyl radical derivative

The behavior was well described by the Bonner–Fisher equation [13] with $J = -15.8\ \text{cm}^{-1}$ and mean molecular field term $\theta = -1.3$ K. The crystallographic analysis also supported the magnetic result; radical **10** is in close contact for the intermolecular interaction along the column. The thermal behavior of the $\chi_m T$ value for **11** in the temperature range of 1.8–320 K is shown in Fig. 10.13. The $\chi_m T$ value at 300 K was 0.313 emu K mol^{-1}, which was a lower value than that of the estimated value of the isolated $S = 1/2$ spin (0.375 emu K mol^{-1}). This result implied the existence of strong intermolecular antiferromagnetic interaction. No change in $\chi_m T$ occurred during cooling until 284 K, then an abrupt decrease occurred and plateaued below 280 K. When heating, the $\chi_m T$ exhibited a behavior like the cooling process and a sharp increase took place near 278 K. As shown in Fig. 10.13, **11** exhibited a fully reversible transition during the heating and cooling processes in the range of 6 K and the transition was completed in a small thermal range *ca.* 280 K. The thermal behavior of the $\chi_m T$ in the range of 284–320 K can be reproduced by the Bleaney–Bowers dimer model [15] with $2J = -122.8\ \text{cm}^{-1}$ and $\theta = -0.71$ K. The thermal behavior of the $\chi_m T$ below 280 K can be roughly reproduced by the Bleaney–Bowers model with $2J = -485\ \text{cm}^{-1}$. To confirm the change in the molecular packing, the DSC measurement was carried out in the range of 250–320 K. The DSC curve exhibited an endothermic peak at 283.9 K, which is the corresponding temperature of the spin transition.

As already described, **11** exhibited the spin transition at *ca.* 284 K. The DSC measurement revealed the possibility of a structural phase transition at the corresponding temperature. Thus, X-ray crystallographic measurements were carried out at 263 and 300 K (Fig. 10.14).

They revealed that **11** can form two polymorphs in response to the temperature and these polymorphs are described as the low-temperature (LT) and high-temperature (HT) phases. Because of the rapid conversion from the HT/LT phase to the LT/HT phase, the reflection data were obtained using the same single crystal at 263 K (LT phase) and 300 K (HT phase). The bond lengths of the amidrazonyl ring are similar to that of the other benzotriazinyl derivatives, which indicates that an unpaired electron

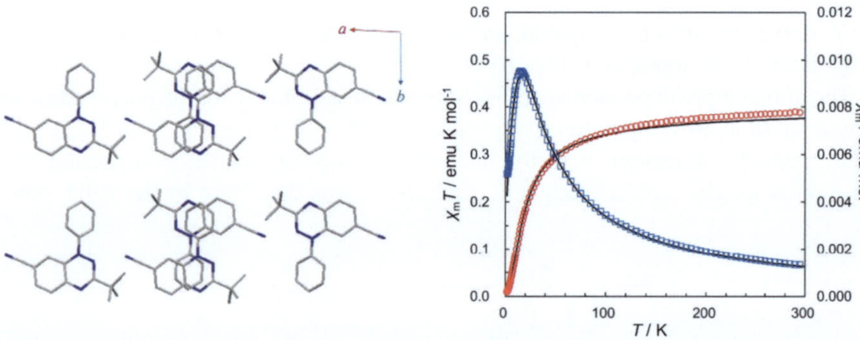

Fig. 10.12 Crystal structure of **10** and its temperature dependence of $\chi_m{}^{-1}$ and $\chi_m T$. Reproduced from Ref. [27] with permission from the Centre National de la Rechercher Scientifique (CNRS) and the Royal Society of Chemistry

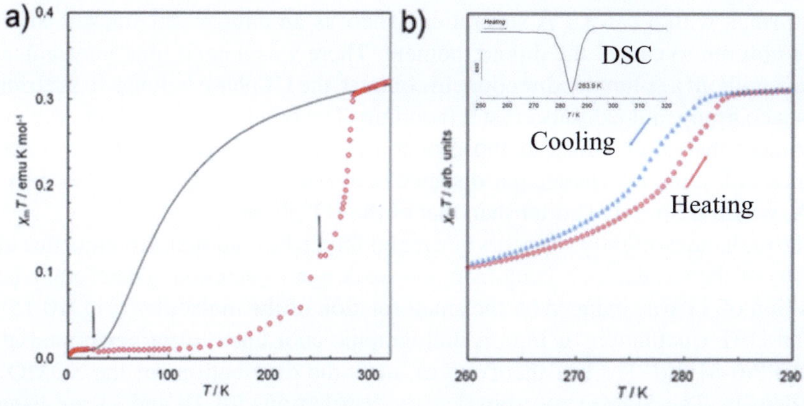

Fig. 10.13 Temperature dependence of c_mT for **11**; **a** The two arrows indicate the temperature of inverting sign of the observed magnetization in the raw data. The solid line represents the best fit to the Bleaney–Bowers model, **b** The measurement was carried out for the sample wrapped in paramagnetic Al foil to avoid any discontinuity of the magnetization in the range of 3–250 K. The red circle and blue triangle represent the heating and cooling processes, respectively. DSC measurement data is shown in the inset. Reproduced from Ref. [27] with permission from the Centre National de la Rechercher Scientifique (CNRS) and the Royal Society of Chemistry

Fig. 10.14 Molecular arrangement of **a** HT and **b** LT phases of **11**. Hydrogen atoms and some *tert*-butyl groups are omitted for clarity. Reproduced from Ref. [27] with permission from the Centre National de la Rechercher Scientifique (CNRS) and the Royal Society of Chemistry

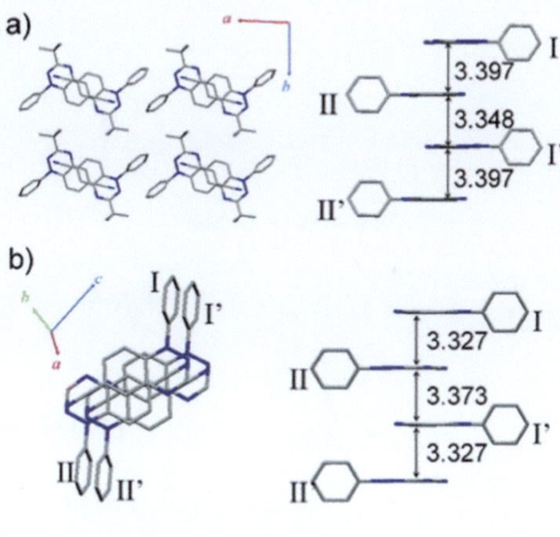

is delocalized over the benzotriazinyl ring. The dihedral angle between the phenyl ring and π-conjugated system of **10** is relatively smaller (45.2°) than that of **11**, **7** (73.8°) and due to the crystal packing effect. The crystal of 10 belongs to the monoclinic *I*2/*a* space group. In the crystal, **10** formed a 1D columnar structure due to the π–π interaction between the phenyl ring and π-conjugated system with an *ca.* 3.6 Å separation. In the columnar assembly, the molecules formed an antiparallel arrangement to cancel the dipole moment. On the other hand, the crystal of the HT phase of **11** belongs to the monoclinic *I*2/*a* space group and formed a 1D columnar structure along the *b* axis due to the weak π-π interaction between the functionalized benzo rings with a *ca.* 3.3 Å separation. There is an antiparallel stacked structure in the column to cancel the dipole moment. There was a negligible intermolecular contact without a columnar direction. In contrast, the LT phase belongs to the triclinic *P*-1 space group and exhibits crystal twinning. The bond lengths and dihedral angle are almost the same values as those of the HT phase. Regarding the contacts of the molecules, the intermolecular distance between the π-conjugated systems is ca. 3.3 Å, which is slightly shorter than that of the HT phase.

The molecules slightly slipped while maintaining the columnar structure due to the rotation of the molecules arising from the weak π–π interaction. Therefore, a phase transition of **11** was induced by the small rotation of the molecules (Fig. 10.15).

The DFT calculations using crystallographic coordinate were performed at the UB3LYP/6-31G(d) level of theory to estimate the distributions of the SOMO and spin density. The SOMO and spin density distributions for **10** and **11** are listed in Fig. 10.16. The SOMO and spin density mainly distributed over the molecules include the substituent and the spin density slightly distributed on the *tert*-butyl group. For **11**, there are negligible differences in the SOMO and spin density distributions between the HT and LT phases. To evaluate the contribution of the intermolecular magnetic interaction, the magnitude of the interaction *J* was estimated at the UB3LYP and UBLYP levels and the 6-31G(d) basis set.

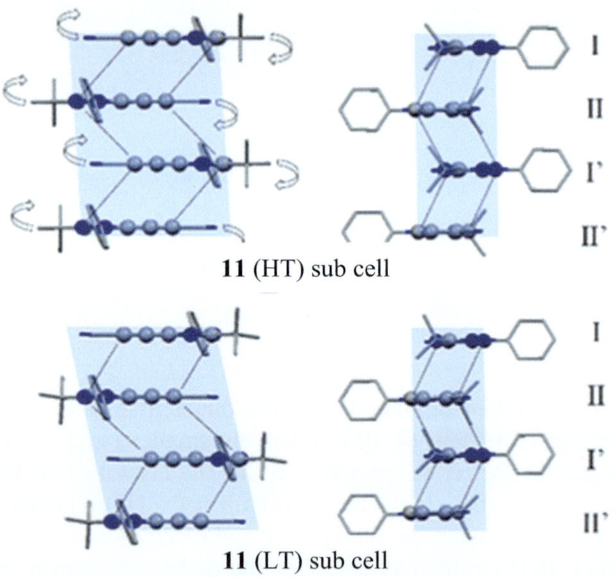

Fig. 10.15 Proposed mechanism of phase transition for **11**. Reproduced from Ref. [27] with permission from the Centre National de la Rechercher Scientifique (CNRS) and the Royal Society of Chemistry

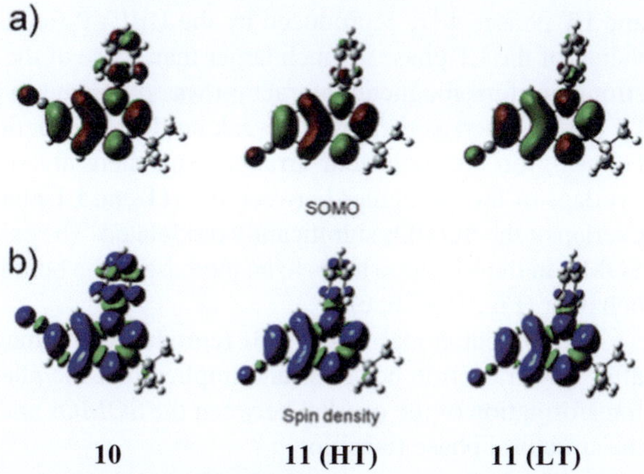

Fig. 10.16 **a** SOMO and **b** spin density distributions of **10** (left) and **11** (HT: middle, LT: right) at the UB3LYP/6-31G(d) level of theory. Reprint with permission from Ref. [27] with permission from the Centre National de la Rechercher Scientifique (CNRS) and the Royal Society of Chemistry

For **10**, the estimated J values within the column using both the UB3LYP ($J = -3.4\ \mathrm{cm}^{-1}$) and UBLYP ($J = -8.7\ \mathrm{cm}^{-1}$) functions were negative, suggesting that there was an antiferromagnetic coupling in the column. The magnitude of the J value of the intermolecular magnetic interaction within and between columns was well reproduced by the UBLYP function. The computation also implied that the antiferromagnetic coupling was mainly derived from the π–π interaction along the column (Fig. 10.17a).

Regarding the radical **11**, there were two kinds of contacts in both the HT and LT phases and the J values were also estimated. The magnetic characteristics of the HT

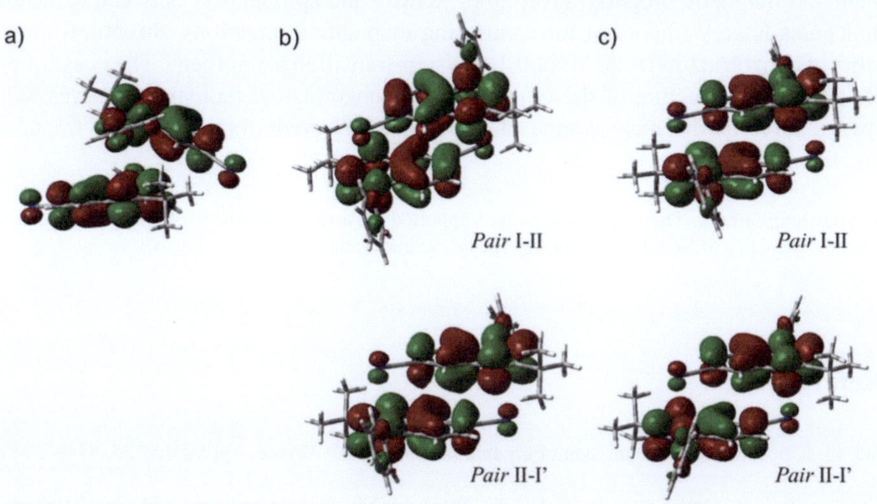

Fig. 10.17 Overlap of SOMOs for **a** **10**, **b** **11**(HT) and **c** **11**(LT). Reprint with permission from Ref. [27] with permission from the Centre National de la Rechercher Scientifique (CNRS) and the Royal Society of Chemistry

and LT phases were reproduced by the UBLYP function. The magnitude of the $2J$ values of the LT phase is much larger than those of the HT phase, indicating that the strong antiferromagnetic interaction was dominant in pair I–II and pair II–I' of the LT phase, whereas there was a weak antiferromagnetic interaction in the HT phase. Focusing on the molecular arrangement, there are small structural changes in the overlaps of the molecules between the HT and LT phases, however, the contact and overlap of the SOMOs significantly modulated. The existence of large overlaps of the SOMO in the LT phase leads to an increase in the bonding character and diamagnetic behavior (Fig. 10.17b, c).

The computational result well reproduced the magnetic interaction before and after spin transition behavior and implied that the phenomena were derived from a transformation of the overlap between the SOMOs and the spin density of **11** due to the structural phase transition.

10.5 Conclusions

We demonstrate that the chemical stability of dihydroquinoline *N*-oxyl based radicals was increased by the introduction of two phenyl groups at 2-position. Introduction of cardo structure is effective for stabilizing aromatic nitroxyl radical derivative by suppression of disproportionation reaction. Chemical modification of quinoline ring by annulation is promising way for designing building block for molecule based magnetic materials. Stable benzotriazinyl derivatives are also important for spin center. Introduction of electron withdrawing group affects the molecular arrangement and magnetic property. Overlap of SOMO and spin density between adjacent molecules is very important for controlling magnetic interactions. Structural flexibility of benzotriazinyl radical contributes spin-transition phenomena. The possibility of varying the structure of these nitroxyl and benzotiazinyl radical derivatives will always serve as the main source of new ideas and goals for exploratory for new functional materials.

Acknowledgements This work was partly supported by a Grant-in-Aid for Scientific Research from the Ministry of Education, Culture, Sports, Science, and Technology, Japan.

References

1. M. Kinoshita, Ferromagnetism of organic radical crystals. Jpn. J. Appl. Phys. **33**, 5718–5733 (1994)
2. O. Piloty, B. Graf Schwerin, Ueber das Nitril der Nitrosoisobuttersäure und seine Derivate. Ber. Deutsch. Chem. Ges. **34**, 1863–1870 (1901)
3. H. Wieland, K. Roth, Weitere Untersuchungen über Derivate des vierwertigen Stickstoffs. Ber. Deutsch. Chem. Ges. **53**, 210–230 (1920)

4. R. Suzuki, R. Tada, Y. Miura, N. Yoshioka, Synthesis and physicochemical properties of methoxy-substituted diphenyldihydroacridine and its Si and Ge bridged analogues and corresponding nitroxide radical derivatives. J. Mol. Struct. **1106**, 399–406 (2016)
5. A. Calder, A.R. Forrester, Nitroxide radicals. Part VI. Stability of *meta-* and *para-*alkyl substituted phenyl-t-butylnitroxides. J. Chem. Soc. C, 1459–1464 (1969)
6. (a) C. Berti, M. Colonna, L. Greci, L. Marchetti, Stable nitroxide radicals from 2-substituted quinoline-N-oxides with organometallic compounds. Tetrahedron **32**, 2147–2151 (1976); (b) R. Benassi, F. Taddei, L. Greci, L. Merchetti, G.D. Andreetti, G. Bocelli, P. Sgarabotto, Crystal and molecular structure of the N-oxyl radicals 1,2-dihydro-3-oxo-2,2-diphenyl-3H-indole 1-oxyl and 1,2-dihydro-2,2-diphenylquinoline 1-oxyl. Attempted calculation of hyperfine coupling constants by the INDO–SCF–MO method. J. Chem. Soc. Perkin II, 786–791 (1980)
7. M. Yao, H. Inoue, N. Yoshioka, Novel aromatic N-oxyl radical based on the benzo[g]quinoline skeleton: synthesis and intermolecular ferromagnetic interaction. Chem. Phys. Lett. **402**, 11–16 (2005)
8. M. Yao, H. Shibuya, T. Kato, H. Inoue, N. Yoshioka, Synthesis and magnetic properties of 2,2,4- and 2,2,6-triphenyl-1,2-dihydroquinoline-N-oxyl derivatives. Polyhedron **24**, 2828–2834 (2005)
9. M. Yao, S. Asakura, M. Abe, H. Inoue, N. Yoshioka, Formation of a one-dimensional stacking structure of π-conjugated nitroxyl radical bearing a 1,2,5-thiadiazole ring and its magnetic property. Cryst. Growth Des. **5**, 413–417 (2005)
10. H.M. McConnell, Ferromagnetism in solid free radicals. J. Chem. Phys. **39**, 1910 (1963)
11. K. Yoshizawa, R. Hoffmann, The role of orbital interactions in determining ferromagnetic coupling in organic molecular assemblies. J. Am. Chem. Soc. **117**, 6921–6926 (1995)
12. T. Suzuki, T. Tsuji, T. Okubo, A. Okada, Y. Obana, T. Fukushima, T. Miyashi, Y. Yamashita, Preparation, structure, and amphoteric redox properties of *p*-phenylenediamine-type dyes fused with a chalcogenadiazole unit. J. Org. Chem. **66**, 8954–8960 (2001)
13. J.C. Bonner, M.E. Fisher, Linear magnetic chains with anisotropic coupling. Phys. Rev. A **135**, 640–658 (1964)
14. V.T. Nguyen, M.L. Scudder, A.N.M.M. Rahman, R. Bishop, D.C. Craig, New aryl edge–edge packing interactions involving the N-oxide functional group. Aust. J. Chem. **52**, 1047–1054 (1999)
15. B. Bleaney, K.D. Bowers, Anomalous paramagnetism of copper acetate. Proc. Roy. Soc. Lon. Ser. A **214**, 451–465 (1952)
16. H.M. Blatter, H. Lukaszewski, A new stable free radical. Tetrahedron Lett. **9**, 2701–2705 (1968)
17. (a) P.A. Koutentis, D. Lo Re, Catalytic oxidation of n-phenylamidrazones to 1,3-diphenyl-1,4-dihydro-1,2,4-benzotriazin-4-yls: an improved synthesis of blatter's radical. Synthesis 2075–2079 (2010); (b) A.A. Berezin, G. Zissimou, C.P. Constantinides, Y. Beldjoudi, J.M. Rawson, P.A. Koutentis, Route to benzo- and pyrido-fused 1,2,4-triazinyl radicals via *N*'-(het)aryl-*N*'-[2-nitro(het)aryl]hydrazides. J. Org. Chem. **79**, 314–327 (2014)
18. Y. Takahashi, Y. Miura, N. Yoshioka, Introduction of three aryl groups to benzotriazinyl radical by Suzuki-Miyaura cross-coupling reaction. Chem. Lett. **43**, 1236–1238 (2014)
19. B. Yan, J. Cramen, R. McDonald, N.L. Frank, Ferromagnetic spin-delocalized electron donors for multifunctional materials: π-conjugated benzotriazinyl radicals. Chem. Commun. **47**, 3201–3203 (2011)
20. (a) C.P. Constantinides, P.A. Koutentis, H. Krassos, J.M. Rawson, A.J. Tasiopoulos, Characterization and magnetic properties of a "super stable" radical 1,3-diphenyl-7-trifluoromethyl-1,4-dihydro-1,2,4-benzotriazin-4-yl. J. Org. Chem. **76**, 2798–2806 (2011), (b) C.P. Constantinides, P.A. Koutentis, J.M. Rawson, Ferromagnetic interactions in a 1D alternating linear chain of π-stacked 1,3-diphenyl-7-(thien-2-yl)-1,4-dihydro-1,2,4-benzotriazin-4-yl radicals. Chem.–Eur. J. **18**, 7109–7116 (2012)
21. Y. Takahashi, Y. Miura, N. Yoshioka, Synthesis and properties of the 3-*tert*-butyl-7-trifluoromethyl-1,4-dihydro-1-phenyl-1,2,4-benzotriazin-4-yl radical. New J. Chem. **39**, 4783–4789 (2015)

22. G.A. Baker Jr., G.S. Rushbrooke, H.E. Gilbert, High-temperature series expansions for the spin-½ heisenberg model by the method of irreducible representations of the symmetric group. Phys. Rev. A **135**, 1272–1277 (1964)
23. Y. Miura, N. Yoshioka, π-Stacked structure of thiadiazolo-fused benzotriazinyl radical: crystal structure and magnetic properties. Chem. Phys. Lett. **626**, 11–14 (2015)
24. (a) O. Kahn, C.J. Martinez, Spin-transition polymers: from molecular materials toward memory devices. Science **279**, 44–48 (1998), (b) C. Janiak, Engineering coordination polymers towards applications. Dalton Trans. 2781–2804 (2003)
25. (a) K. Mukai, Anomalous magnetic properties of stable crystalline phenoxyl radicals. Bull. Chem. Soc. Jpn. **42**, 40–46 (1969); (b) W. Fujita, K. Awaga, Room-temperature magnetic bistability in organic radical crystals. Science **286**, 261–262 (1999); (c) Y. Morita, S. Suzuki, K. Fukui, S. Nakazawa, H. Kitagawa, H. Kishida, H. Okamoto, A. Naito, A. Sekine, Y. Ohashi, M. Shiro, K. Sasaki, D. Shiomi, K. Sato, T. Takui, K. Nakasuji, Nat. Mater. **7**, 48–51 (2008); (d) A. Alberola, D.J. Eisler, L. Harvey, J.M. Rawson, Molecular tailoring of spin-transition materials: preparation, crystal structure and magnetism of trifluoromethyl-pyridyl-1,3,2-dithiazolyl. CrystEngComm **13**, 1794–1796 (2011)
26. C.P. Constantinides, A.A. Berezin, G.A. Zissimou, M. Manoli, G.M. Leitus, M. Bendikov, M.R. Probert, J.M. Rawson, P.A. Koutentis, A magnetostructural investigation of an abrupt spin transition for 1-phenyl-3-trifluoromethyl-1,4-dihydrobenzo[e][1,2,4]triazin-4-yl. J. Am. Chem. Soc. **136**, 11906–11909 (2014)
27. Y. Takahashi, N. Tsuchiya, Y. Miura, N. Yoshioka, Magneto-structural correlation of cyano-substituted 3-*tert*-butyl-1-phenyl-1,2,4-benzotriazin-4-yl: spin transition behaviour observed in a 6-cyano derivative. New J. Chem. **42**, 9949–9955 (2018)

Chapter 11
Construction of Molecule-Based Strongly Isotropic Lattices and Their Physical Properties

Qi Chen, Yoshiaki Shuku, Rie Suizu, and Kunio Awaga

Abstract "Strong isotropy" is a geometrically defined symmetry that involves both atomic positions and bonding configurations. It is mathematically proven that only three specific lattices: honeycomb, diamond, and K_4, possess this property. In this chapter, we introduce a methodology for rationally constructing these lattices using polyhedral π-conjugated molecules and metal–organic frameworks (MOFs). We also explore the various physical properties that emerge from the synergistic interaction between strong isotropy and molecular degrees of freedom.

Keywords Strong isotropy · Line graph · Molecular assembly · Spin frustration · Flat band · Chirality

11.1 Strongly Isotropic Lattices

Crystal structures have been characterized based on the crystallographic symmetry, which is determined by the atomic positions within a unit cell. Recently, a new lattice symmetry named "strong isotropy" has been proposed based on geometry [1–3]. As shown in Scheme 11.2a, the honeycomb lattice consists of three bonds (red, blue and green) that are oriented in different directions. Any bond swapping, e.g., between red and blue, can be explained by a symmetry operation. This is the strong isotropy that preserves the crystal net after any permutation of edges with a common vertex. In contrast, the square lattice does not exhibit this property, as shown in Scheme 11.2b. The crystallographic symmetry is determined only by atomic positions, but the strong isotropy is governed by both atomic and bond positions. Therefore, the regulation of the latter is more rigorous. Furthermore, it has been mathematically proven that only honeycomb, diamond and K_4 lattices satisfy the requirement of strong isotropy. Scheme 11.1 shows the three strongly isotropic lattices that are characterized by the

The original version of the chapter has been revised. A correction to this chapter can be found at https://doi.org/10.1007/978-981-96-4479-7_12

Q. Chen · Y. Shuku · R. Suizu · K. Awaga (✉)
Department of Chemistry and Integrated Research Consortium on Chemical Science, Nagoya University, Nagoya, Japan
e-mail: awagakunio@gmail.com

corrected publication 2026
S. Kobatake and H. Uekusa (eds.), *Advances in Organic Crystal Chemistry*,
https://doi.org/10.1007/978-981-96-4479-7_11

specific angles: cos $\theta = -1/2$, $-1/3$, and $+1/3$, respectively. It is mathematically demonstrated that the inclusion of the numbers 2 and 3 in these expressions is due to these being 2D and 3D lattices with very high isotropy. Since cos θ for the diamond and K_4 lattices are $-1/3$ and $+1/3$, respectively, they are called "mathematical twins". Although the visibility of the K_4 lattice is much less than that of the diamond lattice, the K_4 lattice is a universal material structure which appears in various scales [4]. It is also notable that the K_4 lattice possesses chirality, while the other two are achiral.

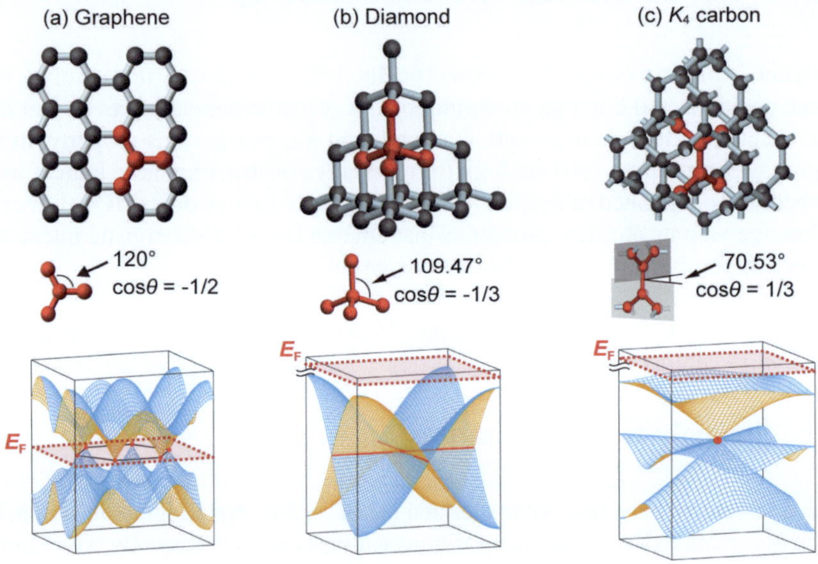

Scheme 11.1 Crystal and band structures of graphene (**a**), diamond (**b**), and K_4 carbon (**c**). Part of the figure was reproduced from Ref. [7] with permission from the Royal Society of Chemistry

Scheme 11.2 Comparison between the honeycomb lattice (**a**) and square lattice (**b**). In the former, bond swapping between the blue and red edges can be expressed by the symmetry operation, but in the latter, this is not the case. Reproduced from Ref. [7] with permission from the Royal Society of Chemistry

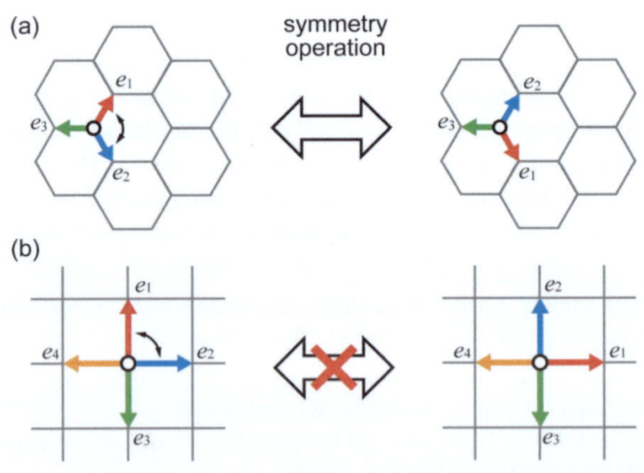

The honeycomb and diamond lattices correspond to the crystal structures of sp^2 and sp^3 carbons, namely graphite (or graphene) and diamond, respectively. Recently, K_4 carbon, in which sp^2 carbons form a K_4 structure, was proposed as a new carbon allotrope based on graph theory [1–3], although it has not been synthesized yet. Therefore, the strongly isotropic lattices can be regarded as the crystal structures of the carbon allotropes. Scheme 11.1 also shows the unique band structures of graphene, diamond and K_4 carbon that possess Dirac cones, triplet Dirac cones, and Dirac nodal lines, respectively [4–6].

Line graphs of strongly isotropic lattices are important for understanding the solid-state properties of these lattices [7], as will be shown later. The first column in Scheme 11.3 shows honeycomb, K_4 and diamond lattices. In the second column, a small blue atom is added on each bond. By drawing bonds between the blue atoms, the original three lattices are transformed into the kagome, hyper-kagome and pyrochlore lattices, respectively, as shown in the third column in Scheme 11.3. This transformation illustrates the concept of line graphs. The line graphs of the honeycomb, K_4 and diamond lattices are the kagome, hyper-kagome and pyrochlore lattices, respectively, which are well recognized as spin frustration lattices.

The basic unit structure of the strongly isotropic lattices is either a triangular structure with three branches or a tetrahedral structure with four branches. The former is used for honeycomb and K_4 lattices, while the latter is used for diamond lattice. To rationally form strongly isotropic lattices using molecules, we have adopted polyhedral π-conjugated molecules, in which the normal vectors of three or four π planes possess C_3 symmetry, expecting that intermolecular π-π overlaps automatically form the strongly isotropic structures, as shown in Scheme 11.4. Regarding the C_3 polyhedral molecules, there are two shapes, namely Δ and Y. In geometry, the relationship between the two shapes is called "dual", because one can be transformed into the other under the rule shown in Scheme 11.5. As shown in Scheme 11.4, metal–organic frameworks (MOFs) and covalent-organic frameworks (COFs) are also promising materials for forming the molecule-based strongly isotropic lattices.

11.2 Molecule-Based Honeycomb Lattices

Molecule-based honeycomb lattices are much more common than other strongly isotropic two lattices and can be found in a wide variety of materials. In other words, molecular honeycomb lattices are easier to construct rationally. In this section, we describe our efforts to construct molecule-based honeycomb lattices, using polyhedral π-conjugated molecules and MOFs, and to reveal their physical properties in terms of molecular degrees of freedom and line graph.

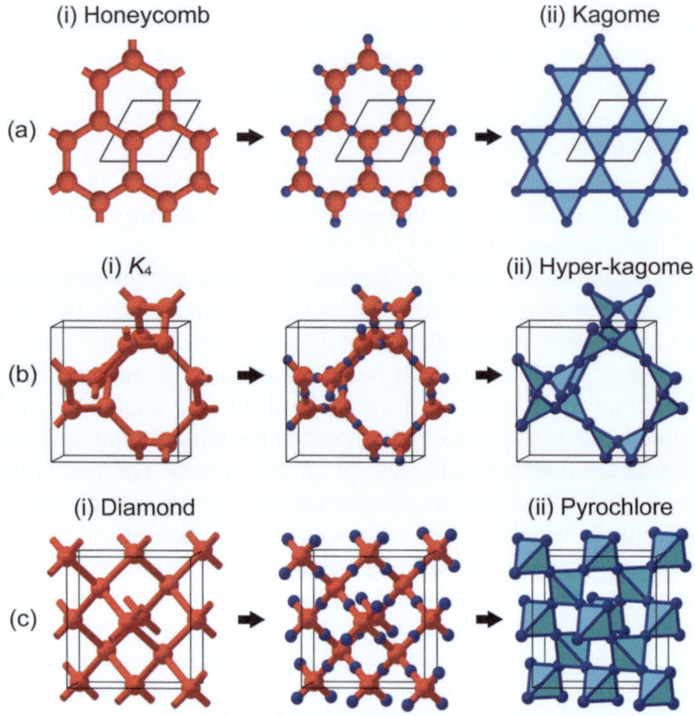

Scheme 11.3 Line graphs of the strongly isotropic lattices, honeycomb, K_4 and diamond. Panels **a–c** show their transformations toward the spin frustration lattices, kagome, hyper-kagome and pyrochlore, respectively. Reproduced from Ref. [7] with permission from the Royal Society of Chemistry

11.2.1 Spin-Liquid Ground State in $Cu_3(HHTP)_2$

Recently, studies on 2D MOFs have rapidly propagated in the field of materials science. Being polymeric molecular materials with π-d conjugated structures, porosities and intriguing electrical properties, they are considered to be promising active materials in various devices, such as organic FET devices, chemiresistive sensors, supercapacitors, electrocatalysts and cathode active materials in batteries. There are many 2D MOFs with honeycomb structures. One typical example is $Cu_3(HHTP)_2$ (HHTP = hexahydroxytriphenylene) (see Fig. 11.1), in which HHTP ligands with C_3 symmetry are connected by square-planar Cu(II) coordination [8]. As explained in Sect. 11.1, the HHTP ligands form a honeycomb lattice, while the Cu(II) ions (S = 1/2) form a kagome lattice, leading to the expectation that this MOF would exhibit geometrical spin frustration. Although the electrical properties of $Cu_3(HHTP)_2$ were investigated by Day et al. [8], no magnetic studies had been conducted from this perspective. Therefore, we investigated the ultralow temperature physical properties of $Cu_3(HHTP)_2$ down to 38 mK [9].

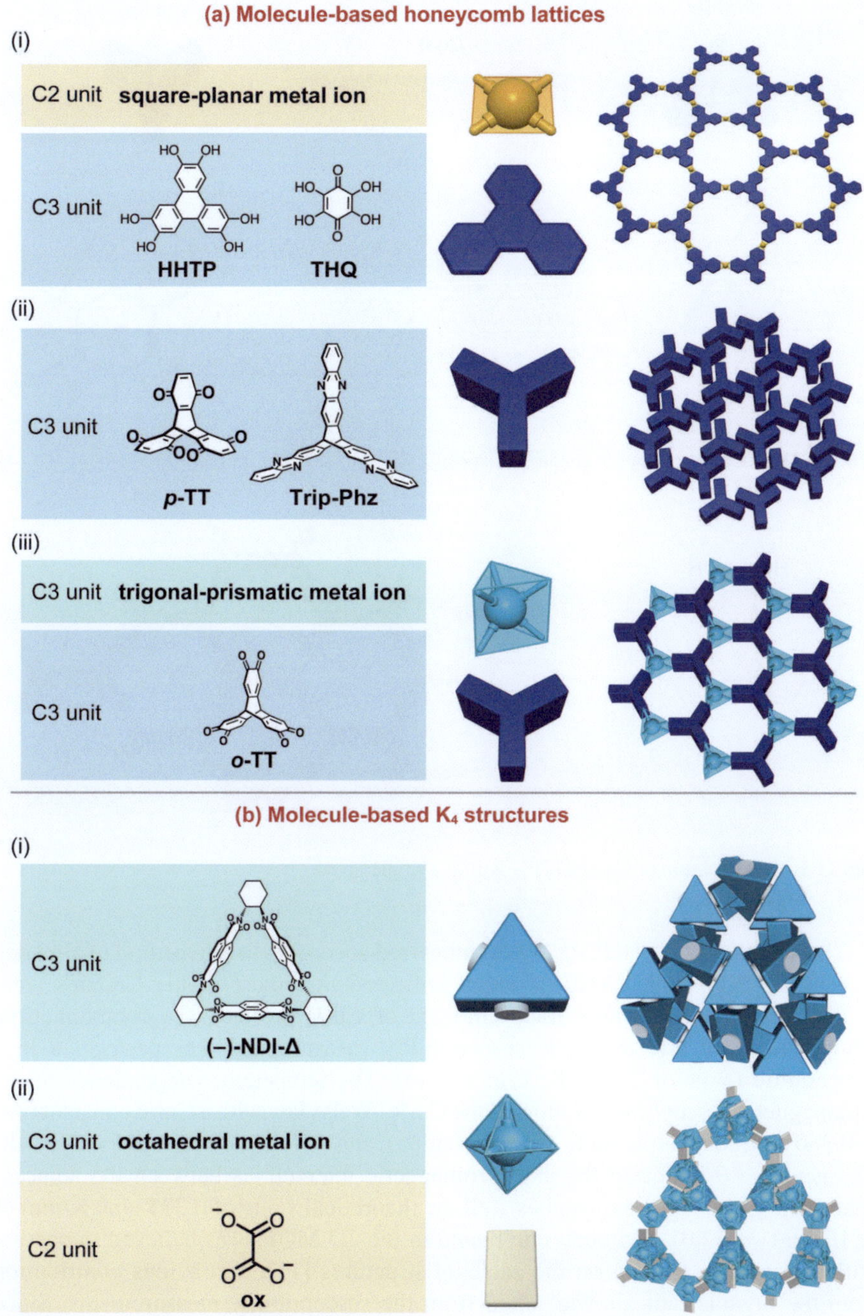

Scheme 11.4 Synthetic strategy to form molecule-based strongly isotropic structures, such as honeycomb lattice (a) and K_4 lattice

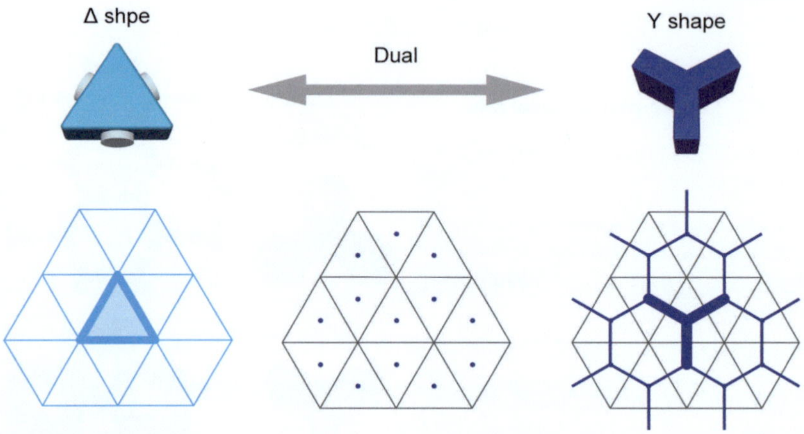

Scheme 11.5 Duality between Δ and Y shapes

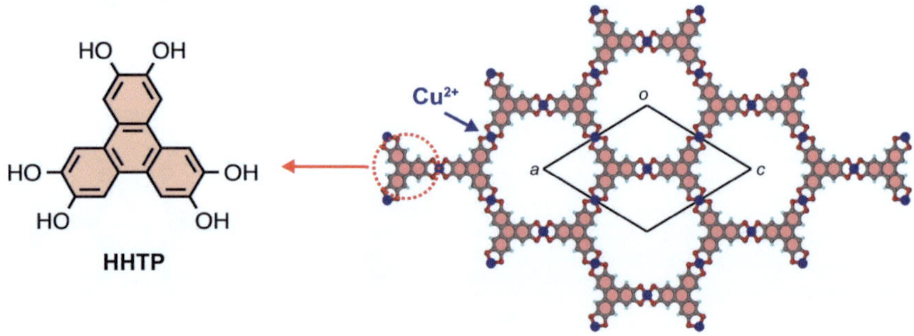

Fig. 11.1 The structure of $Cu_3(HHTP)_2$ within a single layer

The sample of $Cu_3(HHTP)_2$ was synthesized according to the method of Hmadeh et al. [10]. XPS and XANES spectra indicated that most of the Cu ions were in the + 2 state, with only trace amounts of Cu(I) present as a contamination. Temperature variable magnetic susceptibility measurement was performed in a temperature range of 2–300 K (Fig. 11.2a-i). The temperature dependence of the paramagnetic susceptibility followed the Curie–Weiss law with a Curie constant of $C = 0.380$ emu K/mol (per mole of Cu(II) cation), and a Weiss constant of $\theta = -3.4$ K. The negative θ indicated the antiferromagnetic interactions between the adjacent spins. The Curie constant matches well the theoretical value of 0.375 emu K/mol ($S = 1/2$ and $g = 2.0$). Although one ligand in the 2D MOFs was believed to possess multiple radicals located on the catechol structures [11], no obvious contribution from the organic radicals was noted from the susceptibility measurements. Since another possible scenario was the strong antiferromagnetic coupling within the domestic $Cu(II)(catechol)_2$ unit, which may also lead to an effective $S = 1/2$, the existence of organic radicals in this MOF cannot be simply excluded from the susceptibility measurements. Then, we investigated the temperature dependence of the AC magnetic susceptibility of $Cu_3(HHTP)_2$ in the temperature range of 38 mK to 4 K. The real part was normalized and combined with high-temperature data to generate a composite χ_m-T plot (Fig. 11.2a-ii). No decrease, anomaly, or λ-shape was observed in the χ_m-T plot, indicating that magnetic long-range ordering (LRO) was not present. This absence of LRO was attributed to the spin frustration among the Cu(II) $S = 1/2$ spins. In a logarithmic scale plot, a relationship of $\chi \propto T^{-0.31}$ was revealed at ultralow temperatures (38–200 mK). This behavior deviated from the Curie–Weiss law and suggested that the contribution of paramagnetic impurities is negligible. Thus, the observed magnetic behavior and the lack of LRO appear to be intrinsic properties of $Cu_3(HHTP)_2$.

We also investigated the temperature dependence of the heat capacity C_p for $Cu_3(HHTP)_2$ down to 65 mK and found that C_p smoothly and monotonically decreases while lowering the temperature, and no sharp peak could be detected in the experimental temperature range. Therefore, the heat capacity measurements

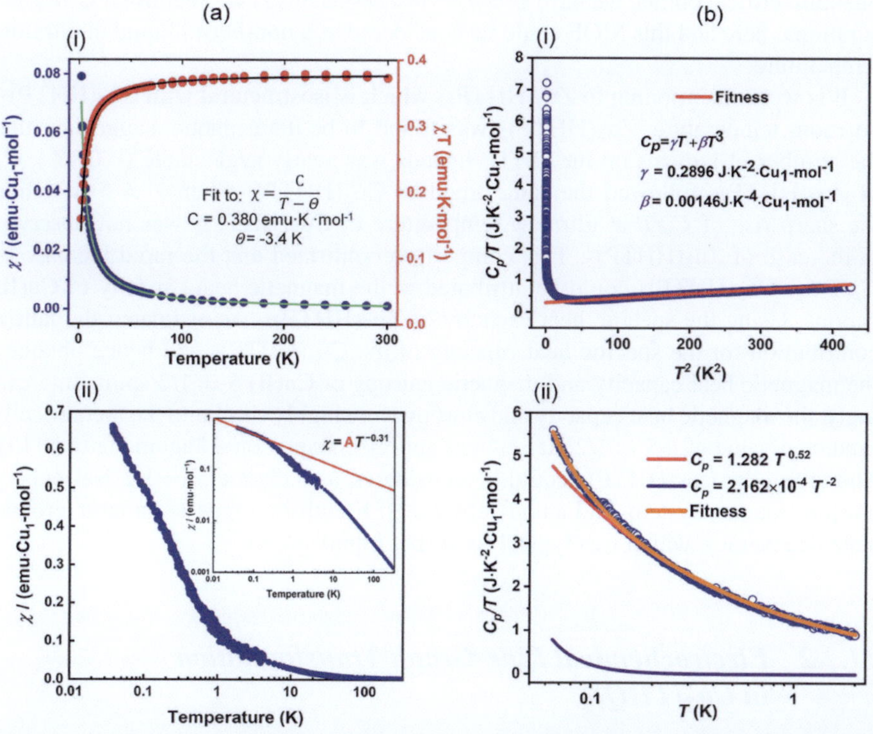

Fig. 11.2 a Magnetic susceptibility of $Cu_3(HHTP)_2$ in the temperature range of 2–300 K (i). Temperature variable magnetic susceptibility of $Cu_3(HHTP)_2$ down to 38 mK (ii). The logarithm scaled plot (inset of (i)) suggested a power law of $\chi_m \propto T^{-0.31}$. (**b**) C_p/T-T^2 plot of $Cu_3(HHTP)_2$ under a temperature range of 0.065–20 K. (i) C_p/T-T plot in the low temperature region (0.065–2 K) (ii). Adapted with permission from Ref. [9]. Copyright 2020, American Chemical Society

also indicated that LRO was not present. Figure 11.2b-i shows the C_p/T-T^2 plots, indicating a rough linear relationship in the range of 7–15 K. This can be expressed as $C_p = \gamma T + \beta T^3$, where the T^3 term apparently represents the heat capacity of lattice vibration, but the physical origin of the linear γT term was not clear. The temperature range of 7–15 K was higher than the Weiss constant of the MOF, and therefore, the linear term here should not be assigned to the fermionic behavior of spins as described in other QSL materials. Nevertheless, after subtracting the lattice vibrational heat capacity, the specific heat capacity in the ultralow temperature region ($T < 2$ K) could be fitted to a function of $C_p = \alpha T^{\varepsilon} + \lambda T^{-2}$, and the best fitness gives $\varepsilon = 0.52$ (Fig. 11.2b-ii). The first term represents the specific heat of the quantum spin liquid state, and the second term arises from Schottky anomaly, which is possibly induced by the nuclear spins. It is worth noting that the spin-contributed heat capacity of herbertsmithite was fitted to $C_p \propto T^{\beta}$ with $\beta = 0.5$–1.0 with different temperature ranges [12]. As described in Sect. 11.3.1, $TBA_{1.5}$[(−)-NDI-Δ] in a 3D hyper-kagome spin-liquid state exhibited the spin-contributed heat capacity of $C \propto T^{0.62}$ [13]. Theoretical approaches suggested that, for the non-Fermi liquid near heavy-fermion quantum critical points, we have $C \propto T^{2/3}$ in 2D systems [14]. Therefore, $C \propto T^{0.52}$ is rational here and this MOF could be considered as a non-Fermi liquid at ultralow temperature.

It is worth mentioning to $Zn_3(HHTP)_2$ which is isostructural with $Cu_3(HHTP)_2$. At room temperature, $Zn_3(HHTP)_2$ was found to be diamagnetic, suggesting that the number of radicals on the HHTP ligands was nearly neglectable (< 0.1%). C_p of $Zn_3(HHTP)_2$ followed the same trend as $Cu_3(HHTP)_2$ when $T > 5$ K, while the sharp rise of C_p/T at ultralow temperature of $Cu_3(HHTP)_2$ was not observed in the case of $Zn_3(HHTP)_2$. This comparison confirmed that the rapid increase of C_p/T for $Cu_3(HHTP)_2$ could be attributed to the magnetic heat capacity of Cu(II) cations. Using the specific heat capacity of $Zn_3(HHTP)_2$, we estimated the lattice contribution for the specific heat capacity of the $Cu_3(HHTP)_2$ and hence obtained the magnetic heat capacity and magnetic entropy of Cu(II) $S = 1/2$ spins. Interestingly, the magnetic heat capacity and entropy were highly similar to the theoretically predicted value of a $S = 1/2$ Heisenberg antiferromagnet on a kagome lattice [15]. Therefore, the $Cu_3(HHTP)_2$ could be considered as a classic $S = 1/2$ Heisenberg antiferromagnet system, and a non-zero $T = 0$ K entropy suggested a large ground state degeneracy, which was typical for a spin liquid system.

11.2.2 Electrochemical Line-Graph Transformation in $Cu_3(THQ)_2$

We have reported that 2D π-d conjugated $Cu(II)_3(THQ)_2$ (THQ = tetrahydroxy-1,4-benzoquinone) MOF demonstrates graphite-like charge storage mechanism in lithium-MOF battery [16]. A new paradigm, called electrochemical magnetometry, was carried out to elucidate the evolution of electronic states during electrochemical process the multi-step redox reactions of $Cu_3(THQ)_2$ in lithium-ion batteries [17]. The pristine $Cu_3(THQ)_2$ exhibits a kagome lattice of Cu(II) ions, characteristic of a frustrated spin-1/2 antiferromagnetic ground state. Upon stepwise reduction, the localized unpaired spins associated with Cu(II) are compensated, and the additional electrons occupy the ligand network, which forms a honeycomb lattice configuration. Notably, the honeycomb lattice is mathematically equivalent to the line graph of the kagome lattice [7]. This suggests that electrochemical reduction of $Cu_3(THQ)_2$ could experimentally realize a line-graph transformation from kagome to honeycomb lattice.

The electrochemical properties of $Cu_3(THQ)_2$ MOF were evaluated by using a coin cell with a lithium metal counter electrode. Galvanostatic charge/discharge profiles (Fig. 11.3a) in the 1.5–4.0 V range show that both cathode types yielded similar ~ 390 mAh/g capacities, suggesting three Li^+ ions insert into the $Cu_3(THQ)_2$ framework. The obtained capacity suggests a three-electron redox reaction in the repeating unit of $[Cu(C_4O_4)]$.

To elucidate the microscopic charge storage mechanism, we plotted differential capacity (dQ/dV) curves against potential (Fig. 11.3b). As dQ/dV reflects the electrochemically active density of states (EA-DOS) [18], it provides insights into the

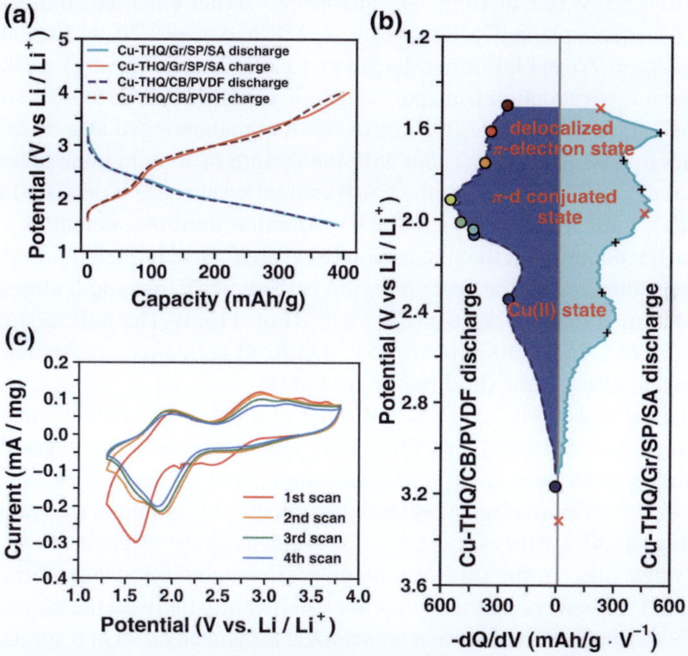

Fig. 11.3 Electrochemical performance of $Cu_3(THQ)_2$ MOF. **a** Initial discharge–charge profiles at 50 mAh/g current density. **b** Differential capacity (dQ/dV) versus potential, with ESR (left) and SQUID (right) detected states marked. **c** Cyclic voltammetry in the first four scans at a scan rate of 1 mV/s. Adapted with permission from Ref. [16]. Copyright 2023, American Chemical Society

electronic band structure of $Cu_3(THQ)_2$. Building on Nishihara et al.'s work [19] on 2D π-d conjugated frameworks, which involves π-d conjugated, M-d related, and linker states, we propose that the Cu(II) state, π-d conjugated state, and delocalized π-electron state (Fig. 11.3b) contribute to three-electron redox reaction of $Cu_3(THQ)_2$. Similar to graphite anodes in Li-ion batteries, where the π-electron band accepts/donates electrons during the charge/discharge process [20], delocalized π-electron state of $Cu_3(THQ)_2$ can accumulate an additional electron. Cyclic voltammetry (CV) was employed to probe the electrochemical kinetics of $Cu_3(THQ)_2$. Figure 11.3c reveals three redox peaks after the first cycle activation, corresponding to the reduction/oxidation of the three states identified in dQ/dV. It should be noted that the delocalized π-electron state exhibits reversible Li^+ intercalation/deintercalation between 1.5 and 2.5 V, while the Cu(II) and π-d conjugated states demonstrate unstable redox behavior during cycling.

To investigate the charge storage mechanism of $Cu_3(THQ)_2$ electrodes, we conducted ex-situ ESR spectra measurements under strict air-free conditions to monitor changes in spin states (Fig. 11.4a). The ESR spectrum of the pristine $Cu_3(THQ)_2$ electrode shows a broad and asymmetric line shape with an isotropic g value of ~ 2.11, due to strong spin–spin exchange interactions [21] and effective microwave absorption by the conductive carbon additive, respectively. After initial discharge to 2.35 V (85 mAh/g), an anisotropic signal emerged, indicating axial symmetry for square planar Cu(II) ions ($g_{\perp} = 2.06$ and $g_{//} = 2.26$, and a peak-to-peak distance $\Delta H_{pp} = 7.7$ mT). Further discharge to 2.07 V (150 mAh/g) weakened this signal, indicating a reduction from paramagnetic Cu(II) ($S = 1/2$) to ESR silent Cu(I) ($S = 0$). At 2.05 V (160 mAh/g), a narrow ESR signal emerged at $g = 2.0047$ with $\Delta H_{pp} = 0.8$ mT, which is consistent with the feature of semiquinone and attributed to a C–O• radical [22]. Although the Cu(II) signal weakened, it persisted to 1.91 V (225 mAh/g), indicating continued Cu(II) reduction until this potential. Given the most of Cu(II) reduction in the first region (blue area), unsaturated quinoidal moiety likely accepts electron in the second region (green area), forming a closed-shell π-conjugated aromatic network, $[Cu^{I}(CO^{-})_4]^{3-}$ (Fig. 11.4b). The radical signal intensified slightly to 1.75 V (300 mAh/g) but stabilized thereafter, suggesting constant radical concentration in the third region (red area).

Quantitative analysis using TEMPOL/KBr revealed a low radical concentration of approximately 1.5% per $[Cu(CO)_4]$ unit. This contradicts the expected radical increase during quinone reduction to semiquinone and subsequent diradical formation. Instead, we propose direct C=O reduction to $C–O^{-}$ without a free radical intermediate (Fig. 11.4b). The observed C–O˙ radicals likely originate from extrinsic defects at edge sites. In the third discharge region, an additional electron is doped into the ligand framework, but this does not significantly increase the radical concentration. We ascribe this electron to a delocalized π-band electron or a spinless singlet pair.

To confirm π-electron delocalization, we compared ESR line shape of $Cu_3(THQ)_2$ discharged to 1.9 V and 1.5 V. The 1.9 V spectrum showed a narrow Lorentzian line ($\Delta H_{pp} = 0.66$ mT) centered at $g = 2.0047$, consistent with localized radical defects. The fully reduced 1.5 V spectrum exhibited an additional broad line ($\Delta H_{pp} = 2.98$

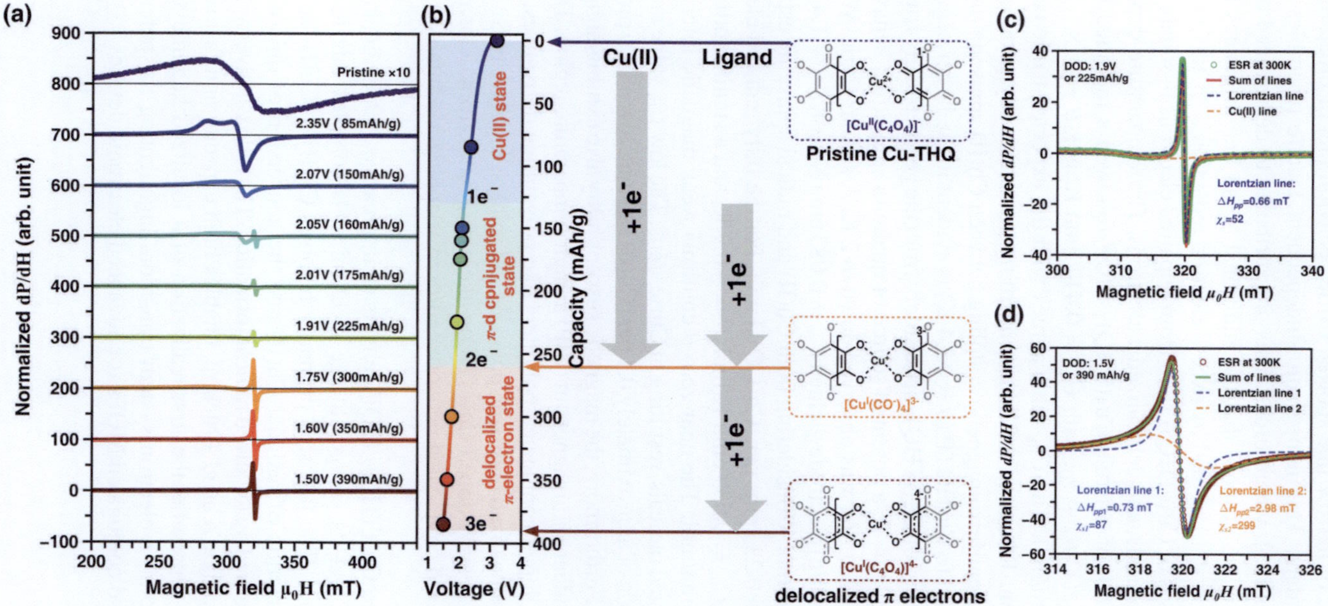

Fig. 11.4 Electronic state evolution during electrochemical reduction. **a** *Ex-situ* X-band ESR spectra of $Cu_3(THQ)_2$ electrodes ($Cu_3(THQ)_2$/CB/PVDF) at the various depth-of-discharge from open circuit voltage to 1.5 V. **b** Proposed charge storage mechanism for $Cu_3(THQ)_2$ MOF. Line shape fitting of ESR spectra for $Cu_3(THQ)_2$ discharged to 1.9 V (**c**) and 1.5 V (**d**). Adapted with permission from Ref. [16]. Copyright 2023, American Chemical Society

mT) with a similar resonant field but distinct temperature dependence. While the spin susceptibility χ_s of narrow line showed Curie behavior, that of broad line was temperature-independent. This temperature-independent feature resembles doped conducting polymers [23], where broad lines are attributed to delocalized carriers (polarons). Therefore, we propose that the additional broad line originates from a delocalized π-electron state formed by excess electron reduction of $[Cu^{I}(CO^{-})_4]^{3-}$.

Magnetic responses of $Cu_3(THQ)_2$ stem from unpaired electrons in square planar Cu(II) ions, ligand radicals, and delocalized π electrons. The temperature-dependent magnetic susceptibility of pristine $Cu_3(THQ)_2$ MOF followed a modified Curie–Weiss law, $\chi_m = C_m/(T–\theta) + \chi_0$, with $C_m = 0.412$ emu K/mol, $\theta = –18.1$ K, and $\chi_0 = 7.7 \times 10^{-5}$ emu/mol. The C_m value agrees with the expected Curie constant for Cu(II) (0.410 emu K/mol, $g = 2.11$), confirming predominantly paramagnetic Cu(II) ions. Stronger antiferromagnetic coupling between Cu(II) spins via the short THQ ligand compared to Cu-HHTP ($\theta = –3.4$ K) is evident in the larger negative θ. The magnetic properties of $Cu_3(THQ)_2$ arise solely from the geometrically frustrated Cu(II) spin (kagome-lattice), a potential source of quantum spin liquid behavior, as no long-range magnetic order was observed down to 2 K. A similar Curie–Weiss fit for the 2.0 V (200 mAh/g) electrode yielded $C_m = 0.085$ emu K/mol, $\theta = -1.4$ K, and $\chi_0 = -4 \times 10^{-6}$ emu/mol. Considering both Cu(II) and radical spins, the estimated spin concentration is 0.23 spins per $[Cu(CO)_4]$ unit. For the fully reduced electrode (1.5 V), the fit gave $C_m = 0.039$ emu K/mol, $\theta = -0.9$ K, and a significantly larger $\chi_0 = 1.02 \times 10^{-3}$ emu/mol. The near-zero θ confirms weak magnetic interactions between these localized radicals, resulting in Curie spins. The low C_m value indicates a radical concentration of approximately 0.1 per $[Cu(CO)_4]$ unit, lower than predicted diradical formation. Furthermore, the magnetic susceptibility also exhibits significant temperature-independent paramagnetism (TIP) at high-temperature region. This TIP behavior resembles doped conjugated polymers like polyaniline derivatives, where TIP is attributed to delocalized high-spin polarons and spinless bipolarons. Similarly, the observed TIP in heavily doped $Cu_3(THQ)_2$ likely originates from delocalized π electrons within the π-d conjugated network and/or overlapping π-orbitals along the *c* direction.

2D π-d conjugated $Cu_3(THQ)_2$ MOF delivers an impressive 390 mAh/g capacity in lithium batteries, yet its charge storage mechanism is complex due to strong electronic correlations. The first step primarily involves Cu(II) to Cu(I) reduction, followed by the reduction of residual Cu(II) and the conversion of quinoid to π-conjugated, accompanied by localized radical formation. The redox-active π-electron state, resembling graphite's band structure, enables high capacity and fast reaction kinetics. This work represents the first realization of a localized kagome spin-to-delocalized honeycomb π-electron system transition, highlighting the potential of heavily doped 2D π-d conjugated MOFs for advanced electronic devices.

11.2.3 Dirac Cones and Flat Bands in the Band Structure of Rb_3(p-TT)

Isostructure and isoelectronicity are crucial concepts in chemistry to understand the chemical and physical properties of atoms, ions, molecules and solid-state materials. Recently, the electronic band structures of carbon allotropes graphene, diamond and theoretically predicted K_4 carbon attract much attention due to their unique structures such as Dirac cone derived from their highly symmetric crystal structures. From this perspective, isostructure of the carbon allotropes formed by supramolecular chemistry [24–26] are promising to realize exotic materials not only with isoelectronic structure of carbon allotropes but also with various functionality of molecule. Triptycene skeleton which has threefold symmetric molecular structure with delocalized π planes is one of the best candidates as isostructure of sp^2 carbon.

In this section, the crystal structure and electronic band structure of triptycene-based electron acceptor, 9,10-dihydro-9,10-[1′,2′]-benzenoanthracene-1,4,5,8,13,16-hexaone (*p*-TT, Fig. 11.5a) will be described. The radical anion salt of *p*-TT forms a supramolecular graphene structure, and the tight-binding band calculation revealed unique features, such as Dirac cones which were "isostructural" to those of graphene and flat bands which is derived from molecular feature [6].

The synthesis of *p*-TT was first reported in 1981 [27]. This molecule has a rigid paddle-wheel shape which consists of three 1,4-benzoquinone (*p*-BQ) moieties. Due to an electron deficient nature of the *p*-BQ skeleton, *p*-TT is known as an electron acceptor. The cyclic voltammogram of *p*-BQ exhibits two reduction peaks at -0.865 and -1.35 V (vs. Fc/Fc$^+$), which are assigned to 0 to -1 and -1 to -2 reductions, respectively. In contrast, the first reduction potential of *p*-BQ is split into three potentials in *p*-TT (0 to -3 state), the first of which appears at -0.690 V (vs. Fc/Fc$^+$). This value is significantly lower than that of *p*-BQ, indicating a stabilization of

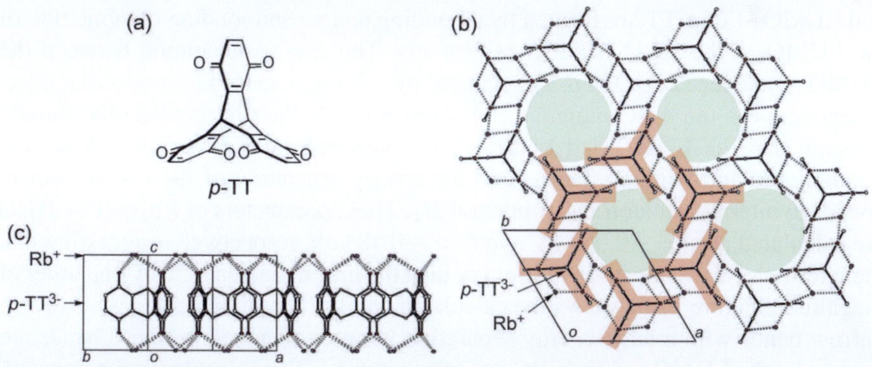

Fig. 11.5 (**a**) Molecular structure of *p*-TT. Honeycomb lattice in Rb_3(*p*-TT)·$2H_2O$: top (**a**) and side (**b**) Hexagonal *p*-TT triradical trianions and cavities formed by them are highlighted in orange and green, respectively. Hydrogen atoms are omitted for clarity. Reproduced from Ref. [6] with permission from the Royal Society of Chemistry

LUMO in *p*-TT. The broad reduction peak of *p*-TT around −1.70 V can be assigned to stepwise reductions from −3 to higher states.

The electron accepting *p*-TT can be reduced by chemically or electrochemically and forms crystals of anion radical salts. Electrochemical crystal growth in a saturated acetonitrile solution of $RbClO_4$ forms single crystals of a rubidium salt, Rb_3(*p*-TT)·$2H_2O$. The chemical stoichiometry of Rb_3(*p*-TT)·$2H_2O$ shows that *p*-TT is in the triradical trianionic state. The crystals of Rb_3(*p*-TT)·$2H_2O$ are found to belong to the hexagonal *P6*/*m* space group, in which one-sixth of the *p*-TT molecule is crystallographically independent. This molecular symmetry indicates that the three negative charges are equally distributed on the three *p*-BQ moieties, the C–O bond lengths of which are consistent with those of 1,4-semiquinones [28]. Figure 11.5b shows a projection of the unit cell of Rb_3(*p*-TT) along the *c* axis. A honeycomb lattice is evident in the figure, and is formed by a so-called intermolecular pancake bond [29], namely a strong π-dimer formation between the *p*-BQ moieties with an interplane distance of 3.02 Å. Due to a strong antiferromagnetic exchange interaction in the inter-molecular π-radical dimers, the magnetic susceptibility measurements for Rb_3(*p*-TT)·$2H_2O$ only exhibited a paramagnetism with a very small Curie constant of 8.4×10^{-3} emu K/mol which is caused by lattice defects (2.2%). Figure 11.5c shows a side view of the honeycomb lattice. The Rb ions and the *p*-TT trianions form alternating flat layers along the *c* axis. Consequently, the 2D honeycomb lattices of *p*-TT are well separated by the Rb ion layers. Since the neighboring *p*-TT layers are related by a translational relation along the *c* axis, there are 1D cavities along this direction, which pass through the centers of the hexagons in the 2D honeycomb lattices.

On the basis of the crystal structure of Rb_3(*p*-TT)·$2H_2O$, electronic band structure was calculated by tight-binding model with the two parameters, t_0 and t_1, which correspond to the intra- and inter-molecular transfer integrals between the *p*-BQ moieties (Fig. 11.6a). Figure 11.6b shows the LUMO of *p*-BQ, and the doubly-degenerated LUMOs and LUMO+1 of *p*-TT, calculated by DFT method. The LUMOs and LUMO+1 of *p*-TT are formed by a bonding and an antibonding combination of the LUMO of the *p*-BQ moiety, respectively. The energy separation between the LUMO+1 and the LUMO of *p*-TT is caused by a through-space intramolecular interaction between the *p*-BQ moieties and corresponds to their intramolecular transfer integral $3t_0$. The LUMO and LUMO+1 of inter-molecular dimer also shows the bonding and anti-bonding nature, and the energy separation of these states corresponds to intermolecular transfer integral $2t_1$. These parameters of Rb_3(*p*-TT)·$2H_2O$ are calculated as $t_0 = -0.063$ eV and $t_1 = -0.484$ eV, respectively, suggesting that the intermolecular interaction is stronger than the intramolecular one by one order of magnitude. Figure 11.6c shows the calculated band structure, which consists of two narrow bands with a large energy separation between them. Since these bands are formed by the LUMOs of six *p*-BQ moieties on two *p*-TT molecules in the unit cell, the lower band is fully occupied by the six electrons in the two $(p\text{-TT})^{3-}$, and the upper band is completely empty. As a result, the Fermi energy lies just at the middle of the gap. The band structure of Rb_3(*p*-TT)·$2H_2O$ clearly involves the Dirac cones at

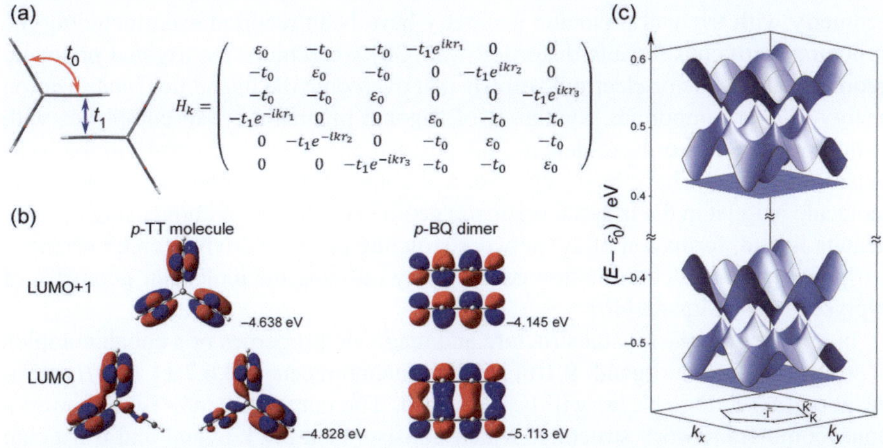

Fig. 11.6 **a** The nearest-neighbor intermolecular arrangement in Rb_3(*p*-TT) · $2H_2O$ and definition of t_0 and t_1, and tight-binding Hamiltonian where r_i ($i = 1$–3) and ε_0 represents the vectors connecting the nearest-neighboring molecules and the onsite energy, respectively. **b** MOs of *p*-TT molecule and intermolecular dimer of *p*-BQ moieties. **c** Electronic band structures of Rb_3(*p*-TT) · $2H_2O$. Part of the figure was reproduced from Ref. [6] with permission from the Royal Society of Chemistry

the $\overline{\text{K}}$ and $\overline{\text{K}'}$ points, as is well known for graphene, though the Fermi level in Rb_3(*p*-TT)·$2H_2O$ does not agree with the crossing level. In addition, the band structure of Rb_3(*p*-TT)·$2H_2O$ exhibits an exotic feature which does not exist in that of graphene. Namely, the band structure of Rb_3(*p*-TT)·$2H_2O$ involves the flat bands that have recently attracted attention [30–34]. Since these flat bands cannot be formed only by the intermolecular interaction t_1, they are caused by the tri-directional intramolecular interaction t_0 of *p*-TT. Rb_3(*p*-TT)·$2H_2O$ can thus provide practical guiding principles for the generation of unique band features, such as Dirac cones and flat bands, by means of supramolecular chemistry.

11.2.4 Ideal Trigonal Prismatic Coordination Geometry of Co(II) in Co(o-TT)

In the previous sections, we described the honeycomb network structures formed by coordinate bonds of metal ions with square planar geometry to threefold symmetric molecules and π-π interactions of threefold symmetric molecules. Here we would like to emphasize that the trigonal prismatic coordination geometry is also allowed as the building block for the 2D honeycomb network structure when the triptycene-based bridging ligand is used (Scheme 11.4). Six-coordinated complexes with octahedral coordination geometry are the most frequently observed for the first-row transition metal ions. In contrast far fewer example of another representative six-coordinated geometry with trigonal prismatic geometry have been reported even including the structures with considerable distortions [35, 36]. To achieve the trigonal prismatic geometry, the steric or electronic effects that overcome the ligand field stabilization energy [37] are important. Syntheses of trigonal prismatic cobalt complexes with multi-dentate (tetra or hexa-dentate) ligands, which can sterically restrain the coordination geometry, have been reported, and some of these complexes show single molecule magnet in the trigonal prismatic geometry [38, 39]. Without using a multi-dentate ligand, formation of 2D network structure can be a driving force to achieve a trigonal prismatic coordination geometry by utilizing the molecular geometry of triptycene-based ligand [40].

In this section, the crystal structure and magnetic properties of a cobalt complex of triptycene-based ligand 9,10-[1,2]benzenoanthracene-2,3,6,7,14,15(9*H*,10*H*)-hexaone (*o*-TT, Fig. 11.7a) will be described. The complex [Co(*o*-TT)] showed a honeycomb framework structure, which consists of a *o*-TT ligand and a trigonal prismatic coordination geometry of the Co ion. The magnetic measurements on [Co(*o*-TT)] clearly revealed a uniaxial magnetic anisotropy of the Co ion [40].

The triptycene derivative *o*-TT which consists of three 1,2-benzoquinone (*o*-BQ) moieties is electron-accepting ligand and the synthesis was first reported in 2018 [41]. The crystals of its cobalt complex [Co(*o*-TT)] · $3CH_3CN$ were prepared by redox reaction between neutral *o*-TT and $Co_2(CO)_8$ in CH_3CN. Figure 11.7b shows the crystal structure of [Co(*o*-TT)] · $3CH_3CN$, which belongs to the orthorhombic *Pbca* space group. The crystal structure consists of a slightly distorted honeycomb framework structure with three acetonitrile molecules in the voids. The *o*-TT molecules and Co ions locates on triangular lattice, respectively. The honeycomb layers are stacked along the *b* axis with an interlayer distance of 5.72 Å. As shown in Fig. 11.7c, the coordination geometry around the Co ion is a trigonal prism. Though the reported

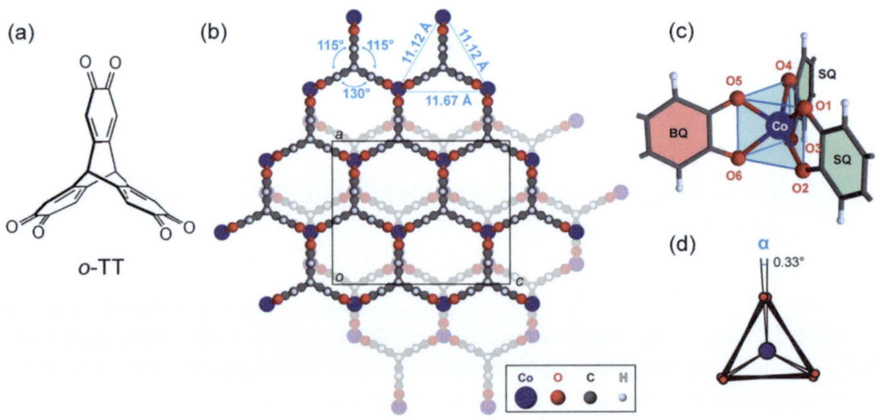

Fig. 11.7 **a** Molecular structure of *o*-TT. Projections of the crystal structure of [Co(*o*-TT)] · $3CH_3CN$ along the *b* axis (**b**). The trigonal prismatic coordination geometry of the Co ions in [Co(*o*-TT)] · $3CH_3CN$ (**c**) and the definition of Bailar twist angle (**d**). Reproduced from Ref. [40] with permission from the Royal Society of Chemistry

trigonal prismatic geometries around the cobalt ions with three-claw ligands are usually distorted with the Bailar twist angles α (Fig. 11.7d) of 1–30° [38, 39], those of [Co(*o*-TT)] · $3CH_3CN$ shows very small values ($\alpha = 0.33$, 0.35 and 0.63°) which indicates an ideal trigonal prismatic geometry with a D_{3h} local symmetry. Since there are no specific interactions between bidentate oxygen atoms such as three-claw ligands, the formation of a 2D honeycomb structure is presumably the driving force underlying the formation of an ideal trigonal prismatic coordination geometry. The Co–O distances are in the range from 2.047 to 2.132 Å, which is comparable to the range of Co–O distances for the octahedral Co(II) ion in a high-spin state ($S = 3/2$) [42]. The analysis of C–O bond distances of *o*-TT in [Co(*o*-TT)] shows a dianionic state, namely $(o\text{-TT})^{2-}$, in which each negative charge is localized on two π-planes as shown in Fig. 11.7c.

Temperature dependence of the $\chi_p T$ for a polycrystalline sample of [Co(*o*-TT)] · $3CH_3CN$ monotonically decreases with a decrease in T from 300 K and can be fitted to the Curie–Weiss law above 50 K with the Curie constant of 3.78 cm^3 K/mol and the Weiss constant of −14.5 K. It is notable that there is no ferrimagnetic behavior in the low temperature region, such as a turnover of Weiss constant from negative to positive with a decrease in temperature down to 2 K. This suggests an $S = 0$ ground state for $(o\text{-TT})^{2-}$. Angular dependence of magnetization M for oriented crystals of [Co(*o*-TT)] · $3CH_3CN$ at 2, 5, 10, 15, 20, and 25 K is shown in Fig. 11.8a. The angle φ is the dihedral angle between the magnetic field and the *ac* plane of the crystal structure, so that the magnetic field is parallel to the *b* axis at $\varphi = 90°$. The magnetization exhibits maxima and minima at $\varphi = 90$ and 0°, respectively. This behavior indicates a uniaxial magnetic anisotropy with the easy axis parallel to the *b* axis. Here the magnetization and *g*-value along *b* axis are defined as M_z and g_z and those along *a* (or *c*) axis are defined as M_x (M_y) and g_x (g_y). The temperature dependence of M_x, M_y and M_z below 150 K are shown in Fig. 11.8b and the ratio of *g* values along easy (g_z) and hard axis ($g_{x,y}$) can be estimated as g_z/g_x (or g_y) = 1.17 (or 1.16), which can be regarded as weak uniaxial magnetic anisotropy.

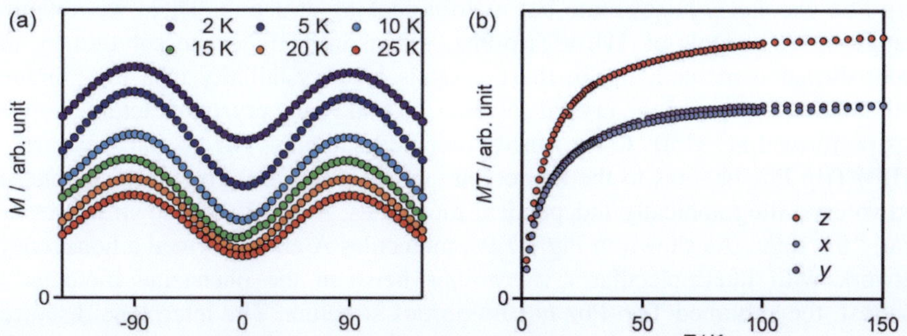

Fig. 11.8 Angular dependence of the magnetization at 2, 5, 10, 15, 20, and 25 K (**a**) and Temperature dependence of M_xT, M_yT and M_zT below 150 K (**b**) for aligned crystals of [Co(*o*-TT)] · $3CH_3CN$. Reproduced from Ref. [40] with permission from the Royal Society of Chemistry

To understand the magnetic anisotropy of trigonal prismatic coordination geometry, the g-value anisotropy caused by the crystal electric field (CEF) was calculated for the Co(II) (d^7) ion in the trigonal prismatic geometry. The ratio of g_z/g_x shows that magnetic anisotropy is significantly dependent on the structural parameters and high-spin states can be (i) the Heisenberg spin with weak anisotropy $g_z > g_x, g_y$ (ii) a strong uniaxial magnetic anisotropy with $g_z \gg g_x, g_y$ and (iii) the XY-model type anisotropy with $g_z \sim 0$.

The anisotropic axis of the trigonal prismatic coordination geometry aligns in the same direction in a 2D network structure and the CEF calculation on a trigonal prismatic coordination geometry model illustrated the potential of various magnetic anisotropy, including Ising, XY and $g_z \gtrsim g_x$. The strategy to form an ideal trigonal prismatic coordination geometry with *o*-TT ligands would be broadly applicable to various triptycene-based ligand and has potential to achieve a strong magnetic anisotropy.

11.2.5 Honeycomb Lattices of Trip-Phz in Bulk Crystals and Monolayers on a Ag(111) Surface

In the previous sections, we introduced molecule-based honeycomb lattices, formed by strong interactions such as exchange interactions between radical moieties and coordination bonds between metals and ligands. In this section, we describe honeycomb lattices of a phenazine analogue of triptycene, Trip-Phz (see the inset of Fig. 11.9a), formed by a van der Waals interaction between the phenazine moieties [43].

This molecule was synthesized by the condensation reaction of 2,3,6,7,14,15-triptycenehexaone and *o*-phenylenediamine in the presence of *N*,*N*-dimethylaminopyridine. The crude product was purified by sublimation under reduced pressure to give pale yellow crystals. The crystal structure of the sublimed Trip-Phz was not a honeycomb but a close-packing structure. Then, we obtained orange block crystals of THF@Trip-Phz, grown in THF/hexane solution by the vapor diffusion method. In air, these crystals lost crystallinity in a few minutes with the evaporation of the crystal solvents, so that X-ray crystal structure analysis was performed at −150 °C for freshly prepared samples. The crystal structure of THF@Trip-Phz belongs to the monoclinic space group, *Cc*. The unit cell consists of two crystallographically independent molecules, **A** and **B**, whose structures are nearly the same. As shown in Fig. 11.9a, molecules **A** and **B** formed a honeycomb network with intermolecular π-π overlaps between the phenazine moieties in contrast, the sublimed Trip-Phz has no porous structure. The interplane distances of 3.44–3.51 Å are comparable to the sum of the van der Waals radii of carbon atoms (3.4 Å). The honeycomb structure possesses a void space with *ca.* 10 Å diameter and stacks along the *c* axis (Fig. 11.9b). Figure 11.9c shows the interlayer contacts between the honeycomb layers. It is found that the phenazine moiety forms

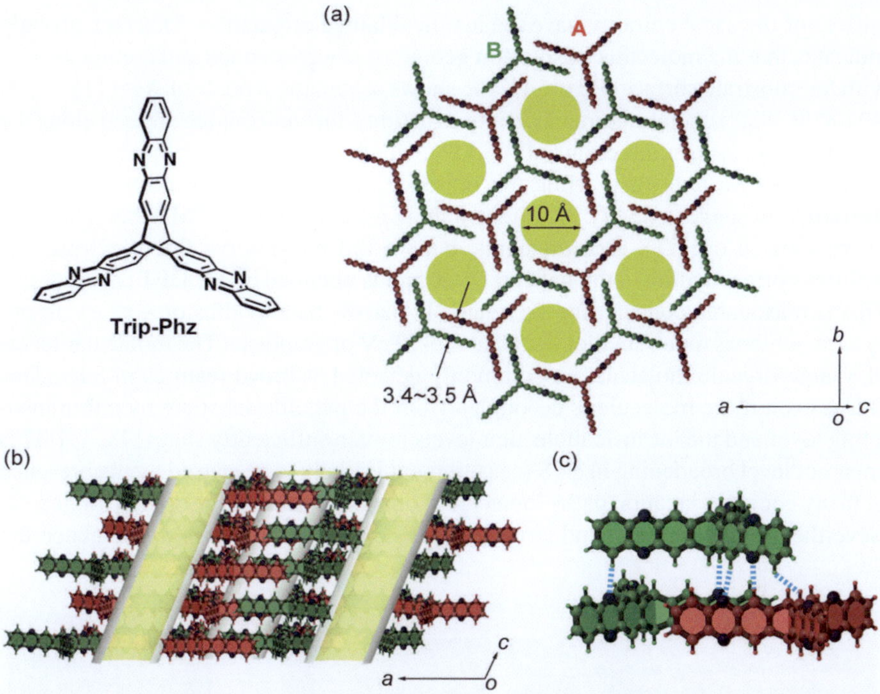

Fig. 11.9 **a** Honeycomb lattice formed by Trip-Phz molecules, **b** 1D channel that runs through the honeycomb layers, and **c** interlayer CH···N hydrogen bonding. The molecules colored in red and green represent molecules A and B, respectively. THF molecules are omitted for clarity. Adapted with permission from Ref. [43]. Copyright 2020, American Chemical Society

interlayer, intermolecular CH···N hydrogen bondings [44, 45]. The sky blue dotted lines in this figure indicate these bonds with C···N distances of 3.23–3.59 Å. They form AA stacking, namely a nearly eclipsed overlap between the neighboring layers, which results in 1D channels, as shown in Fig. 11.9b. It is notable that the interior surfaces of the 1D channels in Trip-Phz are covered by π electrons.

Furthermore, we successfully grew a 2D honeycomb monolayers of Trip-Phz on a substrate and elucidated its intriguing band structure [46]. Figure 11.10a displays an STM image of Trip-Phz molecules on a Ag(111) substrate that was co-deposited with ~0.01 ML Cs and annealed at ~370 K. By this treatment, the surface was preferentially covered by the honeycomb lattices. This image was obtained under a UHV condition with the base pressure less than 2.0×10^{-10} Torr. Although the Trip-Phz itself is achiral, the honeycomb lattice of Trip-Phz molecules possess a chirality due to different stacking sequences of the π-π pancake bondings between the phenazine moieties. Figure 11.10b shows high-resolution STM images with different chirality, which were taken at domains **A** and **B** in Fig. 11.10a, respectively. They also have different lattice orientation as revealed by the unit vectors (also see the dashed lines in Fig. 11.10a). Interestingly, we find that all observed honeycomb lattices of the same chirality have the identical lattice orientation. This fact strongly indicates that the molecular adsorption geometry is well-defined and commensurate with the substrate surface. Figure 11.10c shows schematic models of Ag(111)-(7×7) R21.8°/R38.2° superstructures. For both structures, the moieties are aligned along the $[11\overline{2}]$ and equivalent directions of Ag(111), which should stabilize these adsorption geometries. We note that, while the presence of a chiral honeycomb lattice was theoretically suggested [47], we directly show it here through STM observations.

We carried out STS measurements, but we did not observe fine spectroscopic features corresponding to the flat and Dirac bands obtained by the DFT calculations. This is reasonable considering the relatively narrow band widths of ~0.2 eV in our system, which is much smaller than that of ~10 eV of graphene. The molecular levels of a large organic molecule are commonly detected as broad features in STS. This is true even if the molecule is decoupled from the metallic substrate by a thin insulating layer and the intrinsic molecular levels remain sufficiently sharp [48, 49]. This apparent level broadening in STS measurement is likely to be caused by the presence of many energy relaxation paths for an electronically excited large organic molecule. Nevertheless, the unique band structures predicted here can strongly influence the

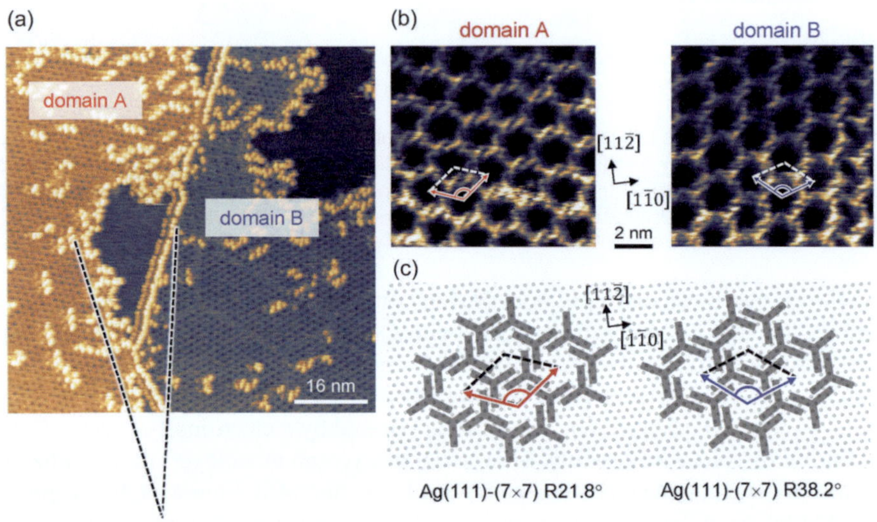

Fig. 11.10 **a** STM image of Trip-Phz molecules adsorbed on a Ag(111) surface that was co-deposited with ~0.01 ML Cs and annealed at ~100 °C ($V_s = -2$ V, $I_t = 10$ pA). Domains A and B have different lattice orientations as indicated by the dashed lines. **b** High resolution STM images of domains **A** and **B** (constant height mode, size: 10 nm × 10 nm, $V_s = -2$ V. The feedback was stabilized at $I_t = 10$ pA). The arrows on the images indicate the unit vectors of honeycomb lattices, while those between the images the crystal orientations of the Ag(111) surface. **c** Structure model of Trip-Phz adsorbed on a Ag(111) surface. Left: Ag(111)-(7×7) R21.8°, Right: Ag(111)-(7×7) R38.2°. The arrows correspond to those in panel (**b**). Adapted with permission from Ref. [46]. Copyright 2024, American Chemical Society

electron transport properties at low temperatures, where the physics is governed by the low-energy electrons.

11.3 Molecule-Based K_4 Structures

In graph theory, the complete graph K_4 consists of four vertices, each pair of which is connected by a unique edge. In 2008, Sunada mathematically converted the graph K_4 into the triply-periodic structure, K_4 crystal, and characterized it [1], while the K_4 structure itself was already known in various research fields, using various names, such as Laves' graph of girth ten, Net 1((10,3)-a), srs net, Y^*, 3/10/c1, hyperoctagon net, and gyroid lattice. Since the K_4 lattice is fundamental, this is ubiquitous in nature, appearing in various materials of different sizes [4]. In this section, we describe a spin liquid state and an enhancement of circularly polarized luminescence (CPL), found in molecule-based K_4 lattices.

11.3.1 Dimer Mott State and Spin-Liquid Ground State in $(TBA)_3[(-)\text{-}NDI\text{-}\Delta]_2$

Recently, S. T. Schneebeli et al. reported a polyhedral π-conjugated molecule with Δ-shaped C_3 symmetry, called NDI-Δ (see the inset of Fig. 11.11), which consists of three planar NDI (naphthalene diimide) moieties connected by three cyclohexane units [50]. This molecule is chiral and exhibits reversible six-step reductions in its CV curves, attributed to the strong electron-accepting ability of the NDI moiety. Although there is no intermolecular π overlap in the crystal structures of neutral NDI-Δ [50, 51], the anionic radical species of NDI-Δ could potentially exhibit three-way branched intermolecular π overlaps due to exchange interactions [13, 52]. Additionally, the bulky terminal cyclohexane units in NDI-Δ might prevent eclipsed π overlap between the NDI moieties, resulting in a twisted π overlap, which is another key factor in forming the K_4 structure [4].

The crystallization of (−)-NDI-Δ radical anion salts was carried out using galvanostatic electrocrystallization. Black block or black needle crystals grew on the cathode under various combinations of solvent and electrolyte these crystals were all isostructural, with the chemical formula $A_3[(-)\text{-NDI-}\Delta]_2$, where A represents a cation. The crystals belonged to the cubic space group $I4_132$ or $P4_132$, where (−)-NDI-Δ formed a K_4 structure. Figure 11.11 shows comparisons between the projections of the K_4 carbon and $(TBA)_3[(-)\text{-NDI-}\Delta]_2$ along the [111] directions. It is clearly seen that the crystal structure of $(TBA)_3[(-)\text{-NDI-}\Delta]_2$ can be reproduced by replacing each sp^2 carbon atom in the K_4 carbon with (−)-NDI-Δ.

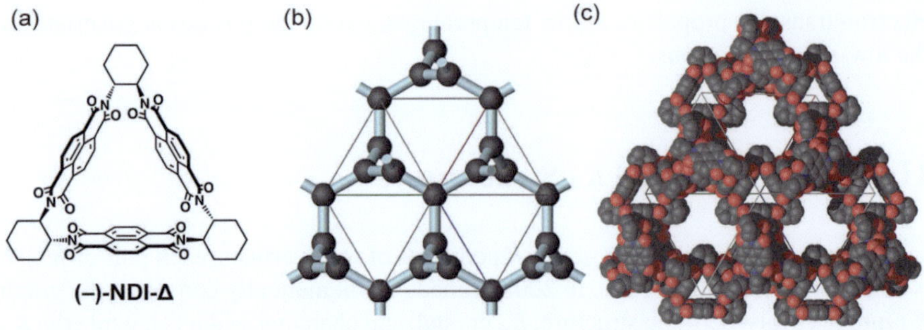

Fig. 11.11 Comparison between the projections of crystal structures of the K_4 carbon and (−)-NDI-Δ in $A_3[(-)\text{-NDI-}\Delta]_2$, along the [111] directions. Adapted with permission from Ref. [13]. Copyright 2017, American Physical Society

Figure 11.12 shows the nearest-neighbor intermolecular arrangements in $(TBA)_3[(-)\text{-NDI-}\Delta]_2$. This chemical stoichiometry indicates that the valence of (−)-NDI-Δ is −1.5, meaning that a negative half-charge exists on each NDI unit, which corresponds to one-third of the total (−)-NDI-Δ. It is indicated that the intermolecular interaction t_1 is much larger than the intramolecular interaction t_0. Therefore, it is reasonable to conclude that one unpaired electron is shared within an intermolecular NDI dimer unit. The weak intramolecular interaction t_0, which corresponds to the interaction between unpaired electrons, would result in electron localization. The dimerized structures and insulating properties of the (−)-NDI-Δ salts closely resemble those of the dimer Mott systems found in the organic radical salts of BEDT-TTF and $Pd(dmit)_2$ (dmit = 1,3-dithio-2-thione-4,5-dithiolate) [53, 54]. The unpaired electrons in $(TBA)_3[(-)\text{-NDI-}\Delta]_2$ exist within the intermolecular pancake dimer, specifically at the midpoint of intermolecular contact. Consequently, it is concluded that the unpaired electrons form a hyper-kagome lattice in the K_4 lattice of (−)-NDI-Δ, as discussed in Scheme 11.3. It is reasonable to expect spin frustration rather than magnetic ordering.

Figure 11.13a shows the temperature dependence of χ_p of $(TBA)_3[(-)\text{-NDI-}\Delta]_2$ in the temperature range of 0.07–300 K. χ_p increases with decreasing temperature and exhibits a small anomaly around 20 K, which is probably caused by the interdimer interaction. After that, χ_p continues to increase down to the lowest temperature. The dependence at high temperatures follows the Curie–Weiss law with a Weiss constant of $\theta = -15$ K, but there is no long-range magnetic ordering down to 70 mK. The purple broken line in Fig. 11.13a shows the temperature dependence of the paramagnetic susceptibility of lattice defects χ_{def}, which follows the Curie–Weiss law with $C_{def} = 2.0 \times 10^{-5}$ emu K/g and $\theta_{def} = -3$ mK. The red curve in this figure shows the intrinsic paramagnetic susceptibility χ_{int}, which is calculated by $\chi_{int} = \chi_p - \chi_{def}$. The results of this analysis clearly suggest the presence of χ_{int} even at 0.5 K despite the 3D magnetic interactions with $\theta = -15$ K.

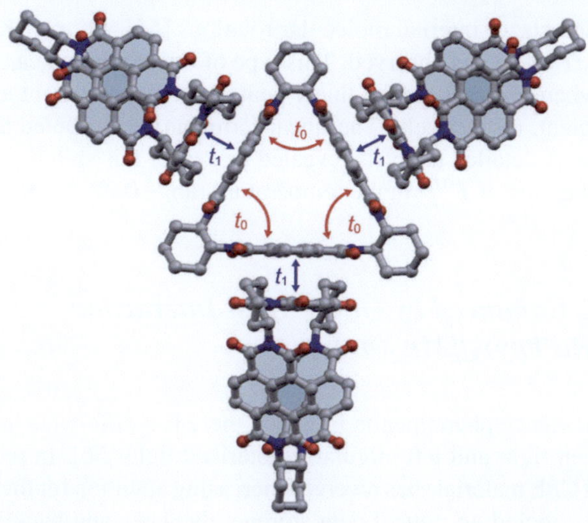

Fig. 11.12 Nearest-neighbor intermolecular arrangements of the (−)-NDI-Δ radical anions in A_3[(−)-NDI-Δ$]_2$

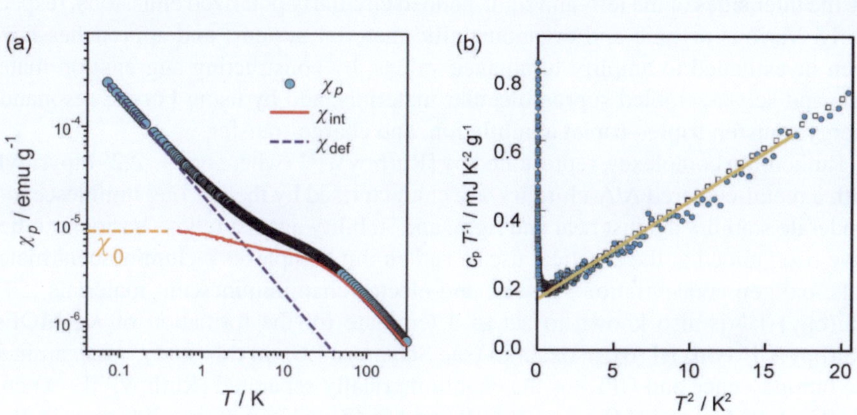

Fig. 11.13 **a** Temperature dependence of the paramagnetic susceptibility χ_p in the temperature range 0.07–300 K for $(TBA)_3$[(−)-NDI-Δ$]_2$. The purple broken line and the red curve indicate those of χ_{def} and χ_{int}, respectively. **b** c_p/T-T^2 plots below 5 K for $(TBA)_3$[(−)-NDI-Δ$]_2$. Adapted with permission from Ref. [13]. Copyright 2017, American Physical Society

Figure 11.13b shows the temperature dependence of the heat capacity c_p for $(TBA)_3$[(−)-NDI-Δ$]_2$ down to 68 mK, with c_p/T plotted against T^2. As the temperature decreases, c_p/T exhibits a linear decrease, followed by a sudden increase below approximately 0.1 K. Such increase at extremely low temperatures has often been observed in organic spin systems and has been attributed to a high-temperature tail of a Schottky peak, formed by nuclear spins, weak magnetic interactions between paramagnetic defects, or internal molecular rotations [55]. Above 1 K, a linear relation between c_p/T and T^2 is observed. This type of linear relation can be fitted by $c_p = \gamma T + \beta T^3$, where γ and β are the linear heat capacity coefficient and lattice heat capacity coefficient, respectively. The linear term can be attributed to the c_p of the spin liquid state. A detailed analysis revealed that c_p of this spin liquid state can be fitted to the power law of $T^{0.62}$ in the temperature range 0.07–4.5 K.

11.3.2 CPL Enhanced by Guest–Host Interaction in $[Ru(bpy)_3][Mn_2(ox)_3]$

CPL is a luminescence phenomenon in which there is a difference in luminescence intensity between right and left circularly polarized light [56]. In recent years, the development of CPL materials has received increasing attention for their many potential applications, including optical data storage, displays, and backlighting for 3D displays [57, 58]. The magnitude of CPL is measured by the luminescence dissymmetry factor, g_{lum}, given by the equation $g_{lum} = 2(I_L–I_R)/(I_L + I_R)$, where I_L and I_R are the intensities of the left- and right-handed circularly polarized emissions, respectively. Various organic and organometallic material systems and approaches have been investigated to amplify luminance values, by constructing aggregation materials and self-assembled supramolecular materials, and by using Förster resonance energy transfer, triplet–triplet annihilation, and charge-transfer.

Ruthenium complexes, represented by $[Ru(bpy)_3]^{2+}$ (where bpy = 2,2′-bipyridyl) with a metal-centered Δ/Λ-chirality, are characterized by their strong luminescence, moderate stability against heat and light, and stability against redox. Various studies have been aimed at the practical use of ruthenium complexes as luminescent materials, oxygen concentration sensors, and electro-chemiluminescent materials [59]. $[Ru(bpy)_3]^{2+}$ is also known to act as a template for the formation of K_4 MOFs, $[Ru(bpy)_3][M_2(ox)_3]$ (ox = oxalate) (see Scheme 11.6). In this study, we examined the luminescence and CPL for the enantiomerically separated $[Ru(bpy)_3]I_2$ (**1**) and $[Ru(bpy)_3][M_2(ox)_3]$ (M = Zn (**2**) [60] and Mn (**3**) [61]) in their solid states [62].

The enantioselective syntheses of **1** ~ **3** were performed in the following procedure. The enantiomers Δ-**1** and Λ-**1** were separated by the diastereomeric recrystallization method from their racemic mixture [63]. Then, using them as templates, single crystals of the enantiomers Δ-(**2/3**) and Λ-(**2/3**) were prepared [60, 61]. X-ray crystal structure analysis of the randomly selected ~10 crystals of **2** and **3** confirmed the MOF structure, in which the chromophore molecule $[Ru(bpy)_3]^{2+}$ was encapsulated in the void spaces of the K_4 skeleton $[M_2(ox)_3]^{2-}$, and confirmed that the chiralities around the metal ions in $[Ru(bpy)_3]^{2+}$ and in $[M_2(ox)_3]^{2-}$ agreed with each other. It is notable that there is no significant difference in the lattice parameters and the atomic positions between the corresponding enantiomers of **2** and **3**.

We carried out solid-state UV–Vis-NIR reflection measurements of the powder samples, which were dispersed into powder $BaSO_4$, and then transformed the

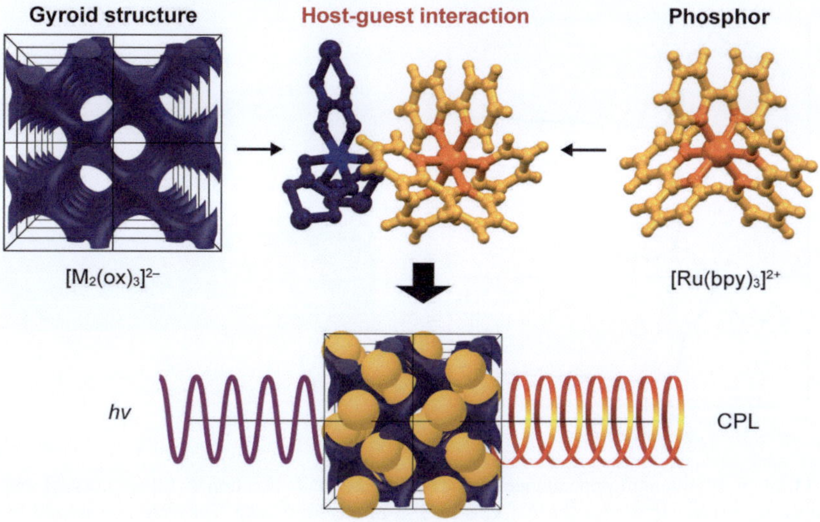

Scheme 11.6 Strategy to enhance CPL in gyroidal MOFs

obtained relative reflectance to $BaSO_4$, r, into the Kubelka–Munk function, $F = (1 - r)^2/(2r)$ [64], which was proportional to absorbance. The black curves in Fig. 11.14 show the wavelength dependence of F for the Δ-forms of **1**–**3**, in which the values of F are normalized by their respective peak values. The F spectra of **2** and **3** are nearly the same, while the shoulder band at 575 nm is enhanced in **1**. The red curves in Fig. 11.14a show the luminescence spectra for the Δ-forms of **1**–**3** in the solid state, which are normalized by their respective peak values. Although their band shapes are similar to each other, the intensities are very different: the luminescence intensities of **1** and **2** are almost the same, while the intensity of **3** is significantly lower than that of **1** and **2**. The decrease in luminescence intensity of **3** can also be easily recognized visually (see Fig. 11.14b). We measured the luminescence quantum yields and luminescence lifetimes for the Δ-forms of **1**–**3** the obtained luminescence quantum yields for excitation at 450 nm are 1.5% for **1**, 1.3% for **2,** and 0.2% for **3**. Much weaker luminescence for **3** was confirmed.

Before the solid-state CPL measurements, we examined the CPL for the solution samples of Λ-**1** and Δ-**1** at room temperature with the excitation wavelength of 365 nm and obtained that $g_{\mathrm{lum}} = +3.0 \times 10^{-4}$ for Λ-**1** and -1.9×10^{-4} for Δ-**1**. These values are in agreement with the reported ones [59, 65]. We then measured the CPL for the enantiomers of **1** ~ **3** in solid state at room temperature with the excitation wavelength of 450 nm. The results are shown in Table 11.1. The CPL dissymmetry factors are calculated as $g_{\mathrm{lum}} = +1.7 \times 10^{-2}$ for Λ-**1** and -2.9×10^{-2} for Δ-**1** they are an order of magnitude larger than those for the solution samples. This is probably a kind of aggregation effect, reviewed in Ref. [66]. While the g_{lum} values for **2** are nearly the same as those of **1**, we found a significantly enhanced CPL effects for **3**, namely $g_{\mathrm{lum}} = +0.73 \times 10^{-1}$ for Λ-**3** and -1.2×10^{-1} for Δ-**3**, which

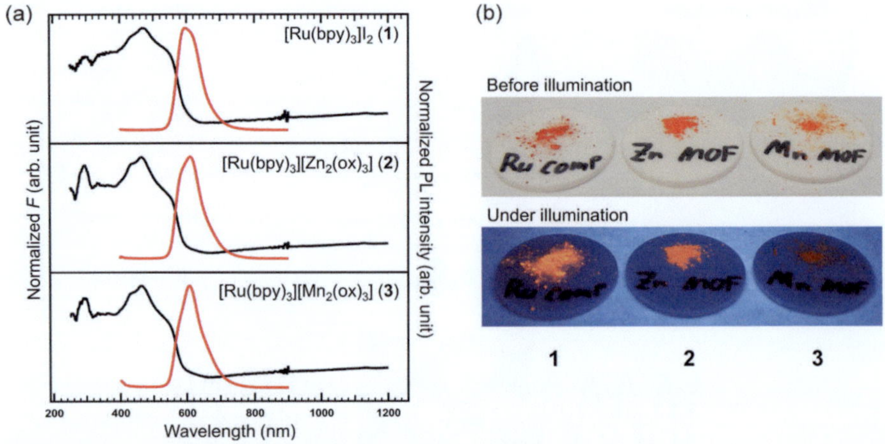

Fig. 11.14 **a** Wavelength dependence of the Kubelka–Munk function F (black curves) and the luminescence (red curves) for the Δ-forms of **1**–**3** in the solid state. They are normalized by the peak values. **b** Photos of the powder samples of **1**–**3** under daylight and UV illumination. Adapted with permission from Ref. [62]. Copyright 2023, American Chemical Society

Table 11.1 Luminescence dissymmetry factor, g_{lum} of **1**–**3**

		1 (solution)	**1** (solid)	**2** (solid)	**3** (solid)
$g_{\mathrm{lum}}{}^{*}$	Λ	$+3.0 \times 10^{-4}$	$+1.7 \times 10^{-2}$	$+6.5 \times 10^{-2}$	$+0.73 \times 10^{-1}$
	Δ	-1.9×10^{-4}	-2.9×10^{-2}	-2.3×10^{-2}	-1.2×10^{-1}

$^{*}g_{\mathrm{lum}} = 2(I_{\mathrm{L}} - I_{\mathrm{R}})/(I_{\mathrm{L}} + I_{\mathrm{R}})$

are an order of magnitude larger than those for **1** and **2**. It is worth noting that the luminescence in **3** is quenched more significantly than that of **2**, because the energy transfer from $[Ru(bpy)_3]^{2+}$ to paramagnetic $[Mn_2(ox)_3]^{2-}$ is more effective than that to diamagnetic $[Zn_2(ox)_3]^{2-}$. It was speculated that a specific energy transfer from $[Mn_2(ox)_3]^{2-}$ to $[Ru(bpy)_3]^{2+}$ would induce an enhancement of the CPL in $[Ru(bpy)_3]^{2+}$. It is concluded that the chiral K_4 skeleton can enhance CPL effects through a homochiral guest–host structure.

11.4 Molecule-Based Diamond Structures

As described in Sect. 11.1, diamond and K_4 lattices are regarded as mathematical twins, and the diamond structure is intuitively easier to understand than K_4. However, although diamond networks have been reported in MOFs and hydrogen-bonded crystals, the diamond lattice is the least common among the three molecule-based strongly isotropic lattices, and it is also the most difficult lattice to construct rationally. In this section, we report unusual phase transitions induced by "bond frustration" in a diamond network of an organic biradical.

11.4.1 Bond Frustration in a Diamond Network of Organic Biradical, bpBDTDA

An organic biradical, bpBDTDA (Fig. 11.15a), crystallizes in the orthorhombic space group *Fdd*2 at temperatures above 359 K (high-temperature phase) [7, 67]. This space group is closely related to that of diamond, specifically the cubic $Fd\bar{3}m$ (see Scheme 11.7). In general, the crystal structure is determined by the lattice and its basis, and the diamond structure (Fig. 11.15d) consists of four layers of basis, **A**-**D**, whose heights are shifted by 1/4 of the lattice length. By reducing the symmetry of the basis, the space group *Fdd*2 for bpBDTDA exists as the least symmetric one. Figure 11.15b shows the *c*-axis projection of the unit cell of bpBDTDA in the high-temperature phase. The fractional coordinates of these **A**-**D** molecules along the *c* axis are shifted by 1/4. The two light blue dashed lines at the ends of the molecules indicate the intermolecular short S···N or S···S contacts. These intermolecular short contacts in all four directions are crystallographically equivalent and form a "diamond network". Additionally, it should be noted that the **A**-**D** molecules independently exhibit uniform stacking along the *c* axis which is formed by a π-π face-to-face overlap between the dithiadiazolyl rings. From this high-temperature phase, bpBDTDA undergoes a unique radical dimerization as the temperature decreases. While organic radicals often exhibit pairwise dimerization from uniform stacking, reflecting their strong electron/spin–lattice interactions, as illustrated in Scheme 11.8a, bpBDTDA with its bulky moiety, exhibits a zigzag dimerization without a centroid shift, as shown in Scheme 11.8b. It was found that the stacking columns **A**-**D** undergo stepwise, spatially inhomogeneous phase transitions toward zigzag dimerization at 359 and 306 K, which can be characterized in terms of "bond frustration", as predicted by the line graph of the diamond lattice.

Figure 11.16a shows the stacking modes in the four columns, **A**, **B**, **C**, and **D**, across the high-temperature (HT), intermediate-temperature (IT), and low-temperature (LT) phases. As described in the previous paragraph, the stacking is uniform in the HT phase. Upon cooling, a two-step first-order phase transition was observed, accompanied by lattice dimerization. In the IT and LT phases, the crystal structures belong to the same monoclinic space group $P2_1$. The unit cell consists of four crystallographically independent molecules, **A**-**D**, which correspond to those in the HT phase. Strangely, in the IT phase, three columns (**A**-**C**) undergo alternating dimerization to form zigzag π-stacking along the monoclinic *b* axis, while column **D** retains uniform stacking. Upon further cooling to the LT phase, the zigzag dimerization pattern of columns **A**-**C** remains unchanged, but the last remaining column, **D**, undergoes zigzag dimerization. The inhomogeneity in the IT phase, with a 3:1 ratio instead of 1:1, is quite unusual. It is also notable that, while there is little positional shift of the

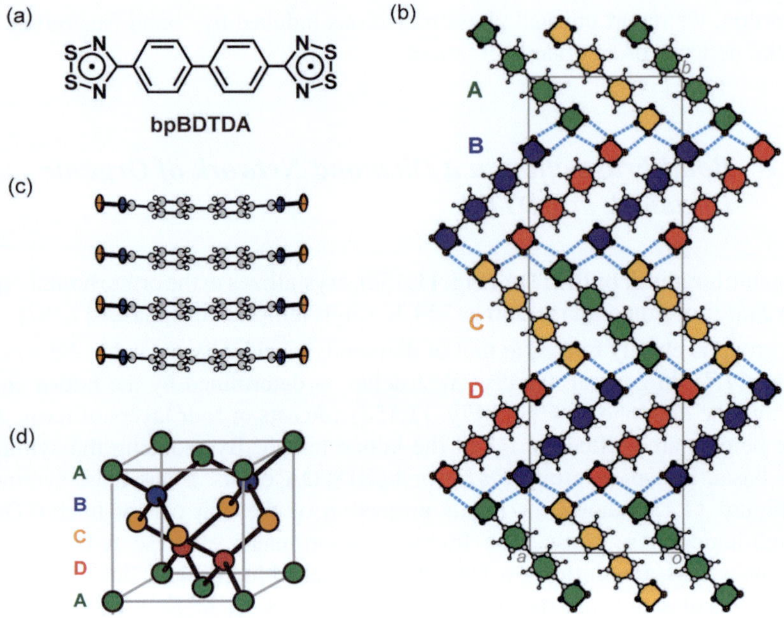

Fig. 11.15 **a** Molecular structures of bpBDTDA. **b** *c* axis projection of the unit cell. **c** The stacking modes in the four columns, **A, B, C,** and **D**. **d** Unit cell of the diamond lattice. Reproduced from Ref. [7] with permission from the Royal Society of Chemistry

Scheme 11.7 Correlation between the symmetry of basis and space groups in the diamond lattice. Reproduced from Ref. [7] with permission from the Royal Society of Chemistry

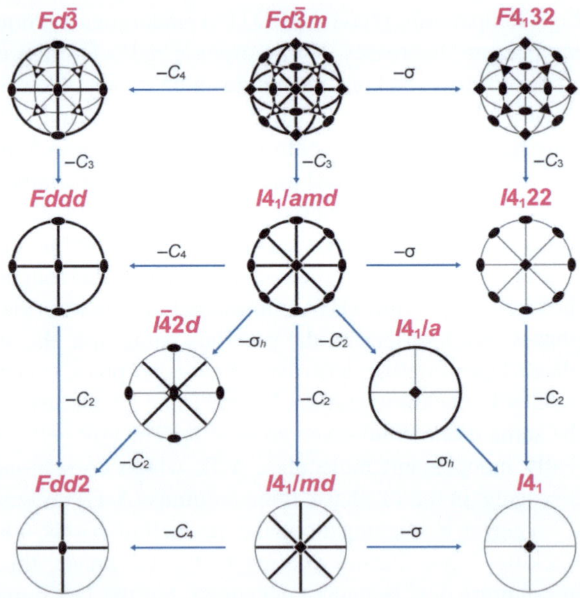

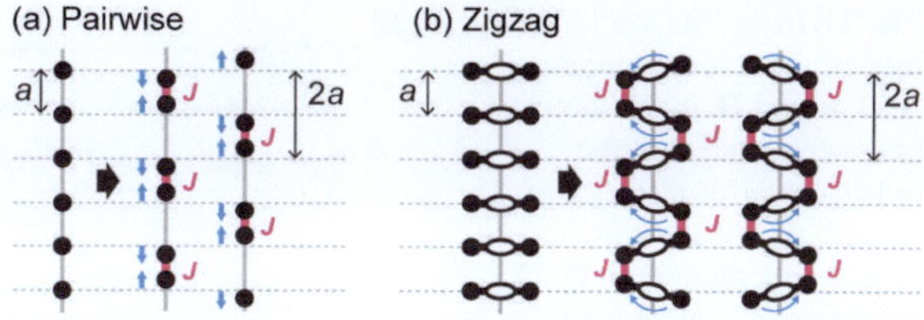

Scheme 11.8 Dimerization patterns in organic radicals **a** pairwise, **b** zigzag. Reproduced from Ref. [7] with permission from the Royal Society of Chemistry

molecular centroids along the π-stacking direction during the first phase transition at 359 K, those of all molecules change significantly in the LT phase below 306 K. This is unusual, as zigzag dimerization does not typically require a shift in molecular centroids (see Scheme 11.8).

The causes of the strange phase transitions in bpBDTDA can be understood by considering the line graph of the diamond structure. Scheme 11.9a shows the structural displacement of column **A** and its correlation to **B** and **D** in the zigzag dimerization. Column **A** is connected to two molecules **B**, which are shifted by +1/4 along the c axis, and also connected to two molecules **D**, which are shifted by −1/4 along the c axis, in the direction perpendicular to **B**-**A**-**B** direction. For clarity, the diagram is divided into two parts, top and bottom. The four diamond bonds in the HT phase are colored white to show that they are equivalent. Note that, as described in Scheme 11.3, the line graph of the diamond lattice is the pyrochlore lattice. Therefore, the lattice points of pyrochlore lattice exist on the bonds in the diamond lattice. After dimerization, there is the zigzag dimerization of columns **B** and **D** associated with the zigzag dimerization of column **A**. As shown in Scheme 11.8b, there are two zigzag dimerization patterns, which are energetically equivalent. However, the zigzag dimerization pattern of columns **B** and **D** can be uniquely determined by the rules, which is to make the molecules as close as possible, that is, shorten the intermolecular interactions. In other words, as column **A** is deformed, columns **B** and **D** are deformed so that the radical moieties are closer together. Scheme 11.9b shows a diamond/pyrochlore lattice, in which one **A**-**B**-**C**-**D**-**A** connection is emphasized with the colors. The zigzag dimerization along the c axis requires an antiphase for the zigzag patterns of tetrahedra **A** at $x = 0$ and tetrahedra **A** at $x = 1$ (Scheme 11.9c), which causes a mismatch in **D**, as shown in Scheme 11.9c. In other words, the diamond structure is not consistent with the zigzag dimerization along the c axis. We believe that this frustration has led to the spatially inhomogeneous stepwise phase transitions observed in this system, as well as to the shifting of the molecular centroids to achieve the zigzag dimerization observed in the LT phase.

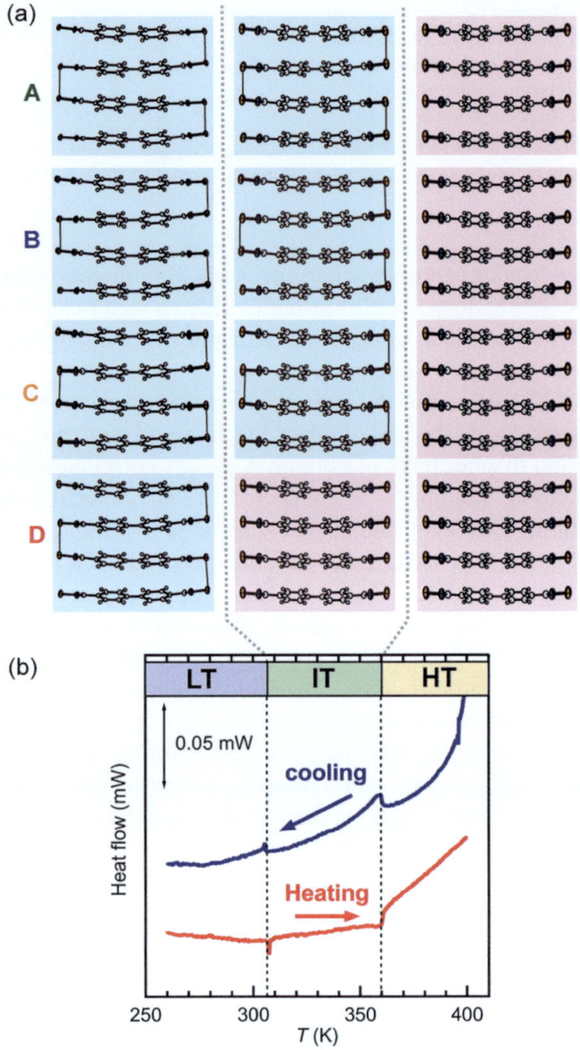

Fig. 11.16 **a** The stacking modes in the four columns, **A**, **B**, **C** and **D**, in the three phases pick background indicates uniform stacking, while blue one indicates zigzag dimerization. **b** DSC curves upon heating (red curve) and cooling (blue curve) in the temperature range between 260 and 400 K. Reproduced from Ref. [7] with permission from the Royal Society of Chemistry

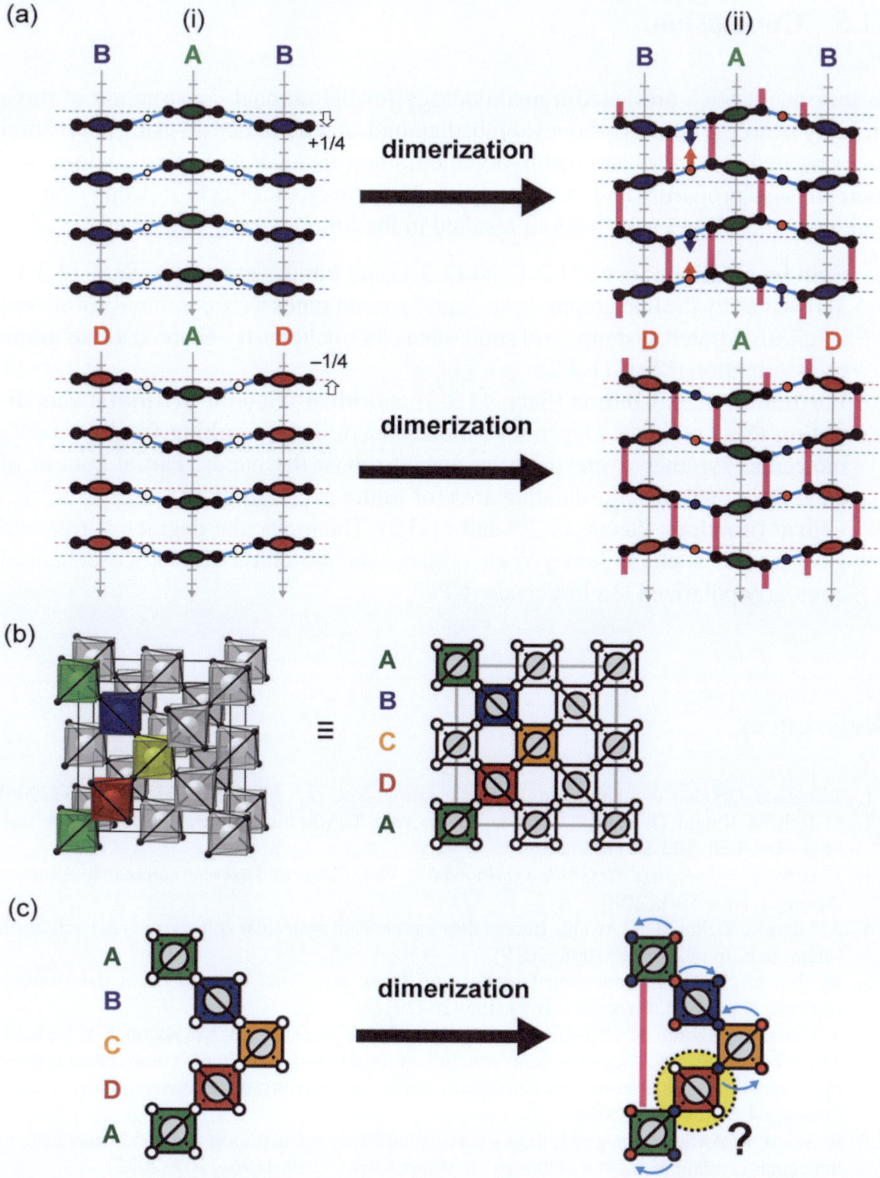

Scheme 11.9 Schematic illustrations for the structural transition from the regular stacking in the HT phase (i) to the zigzag stacking in the IT/LT phases (ii). **a** Side views for the structural relations between the stackings, **A** and **B**, and between **A** and **D**. **b** Schematic view of tetrahedron model for pyrochlore lattice. **c** Tetrahedron model for the bond frustration in Panel (**b**). Reproduced from Ref. [7] with permission from the Royal Society of Chemistry

11.5 Conclusion

In this chapter, we presented a methodology for the rational construction of three strongly isotropic lattices—honeycomb, diamond, and K_4 lattices—using polyhedral π-conjugated molecules and metal–organic frameworks (MOFs). We also discussed their physical properties and functionalities. The synergistic effects of strong isotropy and molecular degrees of freedom resulted in the following key outcomes:

1. **Spin frustration** (Sects. 11.2.1 and 11.3.1) and **bond frustration** (Sect. 11.4.1), as revealed by the line graphs. Spin-liquid ground states were commonly observed in spin-frustrated systems, and solid-state electrochemistry enabled a line-graph transformation (Sect. 11.2.2).
2. **Formation of flat bands** (Sect. 11.2.3) and **ideal trigonal prismatic coordination** (Sect. 11.2.4). Degenerate molecular orbitals, resulting from C_3 or T_d molecular symmetry, inevitably produced robust flat bands. Partial control of band filling will be a fascinating area for future exploration.
3. **Chiral functions** (Sects. 11.2.5 and 11.3.2). The molecular degrees of freedom gave rise to a chiral honeycomb lattice, and the chiral K_4 lattice enhanced circularly polarized luminescence (CPL).

References

1. T. Sunada, Crystals that nature might miss creating. Not. Am. Math. Soc. **55**, 208–215 (2008)
2. M. Itoh, M. Kotani, H. Naito, T. Sunada, Y. Kawazoe, T. Adschiri, New metallic carbon crystal. Phys. Rev. Lett. **102**, 055703 (2009)
3. T. Sunada, *Topological Crystallography with a View Towards Discrete Geometric Analysis* (Springer, New York, 2013)
4. A. Mizuno, Y. Shuku, K. Awaga, Recent developments in molecular spin gyroid research. Bull. Chem. Soc. Jpn **92**, 1068–1093 (2019)
5. M. Tsuchiizu, Three-dimensional higher-spin Dirac and Weyl dispersions in the strongly isotropic K4 crystal. Phys. Rev. B **94**, 195426 (2016)
6. Y. Shuku, A. Mizuno, R. Ushiroguchi, C.S. Hyun, Y.J. Ryu, B.-K. An, J.E. Kwon, S.Y. Park, M. Tsuchiizu, K. Awaga, An exotic band structure of a supramolecular honeycomb lattice formed by a pancake π–π interaction between triradical trianions of triptycene tribenzoquinone. Chem. Comm. **54**, 3815–3818 (2018)
7. R. Suizu, K. Awaga, Line graph theory reveals hidden spin frustration and bond frustration in molecular crystals with strong isotropy. J. Mater. Chem. C **10**, 1196–1203 (2022)
8. R.W. Day, D.K. Bediako, M. Rezaee, L.R. Parent, G. Skoruskii, M.Q. Arguilla, C.H. Hendon, I. Stassen, N.C. Gianneschi, P. Kim, M. Dincă, Single crystals of electrically conductive two-dimensional metal–organic frameworks: Structural and electrical transport properties. ACS Cent. Sci. **5**, 1959–1964 (2019)
9. Y. Misumi, A. Yamaguchi, Z. Zhang, T. Matsushita, N. Wada, M. Tsuchiizu, K. Awaga, Quantum spin liquid state in a two-dimensional semiconductive metal-organic framework. J. Am. Chem. Soc. **142**, 16513–16517 (2020)
10. M. Hmadeh, Z. Lu, Z. Liu, F. Gándara, H. Furukawa, S. Wan, V. Augustyn, R. Chang, L. Liao, F. Zhou, E. Perre, V. Ozolins, K. Suenaga, X. Duan, B. Dunn, Y. Yamamto, O. Terasaki,

O.M. Yaghi, New porous crystals of extended metal-catecholates. Chem. Mater. **24**, 3511–3513 (2012)

11. D. Sherbela, L. Sun, M.A. Blood-Forsythe, S. Er, C.R. Wade, C.K. Brozek, A. Aspuru-Guzik, M. Dincă, High electrical conductivity in Ni_3(2,3,6,7,10,11-hexaiminotriphenylene)$_2$, a semiconducting metal–organic graphene analogue. J. Am. Chem. Soc. **136**, 8859–8862 (2014)
12. J.S. Helton, K. Matan, M.P. Shores, E.A. Nytko, B.M. Bartlett, Y. Yoshida, Y. Takano, A. Suslov, Y. Qiu, J.-H. Chung, D.G. Nocera, Y.S. Lee, Spin dynamics of the spin-1/2 kagome lattice antiferromagnet $ZnCu_3(OH)_6Cl_2$. Phys. Rev. Lett. **98**, 107204 (2007)
13. A. Mizuno, Y. Shuku, M.M. Matsushita, M. Tsuchiizu, Y. Hara, N. Wada, Y. Shimizu, K. Awaga, 3D Spin-liquid state in an organic hyperkagome lattice of mott dimers. Phys. Rev. Lett. **119**, 057201 (2017)
14. T. Senthil, M. Vojta, S. Sachdev, Weak magnetism and non-fermi liquids near heavy-fermion critical points. Phys. Rev. B **69**, 035111 (2004)
15. N. Elstner, A.P. Young, Spin-1/2 heisenberg antiferromagnet on the kagome lattice: high-temperature expansion and exact-diagonalization studies. Phys. Rev. B **50**, 6871 (1994)
16. Q. Chen, O. Adeniran, Z.F. Liu, Z. Zhang, K. Awaga, Graphite-like charge storage mechanism in a 2D π-d conjugated metal-organic frame-work revealed by stepwise magnetic monitoring. J. Am. Chem. Soc. **145**, 1062–1071 (2023)
17. Q. Chen, Z. Zhang, K. Awaga, Magnetometric characterization of inter-mediates in the solid-state electrochemistry of redox-active metal-organic frameworks. J. Vis. Exp. **196**, e65335 (2023)
18. K. Wada, K. Sakaushi, S. Sasaki, H. Nishihara, Multielectron-transfer-based rechargeable energy storage of two-dimensional coordination frameworks with non-innocent ligands. Angew. Chem. Int. Ed. **57**, 8886–8890 (2018)
19. K. Sakaushi, H. Nishihara, Two-dimensional π-conjugated frameworks as a model system to unveil a multielectron-transfer-based energy storage mechanism. Acc. Chem. Res. **54**, 3003–3015 (2021)
20. Y. Li, Y. Lu, P. Adelhelm, M.M. Titirici, Y.S. Hu, Intercalation chemistry of graphite: Alkali metal ions and beyond. Chem. Soc. Rev. **48**, 4655–4687 (2019)
21. J.P. Donoso, C.J. Magon, J.F. Lima, O.R. Nascimento, E. Benavente, M. Moreno, G. Gonzalez, Electron paramagnetic resonance study of copper-ethylenediamine complex ion intercalated in bentonite. J. Phys. Chem. C **117**, 24042–24055 (2013)
22. C.C. Felix, R.C. Sealy, o-Benzosemiquinone and its metal chelates. electron spin resonance investigation of radicals from the photolysis of catechol in the presence of complexing metal ions. J. Am. Chem. Soc. **104**, 1555–1560 (1982)
23. V. Sitaram, A. Sharma, S.V. Bhat, K. Mizoguchi, R. Menon, Electron spin resonance studies in the doped polyaniline paniampsa: evidence for local ordering from linewidth features. Phys. Rev. B **72**, 035209 (2005)
24. O. Ivasenko, D.F. Perepichka, Mastering fundamentals of supramolecular design with carboxylic acids. Common lessons from X-ray crystallography and scanning tunneling microscopy. Chem. Soc. Rev. **40**, 191–206 (2011)
25. M. Yamamoto, R. Suizu, S. Dutta, P. Mishra, T. Nakayama, K. Sakamoto, K. Wakabayashi, T. Uchihashi, K. Awaga, Self-assembled honeycomb lattice in the monolayer of cyclic thiazyl diradical BDTDA (= 4,4'-bis(1,2,3,5-dithiadiazolyl)) on Cu(111) with a zero-bias tunneling spectra anomaly. Sci. Rep. **5**, 18359 (2015)
26. A. Yamamoto, S. Uehara, T. Hamada, M. Miyata, I. Hisaki, N. Tohnai, Diamondoid p rous organic salts toward applicable strategy for construction of versatile porous structures. Cryst. Growth Des. **12**, 4600–4606 (2012)
27. G. Russell, N.K. Suleman, H. Iwamura, O.W. Webster, Radical anions of triptycene bis- and tris(quinones). J. Am. Chem. Soc. **103**, 1560–1561 (1981)
28. J.-M. Lü, S.V. Rosokha, I.S. Neretin, J.K. Kochi, Quinones as electron acceptors. X-ray structures, spectral (EPR, UV−vis) characteristics and electron-transfer reactivities of their reduced anion radicals as separated vs contact ion pairs. J. Am. Chem. Soc. **128**, 16708–16719 (2006)

29. K.E. Preuss, Pancake bonds: π-Stacked dimers of organic and light-atom radicals. Polyhedron **79**, 1–15 (2014)
30. A. Mielke, Ferromagnetism in the Hubbard model on line graphs and further considerations. J. Phys. A **24**, 3311–3321 (1991)
31. H. Tasaki, Ferromagnetism in the Hubbard models with degenerate single-electron ground states. Phys. Rev. Lett. **69**, 1608–1611 (1992)
32. A. Mielke, H. Tasaki, Ferromagnetism in the Hubbard model. Commun. Math. Phys. **158**, 341–371 (1993)
33. H. Tasaki, Ferromagnetism in Hubbard models. Phys. Rev. Lett. **75**, 4678–4681 (1995)
34. T. Misumi, H. Aoki, New class of flat-band models on tetragonal and hexagonal lattices: gapped versus crossing flat bands. Phys. Rev. B **96**, 155137 (2017). (and references therein)
35. E. Cremades, J. Echeverría, S. Alvarez, The trigonal prism in coordination chemistry. Chem. Eur. J. **16**, 10380–10396 (2010)
36. S. Alvarez, Distortion pathways of transition metal coordination polyhedra induced by chelating topology. Chem. Rev. **115**, 13447–13483 (2015)
37. J.C. Knight, S. Alvarez, A.J. Amoroso, P.G. Edwards, N. Singha, A novel bipyridine-based hexadentate tripodal framework with a strong preference for trigonal prismatic co-ordination geometries. Dalton Trans. **39**, 3870–3883 (2010)
38. B. Yao, M.K. Singh, Y.-F. Deng, Y.-N. Wang, K.R. Dunbar, Y.-Z. Zhang, Trigonal prismatic cobalt(ii) single-ion magnets: manipulating the magnetic relaxation through symmetry control. Inorg. Chem. **59**, 8505–8513 (2020). (and references cited therein)
39. A. Landart-Gereka, M.M. Quesada-Moreno, I.F. Díaz-Ortega, H. Nojiri, M. Ozerov, J. Krzystek, M.A. Palacios, E. Colacio, Large easy-axis magnetic anisotropy in a series of trigonal prismatic mononuclear cobalt(ii) complexes with zero-field hidden single-molecule magnet behaviour: the important role of the distortion of the coordination sphere and intermolecular interactions in the slow relaxation. Inorg. Chem. Front. **9**, 2810–2831 (2022). (and references cited therein)
40. Y. Shuku, R. Suizu, M. Tsuchiizu, K. Awaga, Ideal trigonal prismatic coordination geometry of Co(II) in a honeycomb MOF with a triptycene-based ligand. Chem. Commun. **59**, 10105–10108 (2023)
41. S. Langis-Barsetti, T. Maris, J.D. Wuest, Triptycene 1,2-quinones and quinols: permeable crystalline redox-active molecular solids. J. Org. Chem. **83**, 15426–15437 (2018)
42. D.A. Shultz, Valence tautomerism in dioxolene complexes of cobalt, in *Magnetism: Molecules to Materials II*, ed. by J.S. Miller, M. Drillon, ch. 8 (Wiley-VCH, Weinheim, 2001), pp. 281–306
43. R. Ushiroguchi, Y. Shuku, R. Suizu, K. Awaga, Variable host−guest charge-transfer interactions in 1D channels formed in a molecule-based honeycomb lattice of phenazine analogue of triptycene. Cryst. Growth Des. **20**, 7593–7597 (2020)
44. W. Jankowski, M. Gdaniec, The β-polymorph of phenazine. Acta Crystallogr., Sect. C: Cryst. Struct. Commun., **C58**, o181−o182 (2002)
45. Z. Berkovitch-Yellin, L. Leiserowitz, The role played by C-H···O and C-H···N interactions in determining molecular packing and conformation. Acta Crystallogr., Sect. B: Struct. Sci. **B40**, 159−165 (1984)
46. R. Nemoto, R. Arafune, S. Nakano, M. Tsuchiizu, N. Takagi, R. Suizu, T. Uchihashi, K. Awaga, Chiral honeycomb lattices of nonplanar π-conjugated supramolecules with protected dirac and flat bands. ACS Nano **18**, 19663–19671 (2024)
47. S. Das, G. Nascimbeni, R.O. de la Morena, F. Ishiwari, Y. Shoji, T. Fukushima, M. Buck, E. Zojer, M. Zharnikov, Porous honeycomb self-assembled monolayers: tripodal adsorption and hidden chirality of carboxylate anchored triptycenes on Ag. ACS Nano **15**, 11168–11179 (2021)
48. S.H. Kim, H.G. Jeong, S.J. Lim, U.D. Ham, Y.J. Song, J. Yu, Y. Kuk, Geometric and electronic properties of porphyrin molecules on Au(111) and NaCl surfaces. Surf. Sci. **613**, 54–57 (2013)
49. H. Imada, K. Miwa, M. Imai-Imada, S. Kawahara, K. Kimura, Y. Kim, Real-space investigation of energy transfer in heterogeneous molecular dimers. Nature **538**, 364–367 (2016)

50. S.T. Schneebeli, M. Frasconi, Z. Liu, Y. Wu, D.M. Gardner, N.L. Strutt, C. Cheng, R. Carmieli, M.R. Wasielewski, J.F. Stoddart, Electron sharing and anion–π recognition in molecular triangular prisms. Angew. Chem. Int. Ed. **52**, 13100–13104 (2013)
51. Z. Liu, G. Liu, Y. Wu, D. Cao, J. Sun, S.T. Schneebeli, M.S. Nassar, C.A. Mirkin, J.F. Stoddart, Assembly of supramolecular nanotubes from molecular triangles and 1,2-dihalohydrocarbons. J. Am. Chem. Soc. **136**, 16651–16660 (2014)
52. A. Mizuno, Y. Shuku, R. Suizu, M.M. Matsushita, M. Tsuchiizu, D.R. Mañeru, F. Illas, V. Robert, Kunio Awaga, Discovery of the K_4 structure formed by a triangular π radical anion. J. Am. Chem. Soc. **137**, 7612–7615 (2015)
53. G. Saito, Y. Yoshida, Frontiers of organic conductors and superconductors, in *Unimolecular and Supramolecular Electronics I, in Topics in Current Chemistry*, vol. 312 (Springer, Berlin, Heidelberg, 2011)
54. R. Kato, Conducting metal dithiolene complexes: structural and electronic properties. Chem. Rev. **104**, 5319–5346 (2004)
55. M. Sorai, S. Murakawa, T. Kotani, H. Suga, Heat capacity of a cubane-tetramer, $[Ni(OCH_3)(SAL)(CH_3OH)]_4$, between 0.4 and 288 K: spin-spin interactions and tunnel-splitting of the internal rotation of methyl-group. J. Phys. Chem. Solids, **44**, 1133–1139 (1983)
56. J.P. Riehl, F.S. Richardson, Circularly polarized luminescence spectroscopy. Chem. Rev. **86**, 1–16 (1986)
57. T. Zhao, J. Han, P. Duan, M. Liu, New perspectives to trigger and modulate circularly polarized luminescence of complex and aggregated systems: energy transfer, photon upconversion, charge transfer, and organic radical. Acc. Chem. Res. **53**, 1279–1292 (2020)
58. X. Luo, M. Pan, Metal-organic materials with circularly polarized luminescence. Coord. Chem. Rev. **468**, 214640 (2022)
59. K. Takaishi, Y. Nakatsuka, H. Asano, Y. Yamada, T. Ema, Ruthenium complexes bearing axially chiral bipyridyls: the mismatched diastereomer showed red circularly polarized phosphorescence. Chem. Eur. J. **28**, e202104212 (2022). (and references therein)
60. C.A. Kent, D. Liu, A. Ito, T. Zhang, M.K. Brennaman, T.J. Meyer, W. Lin, Rapid energy transfer in non-porous metal–organic frameworks with caged $Ru(bpy)_3{}^{2+}$ chromophores: oxygen trapping and luminescence quenching. J. Mater. Chem. A **1**, 14982 (2013)
61. R. Andrés, M. Gruselle, B. Malézieux, M. Verdaguer, J. Vaissermann, Enantioselective synthesis of optically active polymeric homo- and bimetallic oxalate-bridged networks $[M_2(ox)_3]_n$. Inorg. Chem. **38**, 4637–4646 (1999)
62. K. Nakashima, R. Suizu, S. Morishita, N. Tsurumachi, M. Funahashi, H. Masu, R. Ozawa, K. Nakamura, Kunio Awaga, Enhanced circularly polarized luminescence by a homochiral guest–host interaction in gyroidal MOFs, $[Ru(bpy)_3]$ $[M_2(ox)_3]$ (bpy = 2,2′-bipyridyl, ox = oxalate, M = Zn, Mn). ACS Mater. Au **3**, 201–205 (2023)
63. F.P. Dwyer, E.C. Gyarfas, The chemistry of ruthenium. Part VII. The oxidation of $_\text{D}$ and $_\text{L}$ tris 2:2′ dipyridyl ruthenium II iodide. J. Proc. R. Soc. N. S. W. **83**, 174–176 (1949)
64. W.W. Wendlandt, H.G. Hecht, *Reflectance Spectroscopy* (Interscience Publishers, New York, 1966)
65. A. Gafni, I.Z. Steinberg, Optical activity and lifetime measurements of tris-(2,2′-bipyridine) ruthenium (ii) bromide: evidence for some singlet-singlet emission. Isr. J. Chem. **15**, 102–105 (1976)
66. J. Liu, H. Su, L. Meng, Y. Zhao, C. Deng, J.C.Y. Ng, P. Lu, M. Faisal, J.W.Y. Lam, X. Huang, H. Wu, K.S. Wong, B.Z. Tang, What makes efficient circularly polarised luminescence in the condensed phase: aggregation induced circular dichroism and light emission. Chem. Sci. **3**, 2737–2747 (2012)
67. R. Suizu, A. Iwasaki, Y. Shuku, K. Awaga, Spatially inhomogeneous, stepwise phase transitions in a thiazyl diradical: a structural mismatch induced by lattice transformation. J. Mater. Chem. C **3**, 7968–7977 (2015)

Correction to: Construction of Molecule-Based Strongly Isotropic Lattices and Their Physical Properties

Qi Chen, Yoshiaki Shuku, Rie Suizu, and Kunio Awaga

Correction to:
Chapter 11 in: S. Kobatake and H. Uekusa (eds.), *Advances in Organic Crystal Chemistry*, https://doi.org/10.1007/978-981-96-4479-7_11

In the original version of the book was incorrectly published with minor corrections and Figure 11.10 and Scheme 11.4 has been revised in the Chapter 11. The correction Chapter has been updated with the changes.

The updated version of this chapter can be found at
https://doi.org/10.1007/978-981-96-4479-7_11

S. Kobatake and H. Uekusa (eds.), *Advances in Organic Crystal Chemistry*,
https://doi.org/10.1007/978-981-96-4479-7_12

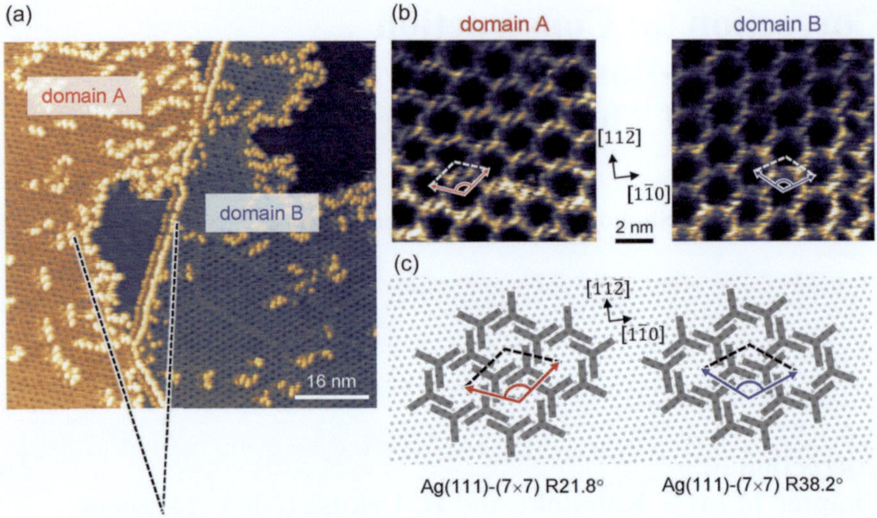

Fig. 11.10 **a** ST image of Trip-Phz molecules adsorbed on a Ag(111) surface that was co-deposited with ~0.01 ML Cs and annealed at ~100 °C (Vs = −2V, It = 10 pA). Domains **A** and **B** have different lattice orientations as indicated by the dashed lines. **b** High resolution STM images of domains **A** and **B** (constant height mode, size: 10 nm × 10 nm, Vs = −2 V. The feedback was stabilized at It = 10 pA). The arrows on the images indicate the unit vectors of honeycomb lattices, while those between the images the crystal orientations of the Ag(111) surface. **c** Structure model of Trip-Phz adsorbed on a Ag(111) surface. Left: Ag(111)-(7×7) R21.8°, Right: Ag(111)-(7×7) R38.2°. The arrows correspond to those in panel (**b**). Adapted with permission from Ref. [46]. Copyright 2024, American Chemical Society

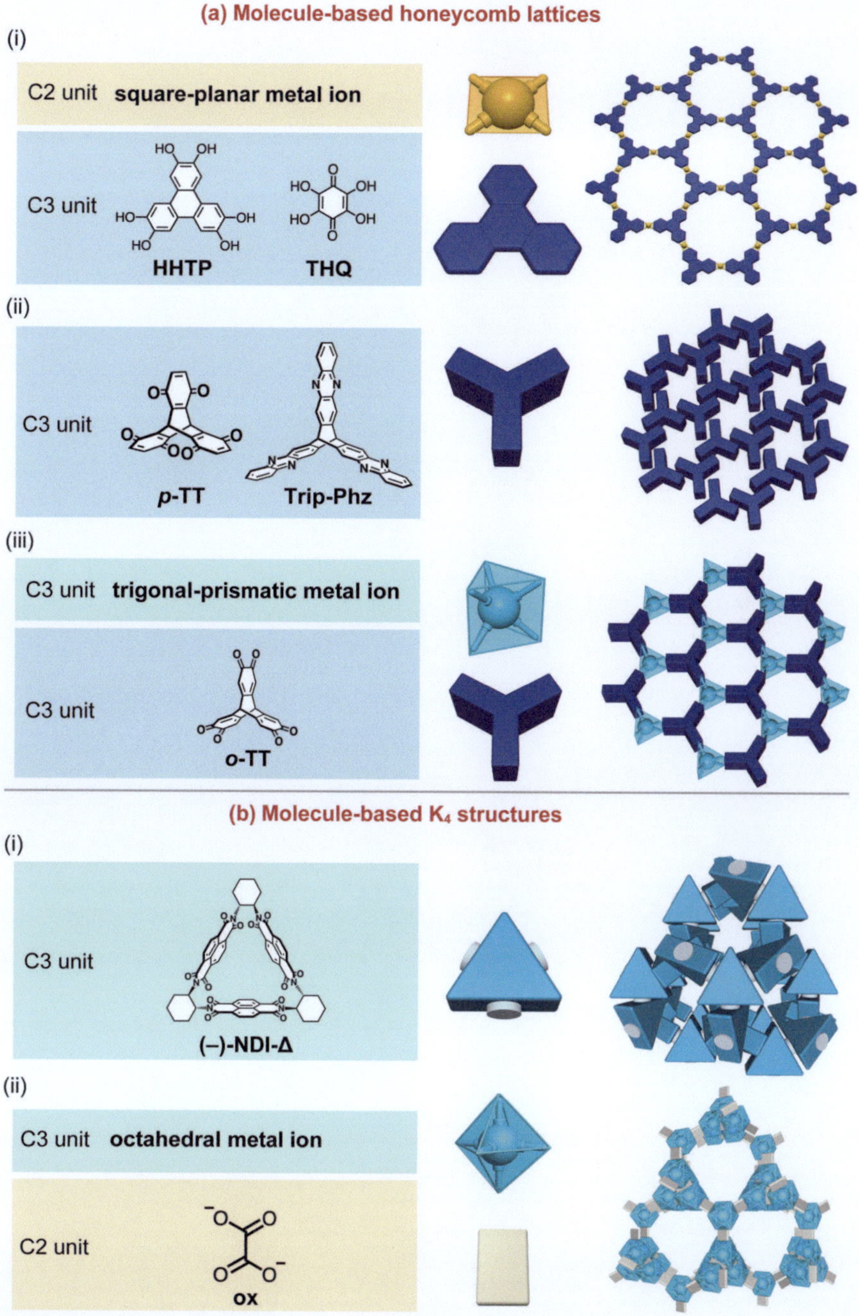

Scheme 11.4 Synthetic strategy to form molecule-based strongly isotropic structures, such as honeycomb lattice (a) and K_4 lattice